Plant Cell Separation and Adhesion

Plant Cell Separation and Adhesion

Edited by

JEREMY A. ROBERTS
Head of Plant Sciences Division
School of Biosciences
University of Nottingham
UK

and

ZINNIA GONZALEZ-CARRANZA
Plant Sciences Division
School of Biosciences
University of Nottingham
UK

Blackwell
Publishing

Blackwell Publishing Editorial Offices:
Blackwell Publishing Ltd, 9600 Garsington Road, Oxford OX4 2DQ, UK
Tel: +44 (0)1865 776868
Blackwell Publishing Professional, 2121 State Avenue, Ames, Iowa 50014-8300, USA
Tel: +1 515 292 0140
Blackwell Publishing Asia Pty Ltd, 550 Swanston Street, Carlton, Victoria 3053, Australia
Tel: +61 (0)3 8359 1011

First published 2007 by Blackwell Publishing Ltd

ISBN-13: 978-14051-3892-5
ISBN-10: 1-4051-3892-0

Library of Congress Cataloging-in-Publication Data
is available

A catalogue record for this title is available from the British Library

Set in 10/12 pt Times
by TechBooks, New Delhi, India
Printed and bound in Singapore
by Markono Print Media Pte Ltd

The publisher's policy is to use permanent paper from mills that operate a sustainable forestry policy, and which has been manufactured from pulp processed using acid-free and elementary chlorine-free practices. Furthermore, the publisher ensures that the text paper and cover board used have met acceptable environmental accreditation standards.

For further information on Blackwell Publishing, visit our website:
www.blackwellpublishing.com

Contents

Contributors

Dr Bernhard Borkhardt Biotechnology Group, Danish Institute of Agricultural Sciences, Thorvaldsensvej 40, DK-1871 Frederiksberg, Denmark
Dr Christopher T. Brett Department of Biomedical and Life Sciences, Wolfson Building, University of Glasgow, Glasgow G12 8QQ, UK
Dr David M. Cavalier MSU-DOE Plant Research Lab, Michigan State University, East Lansing, Michigan 48824, USA
Dr Céline Faugeron Laboratoire de Glycobiologie et Physiologie Végétale, Université d'Artois, Faculté Jean Perrin, SP18 Rue Jean Souvraz, 62307 Lens, France
Dr Hiroo Fukuda Department of Biological Sciences, Graduate School of Science, University of Tokyo, Japan
Dr James Giovannoni Boyce Thompson Institute for Plant Research, Tower Road, Cornell University, Ithaca, NY 14853, USA and US Department of Agriculture – Agricultural Research Service, Tower Road, Cornell University, Ithaca, NY 14853, USA
Dr Zinnia Gonzalez–Carranza Plant Sciences Division, School of Biosciences, University of Nottingham, Sutton Bonington Campus, Loughborough, LE12 5RD, UK
Professor Martha Hawes Department of Plant Sciences, Division of Plant Pathology and Microbiology, University of Arizona, Tucson AZ, USA
Professor Kenneth Keegstra MSU-DOE Plant Research Lab, Michigan State University, East Lansing, Michigan 48824, USA
Dr Hideo Kuriyama Department of Biological Sciences, Graduate School of Science, University of Tokyo, Japan
Professor Marta Laskowski Biology Department, Oberlin College, Oberlin OH, USA
Dr Olivier Lerouxel MSU-DOE Plant Research Lab, Michigan State University, East Lansing, Michigan 48824, USA
Dr Michelle E. Leslie Curriculum in Genetics and Molecular Biology, University of North Carolina, Chapel Hill, NC 27599, USA
Dr Michael W. Lewis Department of Biology, University of North Carolina, Chapel Hill, NC 27599, USA
Dr Aaron H. Liepman MSU-DOE Plant Research Lab, Michigan State University, East Lansing, Michigan 48824, USA
Professor Sarah J. Liljegren Department of Biology and Curriculum in Genetics and Molecular Biology, University of North Carolina, Chapel Hill, NC 27599, USA
Dr Catherine Martel Department of Plant Biology, Cornell University, Ithaca, NY 14853, USA and Boyce Thompson Institute for Plant Research, Tower Road, Cornell University, Ithaca, NY 14853, USA

Dr Jean-Claude Mollet Laboratoire de Glycobiologie et Physiologie Végétale, Université d'Artois, Faculté Jean Perrin, SP18 Rue Jean Souvraz, 62307 Lens, France

Dr Henri Morvan Laboratoire de Glycobiologie et Physiologie Végétale, Université d'Artois, Faculté Jean Perrin, SP18 Rue Jean Souvraz, 62307 Lens, France

Dr Lars Østergaard Department of Crop Genetics, John Innes Centre, Norwich Research Park, Colney, Norwich, Norfolk. NR4 7UH, UK

Professor Jeremy Roberts Plant Sciences Division, School of Biosciences, University of Nottingham, Sutton Bonington Campus, Loughborough, LE12 5RD, UK

Dr Peter Ulvskov Biotechnology Group, Danish Institute of Agricultural Sciences, Thorvaldsensvej 40, DK-1871 Frederiksberg, Denmark

Dr Fushi Wen Department of Plant Sciences, Division of Plant Pathology and Microbiology, University of Arizona, Tucson AZ, USA

Dr Keith W. Waldron The Division of Food Materials Science, Institute of Food Research, Norwich Research Park, Colney, Norwich, NR4 7UA, UK

Preface

The sequencing of plant genomes has provided a new window of opportunity for the study of plant biology. This information has fuelled the development of the 'omic' technologies and it is timely that these approaches, coupled with forward and reverse genetic strategies, should be applied to phenomena such as cell separation and adhesion. This is the first volume to focus exclusively on these processes and to seek to link improvements in our scientific understanding with methods that may allow us to manipulate cell separation and adhesion to the benefit of the agricultural and horticultural industries. As a consequence, the book contains contemporary material of interest to the experimental scientist in addition to information that should be of value to those who seek to apply such knowledge.

The opening chapter outlines those sites where cell separation and adhesion have been documented to take place. Chapter 2 provides a detailed account of cell wall structure and its biosynthesis and assembly, because in order to dismantle cellular components, it is important to first understand how they may be constructed. The third chapter describes vascular cell differentiation in the xylem and discusses the importance of cell–cell contact for the process to occur, whilst Chapter 4 explores the role of cell adhesion during pollen tube guidance. A variety of cell separation phenomena including: root development, organ abscission, pod dehiscence, and fruit ripening, are then examined in detail. In the final chapter, the role of polymer cross-linking in cell-to-cell adhesion is examined and the importance of the process in product quality is evaluated.

The compilation of this book represents the first monograph ever produced that focuses exclusively on the processes of cell separation and adhesion during plant development. We believe that it provides the reader with a detailed overview of the major advances in our knowledge of these phenomena and the commercial significance of understanding how they may be regulated.

During the final stages of editing this volume Daphne Osborne sadly died. During her scientific career Daphne made an outstanding contribution to our understanding of cell separation processes in plants and, in particular, to our knowledge of abscission. This book is dedicated to her memory.

Jerry Roberts
Zinnia Gonzalez-Carranza

Annual Plant Reviews

A series for researchers and postgraduates in the plant sciences. Each volume in this series focuses on a theme of topical importance and emphasis is placed on rapid publication.

Titles in the series:

1. **Arabidopsis**
   Edited by M. Anderson and J.A. Roberts
2. **Biochemistry of Plant Secondary Metabolism**
   Edited by M. Wink
3. **Functions of Plant Secondary Metabolites and their Exploitation in Biotechnology**
   Edited by M. Wink
4. **Molecular Plant Pathology**
   Edited by M. Dickinson and J. Beynon
5. **Vacuolar Compartments**
   Edited by D.G. Robinson and J.C. Rogers
6. **Plant Reproduction**
   Edited by S.D. O'Neill and J.A. Roberts
7. **Protein–Protein Interactions in Plant Biology**
   Edited by M.T. McManus, W.A. Laing and A.C. Allan
8. **The Plant Cell Wall**
   Edited by J.K.C. Rose
9. **The Golgi Apparatus and the Plant Secretory Pathway**
   Edited by D.G. Robinson
10. **The Plant Cytoskeleton in Cell Differentiation and Development**
    Edited by P.J. Hussey
11. **Plant–Pathogen Interactions**
    Edited by N.J. Talbot
12. **Polarity in Plants**
    Edited by K. Lindsey
13. **Plastids**
    Edited by S.G. Moller
14. **Plant Pigments and their Manipulation**
    Edited by K.M. Davies
15. **Membrane Transport in Plants**
    Edited by M.R. Blatt
16. **Intercellular Communication in Plants**
    Edited by A.J. Fleming
17. **Plant Architecture and Its Manipulation**
    Edited by C. Tuurnbull
18. **Plasmodesmata**
    Edited by K.J. Oparka
19. **Plant Epigenetics**
    Edited by P. Meyer

20. **Flowering and Its Manipulation**
Edited by C. Ainsworth
21. **Endogenous Plant Rhythms**
Edited by A. Hall and H. McWatters
22. **Control of Primary Metabolism in Plants**
Edited by W.C. Plaxton and M.T. McManus
23. **Biology of the Plant Cuticle**
Edited by M. Riederer
24. **Plant Hormone Signaling**
Edited by P. Hadden and S.G. Thomas
25. **Plant Cell Separation & Adhesion**
Edited by J.R. Roberts and Z. Gonzalez-Carranza
26. **Senescence Processes in Plants**
Edited by S. Gan
27. **Seed Development, Dormancy and Germination**
Edited by K.J. Bradford and H. Nonogaki
28. **Plant Proteomics**
Edited by C. Finnie
29. **Regulation of Transcription in Plants**
Edited by K. Grasser
30. **Light and Plant Development**
Edited by G. Whitelam
31. **Plant Mitochondria**
Edited by David C. Logan

1 Cell separation and adhesion processes in plants

Jeremy Roberts and Zinnia Gonzalez-Carranza

1.1 Introduction

The plant cell wall plays a critical role both in the protection of the protoplast and in providing a framework to enable plant growth and development. Without such a structure it would not have been possible for an Australian eucalyptus tree to grow to a height of 132 metres or for the roots of a wild fig tree to penetrate 120 metres into the soil. Whilst one of the keys to the successful growth of higher plants has been the capacity to develop an extracellular skeleton, this feature also generates restrictions in terms of the ways that an organism is able to grow and reproduce. As a consequence, mechanisms have been established to enable the wall to be resculptured or removed entirely (Knox, 1992; Walton, 1994). A comprehensive description of cell wall structure, biosynthesis and assembly can be found in Chapter 2.

The process of separating an individual cell from its neighbour has to be highly co-ordinated both in time and space so that neither the internal nor external integrity of the plant is compromised (Roberts *et al.*, 2000). At times only the junction between two adjacent cells may be selectively degraded, for instance during the formation of guard cell within the stomatal complex, while on other occasions a whole organ may undergo wall breakdown as in the softening process that accompanies the ripening of fleshy fruits (Roberts *et al.*, 2002). Wall degradation is largely an autolytic phenomenon catalysed by the interaction between a plant and its environment; however, the process may also be triggered as a result of an interaction with another organism. For instance, pathogenic fungi and bacteria that induce soft rots of plant tissues have their effect by secreting enzymes that degrade the cell wall of their host (Barras *et al.*, 1994; Annis and Goodwin, 1997). Whether these organisms bring about wall dissolution in a similar fashion to that which occurs *in planta* is not clear, but a major distinction is that while hydrolytic enzymes emanating from fungi or bacteria often bring about extensive tissue disintegration and cell death; this does not occur during events such as ripening and abscission.

Some organisms are able to promote cell separation at a predetermined site in a plant. For instance, the pod midge lays its eggs in the developing seeds of crops such as *Brassica napus*. The larva emerges from the egg and begins to consume the contents of the seed where upon growth causes it to become increasingly incarcerated inside the pod. However, once the larva is mature it stimulates cell separation along the dehiscence zone of the silique and once released from the pod falls to the ground where it pupates (Meakin and Roberts, 1991). Whether the pod midge larva is able to secrete enzymes that bring about wall breakdown or triggers prematurely the normal plant's own dehiscence mechanism is unknown.

Just as cell separation is an important phenomenon during plant development so too is the capacity of cells to adhere to one another and formulate a physical and chemical association (Jarvis *et al.*, 2003). Occasionally, during the life cycle of a plant, two distinct cell types may come into contact and develop a bond such as the one that occurs when a pollen grain lands on a receptive stigmatic surface (Zinkl and Preuss, 2000) or during the fusion of floral organs such as carpels. However, in general plant cells develop with a physical association to one another and the adhesion process has to accommodate the co-ordination of growth and development. Adhesion can also take place between two distinct individuals as in the case of a graft union (Jeffree and Yeoman, 1983) or even between a plant and another organism such as *Rhizobium*, which can ultimately lead to the formation of a root nodule or a symbiotic association (Smit *et al.*, 1987; Vicré *et al.*, 2005).

1.2 Cell separation processes

During the life cycle of a higher plant, it is possible to identify many sites where cell separation occurs and these are highlighted in Plate 1. During germination co-ordinated wall breakdown is an important event to facilitate radicle emergence and an increase in the expression of wall degrading enzymes has been identified at the site of root expansion in tomato and Arabidopsis seeds (Sitrit *et al.*, 1999). In situ hybridisation studies have revealed that elevated expression occurs in the cells adjacent to the radicle tip and that this takes place prior to root elongation indicating that some co-ordination of the processes must be taking place (Chen and Bradford, 2000; Chen *et al.*, 2002). This would facilitate root elongation and ensure that growth took place when separation of the endosperm cells had already been facilitated. The nature of the signalling events between the root and adjacent cells is unknown, but it will be important to ascertain whether similar signals are generated at other sites where wall breakdown occurs.

Once root emergence and elongation take place, lateral roots are induced within the pericycle cells that will ultimately penetrate the cortical cells and emerge through the epidermis of the primary root. A role for cell separation in this process has been previously documented and proposed that the emerging lateral root tip secretes cell wall degrading enzymes into the cortex of the root (Peretto *et al.*, 1992). Reporter gene studies indicate that it is the cells in the cortex and epidermis adjacent to the lateral root tip that up-regulate genes involved in cell separation and once more invoke a signalling mechanism between two distinct tissues in co-ordinating wall loosening (Roberts *et al.*, 2002). This process and that of cell separation at the root cap is discussed in detail in Chapter 5.

A process of cell separation that we know little about is that which takes place during the formation of the stoma. An anatomical study of the process leading to aperture formation has been described in some detail and the dissolution of the middle lamella seems analogous to what has been identified during abscission and dehiscence (Stevens and Martin, 1978). The phenomenon is particularly interesting in that only one face of the wall breaks down. Whether this is due to targeted secretion of cell wall degrading enzymes or specific modifications to the other wall faces is yet to be ascertained, but studies of mutants with multiple stomatal complex developments might allow the isolation of some of the signals involved.

Intercellular space formation is another example of targeted wall breakdown and, like stomatal pore development, requires synchronisation between cells adjacent to the site where such an aperture is formed (DeChalain and Berjak, 1979; Jeffree *et al.*, 1986). In tissues such as leaves and hypocotyls where a network of spaces develop the signalling process that co-ordinates these events must operate over some distance (Prat *et al.*, 1997). Xylem formation is another example of co-ordinated cell wall degradation that takes place during organ development and is discussed in detail in Chapter 3.

Environmental stresses have a profound effect on plant development, and those plants that adapt to changes in conditions are able to exploit marginal niches. During waterlogging, plants rapidly experience problems associated with anaerobiosis and those plants that tolerate this condition are frequently able to generate cavities within tissues that can acts as a conduit for oxygenation of the cells under stress (Evans, 2003). The formation of aerenchyma within the roots of plants such as rice and maize is a good example of this development and one which seems to be mediated by elevated ethylene production. Ethylene is known to be an important co-ordinator of cell separation during abscission and the ripening of climacteric fruits and it has the capacity to induce the formation of cell dissolution and aerenchyma development. Although the mechanisms that define which cells will undergo collapse are not known, it is clear that cell wall dissolution is involved probably mediated by wall degrading enzymes (see Chapter 5).

Perhaps the most well-documented examples of cell separation take place when organs are shed and fleshy fruits soften (Rose *et al.*, 2003). These two processes are discussed in detail in Chapters 6 and 8 while the phenomenon that shares many features in common with them, that of organ dehiscence is described in Chapter 7.

1.3 Cell adhesion processes

Whilst the cell wall provides the rigidity for each building block of a plant, it is the adhesion between the cells that provides the overall mechanical strength of the plant. The adhesive properties are not static throughout the life cycle of a plant but change to enable growth and development (see Plate 2). In many ways, cell separation can be considered to be the process whereby the adhesion events between cells are dismantled (Jarvis *et al.*, 2003). In this way by discovering what events maintain cell adhesion we may be able to unravel the converse.

A number of mutants have been isolated, particularly in Arabidopsis, where the cell adhesion process is disrupted in some way. Some of these result in substantial phenotypic changes where cells protrude through the epidermal layers in cotyledons, leaves and hypocotyls (Bouton *et al.*, 2002). Others show enhanced gaps between adjacent cotyledon or hypocotyl cells (Singh *et al.*, 2005). Intriguingly, both these mutants are the consequence of lesions in genes encoding members of the glycosyltransferase family although their functions remain, as yet, unknown. Another group of mutants have been characterised where tissues that normally remain discrete undergo fusion and result in the aberrant formation of leaves and floral organs (Lolle *et al.*, 1992, 1998; Yephremov *et al.*, 1999; Krolikowski *et al.*, 2003). In these mutants, the cuticular structure seems to be substantially compromised or more subtly altered such that the cuticle becomes much more permeable to small molecules, some of which presumably act to promote fusion (Pruitt *et al.*, 2000; Tanaka *et al.*, 2004; Kurdyukov *et al.*, 2006). The role of some of these genes in the recognition processes that take place between pollen grains and the stigmatic surface (see Chapter 4) is open to question.

1.4 Manipulation of cell separation and adhesion in crop plants

It is evident from the previous discussion that cell separation and adhesion phenomena make a major contribution to the growth and development of a plant. Indeed, the processes of cell and tissue *in vitro* culture that have contributed substantially to our scientific knowledge and, in addition, they have been used extensively for the propagation of cultivars of commercial interest, the production of secondary metabolites, and the maintenance of valuable germplasm resources rely on these events.

In the early 1970s, there was considerable interest in being able to manipulate leaf shedding and an ability to co-ordinate the abscission of organs such as leaves, flowers and fruit, which continues to be of commercial interest in enhancing yield or facilitating the harvesting process. Moreover, understanding seed shedding could be of interest to agricultural practices where some weed seeds can lie dormant for many years and the recent discovery of a transcription factor in rice that regulates seed abscission may make an important contribution in this area (Konishi *et al.*, 2006). Attempts to restrict pod shatter in crops such as oilseed rape by breeding strategies have had limited success and whilst there is evidence that genetic manipulation approaches might be more successful, these have not been adopted so far although a number of patents have been submitted in this area (see Chapter 7).

The first genetically modified (GM) plant product launched in Europe used an antisense strategy to reduce cell wall degradation in tomatoes and hence elevate the shelf life of the fruit (Bird *et al.*, 1988). Whilst the processed product proved to be successful commercially, the attitude in Europe to GM products led some retailers choosing to outlaw such material and eventually the product was withdrawn.

The ability to manipulate plant architecture is of increasing interest for the improvement of crop efficiency and yield (Battey, 2005). By changing the way that root development takes place there may be opportunities to improve the efficiency of water and nutrient capture (Lopez-Bucio *et al.*, 2005). One strategy that might achieve this would be to manipulate the cell separation events that facilitate lateral root formation or to use promoters of genes encoding root cell wall loosening agents to attenuate the expression of plant hormone biosynthetic or metabolising genes.

Cell separation and adhesion are also important properties of plants both for the food and non-food industries (see Chapter 9). The texture of a fresh product is dependent on these characteristics and whether a fruit or vegetable is crunchy or mealy correlates with the strength of the adhesion between cells. The degree of degradation of the cell wall can also have a pronounced effect on both the texture and the properties of a processed product – this phenomenon is of interest for the generation of both fabrics and wood-based materials (Yuan and Knauf, 1997).

1.5 Conclusions

The objective of this volume is to explore the process of cell separation and adhesion during the life cycle of a plant and to compare and contrast some of the events that bring them about. In addition, as these phenomena take place in different tissues, we examine whether common events occur during cell wall degradation and assembly at different sites. We hope that the chapters in this volume stimulate the interest of the reader in processes that have been fundamental to the successful growth of plants and have posed some of the biggest challenges for evolution to overcome.

References

Annis, S.L. and Goodwin, P.H. (1997) Recent advances in the molecular genetics of plant cell wall-degrading enzymes produced by plant pathogenic fungi. *European Journal of Plant Pathology* **103**, 1–14.

Barras, F., van Gijsegem, F. and Chatterjee, A.K. (1994) Extracellular enzymes and pathogenesis of soft-rot erwinia. *Annual Review of Phytopathology* **32**, 201–234.

Battey, N.H. (2005) Applications of plant architecture. *Annual Plant Reviews* **17**, 288–314

Bird, C.R., Smith, C.J., Ray, J.A., Moureau, P., Bevan, M.W., Bird, A.S., Hughes, S., Morris, P.C., Grierson, D. and Schuch, W. (1988) The tomato polygalacturonase gene and ripening-specific expression in transgenic plants. *Plant Molecular Biology* **11**, 651–662.

Bouton, S., Leboeuf, E., Mouille, G., Leydecker, M.-T., Talbotec, J., Granier, F., Lahaye, M., Höfte, H. and Truong, H.-N. (2002) QUASIMODO1 Encodes a putative membrane-bound glycosyltransferase required for normal pectin synthesis and cell adhesion in Arabidopsis. *The Plant Cell* **14**, 2577–2590.

Chen, F. and Bradford, K.J. (2000) Expression of an expansin is associated with endosperm weakening during tomato seed germination. *Plant Physiology* **124**, 1265–1274.

Chen, F., Nonogaki, H. and Bradford, K.J. (2002) A gibberellin-regulated xyloglucan endotransglycosylase gene is expressed in the endosperm cap during tomato seed germination. *Journal of Experimental Botany* **53**, 215–223.

De Chalain, T.M.B. and Berjak, P. (1979) Cell death as a functional event in the development of the leaf intercellular spaces in *Avicennia marina* (Forsskal) Vierh. *New Phytologist* **83**, 147–155.

Evans, D.E. (2003) Aerenchyma formation. *New Phytologist* **161**, 35–49.

Jarvis, M.C., Briggs, S.P.H. and Knox, J.P. (2003) Intercellular adhesion and cell separation in plants. *Plant, Cell and Environment* **26**, 977–989.

Jeffree, C.E., Dale, J.E. and Fry, S.C. (1986) The genesis of intercellular spaces in developing leaves of *Phaseolus vulgaris* L. *Protoplasma* **132**, 90–98.

Jeffree, C.E. and Yeoman, M.M. (1983) Development of intercellular connections between opposing cells in graft union. *New Phytologist* **93**, 491–509.

Knox, J.P. (1992) Cell adhesion, cell separation and plant morphogenesis. *The Plant Journal* **2**, 137–141.

Konishi, S., Izawa, T., Lin, S.Y., Ebana, K., Fukuta, Y., Sasaki, T., Yano, M. (2006) An SNP caused loss of seed shattering during rice domestication. *Science* **312**, 1392–1396.

Krolikowski, K.A., Victor, J.L., Wagler, T.N., Lolle, S.J. and Pruitt, R.E. (2003) Isolation and characterization of the Arabidopsis organ fusion gene HOTHEAD. *The Plant Journal* **35**, 501–511.

Kurdyukov, S., Faust, A., Trenkamp, S., Bar, S., Franke, R., Efremova, N., Tietjen, K., Schreiber, L., Saedler, H. and Yephremov, A. (2006) Genetic and biochemical evidence for involvement of HOTHEAD in the biosynthesis of long-chain alpha-,omega-dicarboxylic fatty acids and formation of extracellular matrix. *Planta* **324**, 315–329.

Lolle, S.J., Hsu, W. and Pruitt, R.E. (1998) Genetic analysis of organ fusion in *Arabidopsis thaliana. Genetics* **149**, 607–619.

Lolle, S.J., Cheung, A.Y. and Sussex, I. M. (1992) FIDDLEHEAD: an Arabidopsis mutant constitutively expressing an organ fusion program that involves interactions between epidermal cells. *Developmental Biology* **152**, 383–392.

Lopez-Bucio, J., Cruz-Ramirez, A., Perez-Torres, A., Ramirez-Pimentel, J.G., Sanchez-Calderon, L. and Herrera-Estrella, L. (2005) Root architecture. *Annual Plant Reviews* **17**, 182–208.

Meakin, P.J. and Roberts, J.A. (1991) Anatomical and biochemical changes associated with the induction of oilseed rape (*Brassica napus*) pod dehiscence by *Dasineura brassicae* (Winn.). *Annals of Botany* **67**, 193–197.

Peretto, R., Favaron, F., Bettini, V., DeLorenzo, G., Marini, S., Alghisi, P., Cervone, F. and Bonfante, P. (1992) Expression and localization of polygalacturonase during the outgrowth of lateral roots in *Allium porrum* L. *Planta* **188**, 164–172.

Prat, R., Andre, J.P., Mutaftschiev, S. and Catesson, A.M. (1997) Three-dimensional study of the intercellular gas space in *Vigna radiata* hypocotyls. *Protoplasma* **196**, 69–77.

Pruitt, R.E., Vielle-Calzada, J.-P., Ploense, S.E., Grossniklaus, U. and Lolle, S.J. (2000) *FIDDLEHEAD*, a gene required to suppress epidermal cell interactions in *Arabidopsis*, encodes a putative lipid biosynthetic enzyme. *Proceedings of the National Academy of Sciences of the United States of America* **97**, 1311–1316.

Roberts, J.A., Whitelaw, C.A., Gonzalez-Carranza, Z.H. and McManus, M. (2000) Cell separation processes in plants – models, mechanisms and manipulation. *Annals of Botany* **86**, 223–235.

Roberts, J.A., Elliott, K.A. and Gonzalez-Carranza, Z.H. (2002) Abscission, dehiscence and other cell separation processes. *Annual Review of Plant Biology* **53**, 131–158.

Rose, J.K.C., Catala, C., Gonzalez-Carranza, Z.H. and Roberts, J.A. (2003) Cell wall disassembly. *Annual Plant Reviews* **8**, 264–324.

Singh, S.K., Eland. C., Harholt, J., Vibe Scheller, H. and Marchant, A. (2005) Cell adhesion in *Arabidopsis thaliana* is mediated by ECTOPICALLY PARTING CELLS 1 – a glycosyltransferase (GT64) related to the animal exostosins. *The Plant Journal* **43** (3), 384–397.

Sitrit, Y., Hadfield, K.A., Bennett, A.B., Bradford, K.J. and Downie, B. (1999) Expression of a polygalacturonase associated with tomato seed germination. *Plant Physiology* **121**, 419–428.

Smit, G., Kijne, J.W. and Lugtenberg, B.J. (1987) Involvement of both cellulose fibrils and a Ca2+-dependent adhesin in the attachment of *Rhizobium leguminosarum* to pea root hair tips. *Journal of Bacteriology* **169**, 4294–4301.

Stevens, R.A. and Martin, E.S. (1978) Structural and functional aspects of stomata I. Developmental studies in *Polypodium vulgare. Planta* **142**, 307–316.

Tanaka, T., Tanaka, H., Machida, C.,Watanabe, M. and Machida, Y. (2004) A new method for rapid visualization of defects in leaf cuticle reveals five intrinsic patterns of surface defects in *Arabidopsis. The Plant Journal* **37**, 139–146.

Vicré, M., Santaella, C., Blanchet, S., Gateau, A and Driouich, A. (2005) Root border-like cells of Arabidopsis. Microscopical characterization and role in the interaction with rhizobacteria. *Plant Physiology* **138**, 998–1008.

Walton, J.D. (1994) Deconstructing the cell wall. *Plant Physiology* **104**, 1113–1118.

Yephremov, A., Wisman, E., Huijser, P., Huijser, C., Wellesen, K. and Saedler, H. (1999) Characterization of the FIDDLEHEAD gene of Arabidopsis reveals a link between adhesion response and cell differentiation in the epidermis. *Plant Cell* **11**, 2187–2202.

Yuan, L and Knauf, V.C. (1997) Modification of plant components. *Current Opinion in Biotechnology* **8**, 227–233.

Zinkl, G.M. and Preuss, D. (2000) Dissecting Arabidopsis pollen-stigma interactions reveals novel mechanisms that confer mating specificity. *Annals of Botany* **85**, 15–21.

2 Cell wall structure, biosynthesis and assembly

Aaron H. Liepman, David M. Cavalier, Olivier Lerouxel
and Kenneth Keegstra

2.1 Introduction

The primary walls of plant cells are composed mainly of polysaccharides, but they also contain structural proteins and many different enzymes (Carpita and McCann, 2000). Among the polysaccharides, cellulose is the most important and normally the most abundant. Cellulose microfibrils are embedded in a matrix of other polysaccharides, glycoproteins and proteins (Plate 3). While the molecular details of this complex network are still unclear, it is clear that the organization of components within cell walls is not random and that the plant cell wall plays important roles in defining the shape of plant cells as well as the ways the cells grow, divide and differentiate to yield the complex forms found in higher plants.

Significant progress has been made in defining the structure of the various components that make up plant cell walls. Considerable information is available on the structure of the individual polysaccharides present in plant cell walls, and with the recent availability of genome sequences, we know the amino acid sequence of most cell wall proteins (see, for example, Schultz *et al.*, 2002). However, large gaps still exist in our understanding of the biosynthesis of wall polysaccharides. It is known that cellulose is made at the plasma membrane and deposited directly into the wall (Doblin *et al.*, 2003), while most other matrix components are made in the Golgi and delivered to the wall in secretory vesicles (Carpita and McCann, 2000). Because wall components are synthesized in different locations, they must be assembled into a functional wall following their biosynthesis. Much remains to be learned about the details of this assembly process, as well as the rearrangements and selective degradation events that are needed to allow cell growth and differentiation.

One reason for the lack of information regarding cell wall polysaccharide biosynthesis is that the process is difficult to study using traditional biochemical techniques. The Golgi and plasma membrane enzymes that produce wall polysaccharides are low abundance, integral membrane proteins that often lose activity after solubilization with detergents, making traditional biochemical purification difficult. Another reason for the slow progress in understanding wall biosynthesis is that the enzymes that synthesize wall polysaccharides are highly specific. In contrast to nucleic acids and proteins, where templates are used for determining the sequence of monomers in the polymer, the structure of polysaccharides is determined by the specificity of the enzymes that synthesize the polymers. In many cases, the precise specificity of the biosynthetic enzymes is still not known, in part because the oligosaccharides that serve as acceptors are not available for establishing biochemical assays.

Only in recent years have some of the proteins that synthesize wall polysaccharides been identified. This has occurred primarily through the application of genetic and genomic strategies. One important advancement is the availability of whole genome sequences, first from Arabidopsis (AGI, 2000) and more recently from other species. Sequence analysis has led to the identification of many candidates for enzymes that may be involved in wall polysaccharide biosynthesis (Coutinho *et al.*, 2003; Carbohydrate-Active Enzymes server at http://afmb.cnrs-mrs.fr/CAZY/). The current challenge is to identify both the biochemical function and the biological role of the various candidate genes.

Several different strategies are being used to define the roles of candidate genes. One is the application of genetics, both forward genetic screens and reverse genetic methods. As discussed in more detail below, genetic strategies have been very successful in identifying the genes and proteins involved in the biosynthesis of cellulose (Doblin *et al.*, 2003) and matrix polysaccharides (see, for example, Madson *et al.*, 2003). Another successful strategy has been the expression of candidate genes in various heterologous expression systems (for example, see Faik *et al.*, 2002; Liepman *et al.*, 2005). In this chapter, we first describe the structure, function and biosynthesis of the major components of primary plant cell walls. We then consider various models of how the components are organized within the primary wall followed by a brief consideration of the metabolism and reorganization of these components during growth and development.

2.2 Primary cell walls: composition and biosynthesis

2.2.1 Cellulose

Cellulose is ubiquitous among plants where it constitutes the major polysaccharide of cell walls; it is also considered the most abundant biopolymer on Earth (Saxena and Brown, 2005). Cellulose is at the core of plant cell walls, where it serves as a scaffold for the binding of other cell wall components. Because an exhaustive summary of the large body of literature relating to cellulose and its biosynthesis is beyond the scope of this chapter, the reader is referred to excellent recent discussions on these topics (Doblin *et al.*, 2002, 2003; Saxena and Brown, 2005). This section highlights some recent developments that have improved our understanding of cellulose biosynthesis.

Cellulose, crystalline 1,4-β-D-glucan is synthesized at the plant plasma membrane. In plants, cellulose microfibrils are about 3 nm in diameter and generally consist of parallel arrangements of thirty-six 1,4-β-D-glucan chains (Doblin *et al.*, 2003). Membrane-bound cellulose synthase enzyme complexes, the most prolific biomachines in nature, have been implicated in the biosynthesis of cellulose. These complexes are visible as hexameric rosettes approximately 25–30 nm in diameter when plant cells are examined using freeze-fracture electron microscopy (Doblin *et al.*, 2002; Saxena and Brown, 2005). Using cytosolic uridine-diphosphoglucose

(UDP-glucose) as substrate, each rosette subunit, is thought to extrude multiple 1,4-β-D-glucan chains that coalesce as microfibrils outside of the plasma membrane (Doblin *et al.*, 2002; Read and Bacic, 2002).

In recent years, significant insight into the molecular details of cellulose biosynthesis has been gained using forward and reverse genetic analyses coupled with advances in plant genomics. The first compelling candidates thought to encode plant cellulose synthase catalytic subunits (CESA proteins) were identified in developing cotton (*Gossypium hirsutum*) fiber cDNA libraries using transcriptional profiling (Pear *et al.*, 1996). Genetic studies have since implicated the CESA proteins in cellulose biosynthesis; when homologs of *GhCESA* sequences are mutated in *Arabidopsis thaliana*, cellulose deficiencies result (Arioli *et al.*, 1998; Taylor *et al.*, 1999; Fagard *et al.*, 2000; Williamson *et al.*, 2001; Tanaka *et al.*, 2003; Scheible and Pauly, 2004). CESA proteins are large ($\sim$1000 amino acids) integral membrane proteins with eight predicted transmembrane domains; they have all been classified as members of the CAZy family GT2 (Richmond, 2000; Coutinho *et al.*, 2003). The catalytic domain, containing D, DXD and QXXRW residues characteristic of processive β-glycosyltransferases (GTs), of each CESA protein occupies the central portion of the protein sequence and is thought to face the cytosol. Another feature common to all CESA proteins is a pair of amino-terminal zinc finger motifs (Doblin *et al.*, 2002; Kurek *et al.*, 2002). CESA proteins have been shown to form higher order structures *in vitro* and *in planta* (Kurek *et al.*, 2002; Taylor *et al.*, 2004). The conserved zinc finger sequences are thought to act as redox-regulated multimerization domains involved in the assembly of CESA monomers into rosette complexes (Kurek *et al.*, 2002). The redox state of a cell may thus regulate cellulose biosynthesis at the level of assembly of CESA proteins into rosettes.

Bioinformatics analysis of plant DNA sequences indicates that plant genomes typically contain many *CESA* genes. An ongoing tally of plant *CESA* sequences indicates that 10 *CESA* genes are present in Arabidopsis, 13 in rice (*Oryza sativa*) and more than 50 in wheat (*Triticum aestivum*) (http://cellwall.stanford.edu/php/summary.php). Why do plant genomes contain so many *CESA* genes? The need for multiple *CESA* genes may be partially explained by the divergence in function and expression patterns among CESA isoforms. Hints of such divergence are evident at the transcriptional level, where transcripts of certain *CESA* genes are present specifically in cells and tissues undergoing deposition of primary cell walls, while distinct *CESA* transcripts are present in cells depositing secondary walls (Hamann *et al.*, 2004; Brown *et al.*, 2005). Phylogenetic analyses indicate that the divergence of genes encoding CESA proteins involved in primary cell wall biosynthesis from those involved in secondary cell wall biosynthesis probably occurred prior to the origin of flowering plants (Doblin *et al.*, 2003; Nairn and Haselkorn, 2005). The expression of particular combinations of *CESA* genes is coordinated and suggests that several different CESA proteins are needed to assemble active rosette complexes. Within the cellulose synthase complex, different CESA proteins may be needed to initiate and extend glucan chains (Read and Bacic, 2002), or for proper assembly of rosette subunits into functional complexes (Doblin *et al.*, 2002).

Despite its critical role in cell wall architecture, mutants with defective cellulose have been reported and these mutants have provided significant insight about cellulose biosynthesis. Cellulose mutants have been isolated by screening of plants with defective cell elongation (e.g. Fagard *et al.*, 2000), or with deficiencies in cell wall strength (e.g. Taylor *et al.*, 1999). Two types of cellulose mutants have been identified with such screens: (1) those with lesions in *CESA* genes, and (2) those affecting genes, the products of which have ancillary functions in cellulose biosynthesis. Mutants in six of the ten *CESA* genes of Arabidopsis have been described and these *CESA* mutants provide additional evidence of functional specialization among these gene products. In addition to aberrant cellulose deposition, *CESA* single mutants exhibit various phenotypes including decreased cell length, weakened stems, collapsed xylem and perturbations in the levels of other cell wall components (Arioli *et al.*, 1998; Taylor *et al.*, 1999; Fagard *et al.*, 2000; Williamson *et al.*, 2001; Tanaka *et al.*, 2003; Scheible and Pauly, 2004). Some *CESA* mutants specifically affect cellulose in primary cell walls (e.g. mutations in *AtCESA1*, *AtCESA3* and *AtCESA6*), while others affect cellulose in secondary cell walls (e.g. *AtCESA4*, *AtCESA7* and *AtCESA8*). Mutants with lesions in any of the other four *AtCESA* genes (*AtCESA2*, *AtCESA5*, *AtCESA9* and *AtCESA10*) have not yet been described, possibly because they are lethal. The close evolutionary relationship between the *AtCESA2*, *AtCESA5*, *AtCESA9* and *AtCESA10* genes with primary wall *CESA* sequences suggests that they may play roles in cellulose deposition in primary walls.

Mutations in a number of non-*CESA* genes also result in cellulose defects. These genes include the Arabidopsis *KORRIGAN* gene encoding a 1,4-β-glucanase (Lane *et al.*, 2001; Szyjanowicz *et al.*, 2004), *CYT1*, a gene whose product is involved in the biosynthesis of GDP-mannose (Lukowitz *et al.*, 2001), several *PEANUT* genes encoding enzymes involved in the biosynthesis of glycosylphosphatidylinositol membrane anchors (Gillmor *et al.*, 2005) and *KOBITO1* – a plant-specific gene of unknown function (Pagant *et al.*, 2002).

Genetic studies suggest that in addition to the CESA proteins, it is likely that cellulose synthase complexes contain additional proteins, though so far no such proteins have been localized to the rosette. Much work remains to elucidate the composition and function of proteins present in plant cellulose synthase complexes. Biochemical experiments aimed to reconstitute cellulose biosynthesis *in vitro* may help shed light on this question. Several groups have recently reported progress in synthesizing cellulose *in vitro* using cell-free extracts from plants (Kudlicka and Brown, 1997; Lai-Kee-Him *et al.*, 2002; Colombani *et al.*, 2004). Significant improvements in the yield of *in vitro* cellulose have been achieved by careful choices of plant material, detergents used for solubilization and additives in the reaction mixture. In addition to synthesizing callose, extracts prepared from suspension cultures of blackberry (*Rubus fruticosus*) cells yielded 20%, and from suspension cultures of aspen (*Populus tremula x tremuloides*) yielded 50%, of *de novo* cellulose exhibiting characteristics quite similar to cellulose microfibrils found in primary cell walls (Lai-Kee-Him *et al.*, 2002; Colombani *et al.*, 2004). An alternative strategy to identify constituents of rosette complexes is the use of affinity chromatography. An epitope-tagged AtCESA7 protein has been used to complement a mutation in this gene, and efforts are being made to purify the protein complex containing this protein (Taylor *et al.*, 2004). These efforts may pave the way toward the purification of cellulose synthase complexes, enabling the identification of constituent polypeptides using sensitive mass spectroscopy-based proteomics technologies. Eventually, it may be possible to reconstitute cellulose biosynthesis *in vitro*, using proteins purified from plants or expressed in heterologous systems and to better understand the function of each rosette constituent.

2.2.2 Callose

Callose, a 1,3-β-D-glucan another polysaccharide produced at the plant plasma membrane. Microscopic examinations of plant tissues stained with the callose-binding fluorochrome aniline blue have localized this polysaccharide to the cell plate during cytokinesis. Callose is also present in the walls of pollen mother cells, pollen tubes and the walls of plasmodesmata (Stone and Clarke, 1992; Jacobs *et al.*, 2003). In addition to its structural roles, callose deposition is induced by a variety of biotic and abiotic stresses. Plants often synthesize callose plugs, or papillae, at the site of pathogen ingress. These papillae are thought to impede or ensnare potential pathogen invaders, allowing the plant to mount a more effective defense response (Jacobs *et al.*, 2003). Like cellulose, callose is synthesized from cytosolic pools of the nucleotide sugar precursor UDP-glucose. Because these polymers are synthesized at the same location in the cell using the same substrate, and because cellulose synthase and callose synthase (CALS) activities often co-purify, it has long been thought that callose synthesis results from the activity of disrupted cellulose synthase complexes (Verma and Hong, 2001). Only recently have CALS enzymes been identified and shown to be distinct from the CESA proteins.

Arabidopsis contains 12 *CALS* genes, all members of the CAZy family GT48, and related to the *FKS1* gene of yeast, a β-1,3-glucan synthase. Similar to the CESA proteins, CALS proteins are polytopic integral membrane proteins, with large hydrophilic regions predicted to contain catalytic residues facing the cytosol. The CALS proteins are among the largest known in plants, at nearly 2000 amino acids in length. Interestingly, CALS proteins appear to be missing UDP-glucose binding sequences, implying that these proteins may not directly bind the substrate. Furthermore, a UDP-glucose binding protein (UGT1), and Rho GTPase interact with the CALS1 protein. The activity of CALS proteins may thus be regulated by these accessory proteins. For more detailed discussions of callose and CALS complexes in plants, the reader is referred to more comprehensive reviews (Stone and Clarke, 1992; Verma and Hong, 2001).

2.2.3 Hemicelluloses

Hemicelluloses are defined as non-cellulosic polysaccharides, excluding pectins that associate non-covalently with cellulose and are extracted from cell walls with 1–4 M alkali solutions. They include xyloglucans, arabinoxylans, galacto(gluco)mannans and mixed-linkage glucans (MLGs) (Bauer *et al.*, 1973; Bacic *et al.*, 1988; Fry, 1989; Hayashi, 1989; McCann and Roberts, 1991; Carpita and McCann, 2000). The main function of the primary cell wall hemicelluloses is assumed to be cross-linking adjacent cellulose microfibrils to form a three-dimensional, load-bearing structure that can resist the turgor pressure of the cell (Keegstra *et al.*, 1973; Valent and Albersheim, 1974; Hayashi *et al.*, 1994a,b; Levy *et al.*, 1997). In general, hemicelluloses found in type I primary cell walls, which are characteristic of most flowering plant cell walls, are composed of xyloglucan (10–20% dry weight), glucomannan (5–10%) and xylan (~5%) (Carpita and Gibeaut, 1993; O'Neill and York, 2003). Alternatively, hemicelluloses found in type II primary cell walls, which are characteristic of the family Poaceae, are composed of xyloglucan (1–5%), xylan (20–40%) and MLG (10–30%) (Carpita and Gibeaut, 1993; Carpita, 1996; O'Neill and York, 2003).

Hemicelluloses are synthesized by an array of glycan synthases and glycosyltransferases. Glycan synthases are defined as processive enzymes that are responsible for the biosynthesis of the backbone of a polysaccharide (Keegstra and Raikhel, 2001). While biochemical studies have shown that hemicelluloses are synthesized in Golgi-enriched fractions (Ray *et al.*, 1969; Ray, 1980; Hayashi and Matsuda, 1981; Gibeaut and Carpita, 1993; White *et al.*, 1993; Porchia *et al.*, 2002), attempts to isolate hemicellulose glycan synthases using traditional biochemical protein purification techniques have been unsuccessful. However, significant progress has been made recently with the identification of a diverse set of cellulose synthase-like (*CSL*) genes in Arabidopsis and rice that belong to the cellulose synthase super family of genes (Richmond, 2000; Hazen *et al.*, 2002). The Arabidopsis and rice genomes contain 29 and 37 *CSL* genes, respectively, which are distributed among eight families (*CSLA–H*) and are predicted to encode polytopic integral membrane proteins. Based upon the sequence similarity to the *CESA* genes, it has been proposed that the *CSL* genes encode hemicellulose glycan synthases (Richmond and Somerville, 2000, 2001; Hazen *et al.*, 2002). Indeed, studies involving heterologous expression of candidate genes by Dhugga *et al.* (2004) and Liepman *et al.* (2005) provided the first direct evidence for the *CSL* hypothesis in mannan and glucomannan biosynthesis (discussed below).

Glycosyltransferases are defined as non-processive enzymes that are responsible for the addition of specific sugar residues to either the backbone of a particular polysaccharide or a nascent side chain of a polysaccharide (Keegstra and Raikhel, 2001). Traditional biochemical purification techniques, forward and reverse genetics and heterologous expression of candidate genes have been employed successfully in studying hemicellulose glycosyltransferases. All hemicellulose glycosyltransferases characterized to date require an acceptor substrate and are type-II integral membrane proteins that contain a short N-terminal sequence residing in the cytoplasm, a single transmembrane domain and a C-terminal catalytic domain residing in the lumen of the Golgi apparatus (Edwards *et al.*, 1999; Perrin *et al.*, 1999; Faik *et al.*, 2000, 2002; Madson *et al.*, 2003).

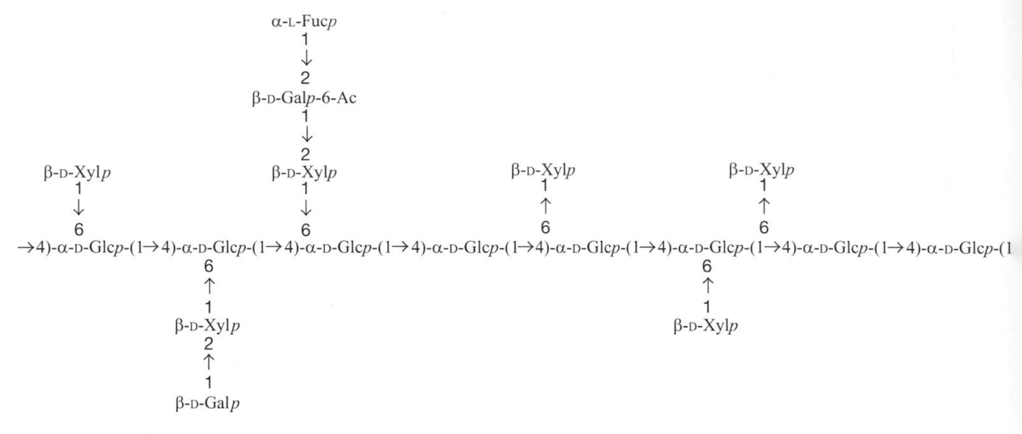

Figure 2.1 Representative dicot xyloglucan structure. A naming convention was developed to describe the structure of xyloglucan by listing, from the non-reducing end to the reducing end of the molecule, a letter code that represents a discrete glycosyl residue substitution pattern (Fry *et al.*, 1993). Single letter codes for common xyloglucan sidechains include 'G' (for and unsubstituted back bone glucosyl residue), 'X' (α-D-Xyl*p*→), 'L' (*β*-D-Gal*p*-(1→2)-*α*-D-Xyl*p*→) and 'F' (*α*-L-Fuc*p*-(1→6)-*β*-D-Gal*p*-(1→2)-*α*-D-Xyl*p*→). Therefore, the depicted xyloglucan fragment would be named XLFGXXXG and is classified as XXXG-type xyloglucan.

2.2.3.1 Xyloglucan

Xyloglucan is composed of a 1,4-*β*-D-glucan backbone that is regularly substituted at the *O*-6 position with either *α*-D-xylosyl residues or side chains composed of discrete sequences of *α*-D-xylosyl, *β*-D-galactosyl and *α*-L-fucosyl residues, and in some species *α*-L-arabinose (Figure 2.1). Xyloglucan can be categorized into two general classes based upon the xylose substitution pattern of endo-(1,4)-*β*-glucanase hydrolyzed xyloglucan: XXXG-type and XXGG-type xyloglucans (Vincken *et al.*, 1997). XXXG-type xyloglucan is the most prevalent type of xyloglucan found in the majority of dicots, monocots and conifers studied to date, while XXGG-type xyloglucan is found primarily in the families Poaceae and Solanaceae (for an extensive comparison of species-specific xyloglucan structure, see O'Neill and York, 2003; Hoffman *et al.*, 2005). XXXG-type xyloglucan is composed predominantly of repeating subunits of XXXG, XXFG, XLFG and XLLG; although the ratios of each oligosaccharide subunit can vary between species. Alternatively, there is a considerable variation in the structure of the repeating oligosaccharide subunits of plants containing XXGG-type xyloglucan (O'Neill and York, 2003; Hoffman *et al.*, 2005). Finally, xyloglucan can be acetylated, with XXXG-type xyloglucan containing 6-*O*-acetyl galactosyl residues (Sims *et al.*, 1996) and XXGG-type xyloglucan of the family Solanaceae containing 5-*O*-acetyl arabinosyl and 6-*O*-acetyl glucosyl residues (York *et al.*, 1988; Kiefer *et al.*, 1989; Pauly and Scheller, 2000). However, neither the identity of the acetyltransferases nor the biological significance of xyloglucan acetylation has been determined.

Xyloglucan fucosyltransferase from pea epicotyls (PsFUT1) is the only xyloglucan biosynthetic enzyme that has been purified with traditional protein purification techniques (Perrin *et al.*, 1999). Amino acid sequences derived from purified PsFUT1 were used to identify an Arabidopsis gene (*AtFUT1*) that encodes a protein with xyloglucan-specific fucosyltransferase activity. AtFUT1 expressed in COS cells utilizes GDP-L-fucose to fucosylate both tamarind (*Tamarindus indica*) seed and nasturtium (*Tropaeolum majus*) seed xyloglucan (both lack fucosylated xyloglucan) (Perrin *et al.*, 1999). Disruption of *AtFUT1* with a T-DNA insertion produced phenotypically normal plants; however, xyloglucan isolated from the leaves and roots lacked fucose (Perrin *et al.*, 2003). Sarria *et al.* (2001) identified nine other putative fucosyltransferase genes (*AtFUT2–10*) that were classified, along with *AtFUT1*, as being members of the CAZy family GT37. Results from heterologous expression and overexpression studies involving AtFUT3–5 indicated that these enzymes were involved in the fucosylation of cell wall carbohydrates other than xyloglucan (Sarria *et al.*, 2001).

Reiter *et al.* (1993) employed a forward genetics strategy whereby 5200 ethylmethane sulfonate (EMS)-mutagenized Arabidopsis plants were screened for altered cell wall monosaccharide composition; the screen resulted in the identification of 23 mutant lines that corresponded to 11 loci designated as *mur1–11*. The distinguishing characteristic of plants that were mutated at the *MUR1*, *MUR2* or *MUR3* locus is a significant decrease (>50%) in the fucose content of cell wall polysaccharides (Reiter *et al.*, 1993, 1997). *MUR1* encodes GDP-D-mannose-4,6-dehydratase, which is required for the biosynthesis of GDP-L-fucose (Bonin *et al.*, 1997). The *mur2* mutant was shown to contain a lesion in the *AtFUT1* gene that resulted in a greater than 98% decrease in fucosylated xyloglucan; however, *mur2* plants did not have any gross morphological phenotype, nor was the tensile strength of the *mur2* cell walls different from the wild type (Vanzin *et al.*, 2002).

The *mur3* locus was initially identified because of a significant decrease in cell wall fucose content (Madson *et al.*, 2003). Apart from collapsed trichome papillae, the gross morphology of the *mur3* plants was indistinguishable from wild-type plants. Further biochemical analysis showed that endoglucanase-digested xyloglucan from *mur3* plants lacked XLFG and XLLG fragments and they had a significant increase in the amount of XLXG fragments. Extracts of *Pichia pastoris* expressing the MUR3 protein utilized UDP-D-galactose as a nucleotide sugar substrate to add an α-(1,2)-galactosyl residue to the third xylosyl residue from the non-reducing end of XXXG to form XXLG; however, neither XLXG nor XLLG were formed *in vitro* (Madson *et al.*, 2003). *MUR3* belongs to CAZy family GT47, which contains 38 other putative Arabidopsis glycosyltransferase genes, 10 of which (*AtGT11–20*) are closely related to *MUR3* (Li *et al.*, 2004). Initial sugar composition analysis of *atgt13* and *atgt18* T-DNA insertion lines showed 10.3 and 13.5% reduction of cell wall galactose, respectively; however, further characterization of these lines is needed (Li *et al.*, 2004). It is assumed that there is at least one other galactosyltransferase that specifically adds a galactosyl residue to the second xylosyl residue from the non-reducing end of XXXG or XXLG to form XLXG or XLLG, respectively.

Based upon similarities between the pea (*Pisum sativum*) xyloglucan xylosyltransferase and the fenugreek (*Trigonella foenum-graecum*) galactomannan 1,6-α-galactosyltransferase (GMGT) (both require a β-glycan acceptor and synthesize products with 1,6-α-linkages), the fenugreek GMGT sequence was used to identify a family of seven putative xyloglucan xylosyltransferase genes in Arabidopsis (Edwards *et al.*, 1999; Faik *et al.*, 2002). The fenugreek 1,6-α-galactosyltransferase and the seven putative Arabidopsis xyloglucan xylosyltransferase genes belong to the CAZy family GT34. Heterologous expression of the seven putative xylosyltransferase genes in *Pichia pastoris* indicated that one gene, *AtXT1*, encodes a protein with xylosyltransferase activity, while none of the other six genes, designated *GT2–7*, appeared to have xylosyltransferase activity. Biochemical analyses indicated that AtXT1 utilizes UDP-D-xylose to add an α-1,6-xylosyl residue to the penultimate glucosyl residue from the reducing end of cellopentaose. While AtXT1 has xylosyltransferase activity and produces a nascent xyloglucan oligosaccharide, further studies using reverse genetics and heterologous expression should determine if *AtXT1* and *AtGT2–7* encode proteins that are involved in xyloglucan biosynthesis.

2.2.3.2 Arabinoxylan

Arabinoxylan is the major hemicellulose in the primary cell walls of graminaceous monocotyledonous plants (Carpita, 1996), and it is composed of a backbone of 1,4-β-linked xylosyl residues that is regularly substituted at the *O*-2 and/or *O*-3 position with α-L-arabinose (O'Neill and York, 2003). In addition, arabinoxylan of the primary cell wall can contain small amounts of α-D-glucuronosyl residues at *O*-2 (Ebringerová and Heinze, 2000) and *O*-acetyl groups at *O*-2 and/or *O*-3 (O'Neill and York, 2003).

The backbone of arabinoxylan is synthesized by 1,4-β-xylan synthase, which has been partially characterized in a diverse group of economically important plants such as maize (*Zea mays*) (Bailey and Hassid, 1966), oats (*Avena sativa*) (Ben-Arie *et al.*, 1973), bean (*Phaseolus vulgaris*) (Bolwell and Northcote, 1983), pea (Baydoun *et al.*, 1989) and wheat (Porchia and Scheller, 2000; Kuroyama and Tsumuraya, 2001). There are two reports available for the identification and partial characterization of the arabinoxylan arabinosyltransferase in bean (Bolwell, 1986) and wheat (Porchia *et al.*, 2002). However, none of the genes encoding xylan biosynthetic enzymes have yet been identified.

2.2.3.3 Galacto(gluco)mannan

Galacto(gluco)mannans are composed of backbone of 1,4-β-D-mannosyl and 1,4-β-D-glucosyl residues that may be substituted at the *O*-6 position of the mannosyl residues with either β-D-galactosyl residues or side chains of β-D-Gal-1,2-β-D-Gal- (O'Neill and York, 2003).

Galacto(gluco)mannans comprise only 5–10% of the dry weight of primary cell wall carbohydrates in most plants but up to 50% of the dry weight of seeds from legumes such as fenugreek and guar (*Cyamopsis tetragonolobus*) (O'Neill and York, 2003). Consequently, the secondary cell walls of fenugreek and guar endosperms have been used as model systems to study galacto(gluco)mannan biosynthesis.

The fenugreek *GMGT* gene was the first plant cell wall glycosyltransferase gene to be cloned and characterized (Edwards *et al.*, 1999). The fenugreek GMGT protein catalyzes the addition of a 1,6-α-galactosyl residue from UDP-D-galactose to primarily the third mannosyl residue from the non-reducing end of a mannohexaose (1,4-β-mannan) acceptor substrate (Edwards *et al.*, 2002). Results from research to determine the feasibility of modifying the man/gal ratio, which is important to the functional properties of galactomannan used in industry, look promising. The man/gal ratio of tobacco galactomannan could be decreased with the overexpression of the fenugreek GMGT (Reid *et al.*, 2003), while *Lotus japonicus* plants transformed with sense, antisense and sense/antisense GMGT constructs produced galactomannans with an increase in the man/gal ratio (Edwards *et al.*, 2004).

In the first study that provided persuasive evidence for the *CSL* hypothesis, Dhugga *et al.* (2004) used transcriptional profiling to identify a candidate gene in guar that is closely related to the *CSLA* family of Arabidopsis and rice. The guar candidate gene encodes an enzyme with mannan synthase activity that localized to the Golgi apparatus when expressed in soybean somatic embryo cells (Dhugga *et al.*, 2004). Heterologous expression of a variety of Arabidopsis *CSL* genes in *Drosophila melanogaster* Schneider 2 cells indicated that *AtCSLA2*, *AtCSLA7* and *AtCSLA9* encode proteins with β-glycan synthase activity. These proteins produced either a mannan homopolymer when supplied with GDP-D-mannose, or a glucomannan heteropolymer when supplied with GDP-D-mannose and GDP-D-glucose (Liepman *et al.*, 2005). Indeed, Liepman *et al.* (2005) have shown for the first time that a single *CSLA* gene encodes a glycan synthase that can produce either a mannan homopolymer or glucomannan heteropolymer depending upon the availability of GDP-D-mannose or GDP-D-glucose. While guar mannan synthase is involved in the deposition of the secondary cell wall thickenings in the guar seed endosperm cell walls (Dhugga *et al.*, 2004), the physiological function of *AtCslA2*, *AtCslA7* and *AtCslA9* remains to be determined.

2.2.3.4 *Mixed-linkage glucan*

MLG is an unbranched hemicellulose composed of 1,3-β-glucosyl and 1,4-β-glucosyl residues in ratios between 1:2 and 1:3 (Carpita, 1996). Using a *Bacillus subtilus* β-D-glucanohydrolase (Staudte *et al.*, 1983) and a maize seedling β-D-glucanase (Huber and Nevins, 1981; Hatfield and Nevins, 1987), the sequence of MLG from elongating tissues and endosperm walls of cereals was shown to be composed of a 3:1 ratio of 1,4-β-linked cellotriose and cellotetraose units that are connected with a 1,3-β-linkage. In addition, interspersed within the cellotriose and cellotetraose repeating units are longer cellodextrins (up to 50 glucosyl residues in length) separated by a single 1,3-β-D-glucosyl residues (Wood *et al.*, 1994; Carpita, 1996). While much research has been conducted on the biosynthesis of MLG in maize (Gibeaut and Carpita, 1993; Buckeridge *et al.*, 1999, 2001; Urbanowicz *et al.*, 2004), the genes that encode the MLG synthases have not been identified.

2.2.4 *Pectic polymers*

Pectic polymers are a major component of the plant primary cell wall. They generally represent one-third of the polysaccharide content and have structural and other physiological functions. Pectins are preponderant in the middle lamella (which serves as the interface of two plant cells), and are involved in processes such as abscission (see Chapter 6) and defense against pathogen attacks, by ensuring cellular adhesion and regulating cell wall porosity (O'Neill *et al.*, 2004; Vorwerk *et al.*, 2004). Pectins are defined as a group of complex polysaccharides all containing 1,4-α-D-linked galacturonic acid (GalA). Three main categories of pectic polysaccharides, homogalacturonan (HG), rhamnogalacturonan I (RG I) and a highly substituted HG named rhamnogalacturonan II (RG II), have been characterized and recently reviewed (Ridley *et al.*, 2001; Willats *et al.*, 2001; O'Neill *et al.*, 2004). Structurally, pectins are subdivided into two groups, related to their backbone characteristics: HG and RG II have in common a 1,4-D-GalA backbone, whereas the RG I backbone contains a 1,4-α-D-GalA-1,2-α-L-Rha disaccharide repeating unit (Figure 2.2).

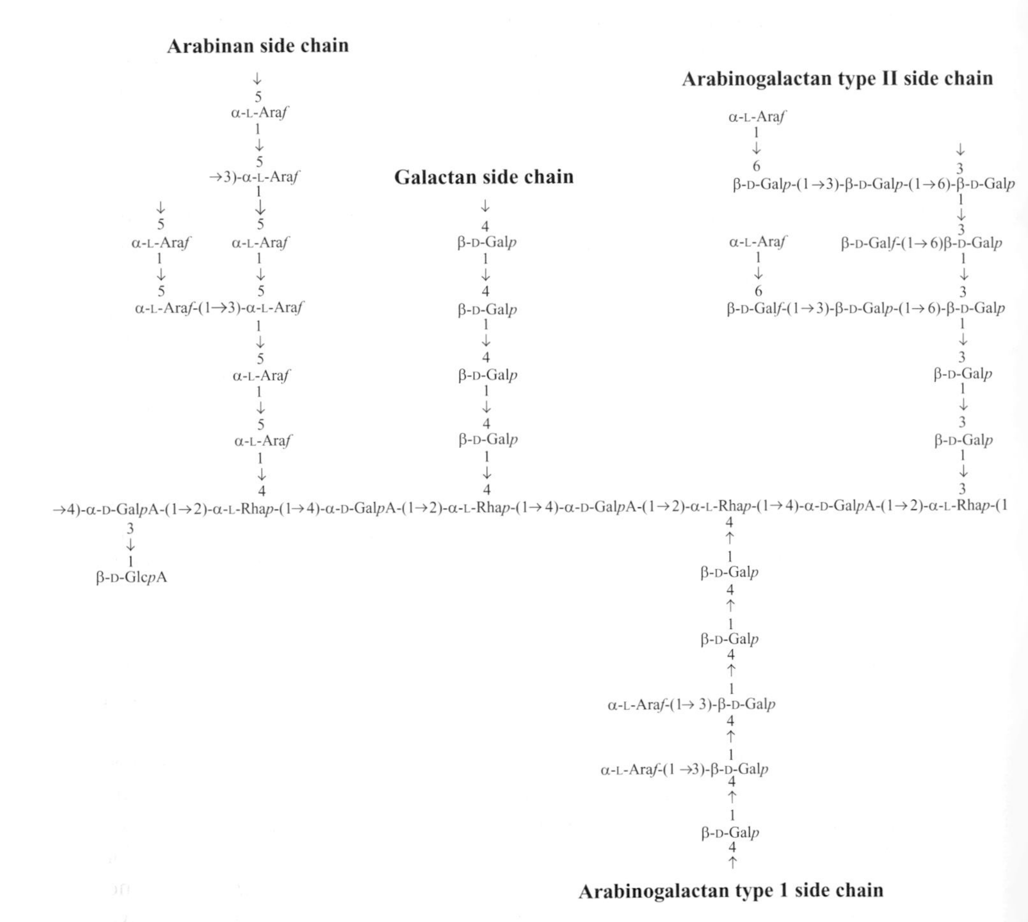

Figure 2.2 Composition and structure of RG I. A backbone of (→4)-α-D-Gal*p*A-(1→2)-α-L-Rha*p*→) is substituted at rhamnosyl residues with various oligosaccharide side chains. Backbone GalA residue can also be decorated by β-D-Glc*p*A, as shown for beet pectins. Antibodies specific to arabinan or galactan side chains have been useful to study heterogeneity of the RG I domain among plants (Willats *et al.*, 2001).

Although these subdivisions are made based on structural characteristics, evidence is available to support the hypothesis that these three subdomains are part of the same macromolecule (Ridley *et al.*, 2001). Immunocytochemical and biochemical studies have shown that, like hemicelluloses, pectins are synthesized in the Golgi apparatus (Moore *et al.*, 1991; Sterling *et al.*, 2001). Because there is a considerable diversity of monosaccharide units and glycosidic linkages that constitute pectic polysaccharides, at least 53 GTs are required for pectin biosynthesis (Mohnen, 1999); however, only three of them have been identified and partially characterized to date (Bouton *et al.*, 2002; Iwai *et al.*, 2002; Egelund *et al.*, 2004).

2.2.4.1 Homogalacturonans

HG is a linear homopolymer composed of 100–200 1,4-α-D-GalA residues (Thibault *et al.*, 1993) and constitutes a large portion of the cell wall pectins. Two different monosaccharide substitutions on HG chains have been described: (1) a xylosyl residue at C-3, thus forming a xylogalacturonan domain, which has been found in low abundance, but is widespread among different plant species (Schols *et al.*, 1995; Renard *et al.*, 1997; Le Goff *et al.*, 2001); (2) an apiose residue at C-2 or C-3 forming an apiogalacturonan domain, restricted to aquatic plants (Hart and Kindel, 1970). A major modification that occurs during the synthesis of HG is the methylesterification of the carboxyl group in 60–70% of the GalA residues (O'Neill *et al.*, 1990). Methylesterification of the HG chains prevents the interaction of two HG molecules via Ca^{2+} cross bridges, thereby providing a control mechanism for cell expansion. A second, less common modification of HG is *O*-acetylation at the C-3 (predominantly) or C-2 position of the GalA residue (Ishii, 1997). While HG methylesterification has been proposed to regulate cell expansion, the role of *O*-acetylation is still unclear.

Recently, HG 1,4-α-D-galacturonyltransferases (HG GalAT) from tobacco (*Nicotiana tabacum*), pea and petunia membranes have been solubilized and characterized enzymatically (Doong and Mohnen, 1998; Scheller *et al.*, 1999, 2001; Akita *et al.*, 2002). *In vitro*, tobacco HG GalAT adds a single GalA residue onto the non-reducing end of acceptor oligogalacturonides with a DP $\geq$ 10; however, this enzyme was not processive as would be expected of a polysaccharide synthase (Scheller *et al.*, 1999). The reason for this observation remains unclear, but it could be a function of the *in vitro* assay (Ridley *et al.*, 2001). However, Akita *et al.* (2002) reported a processive HG GalAT activity present in *Petunia axillaris* pollen tubes. HG GalAT from pea is a Golgi-localized enzyme, with its catalytic site facing the lumen (Sterling *et al.*, 2001), supporting the immunocytochemical studies to localize HG in the Golgi (Staehelin and Moore, 1995). A good candidate gene for HG GalAT has been identified by reverse genetics in *Arabidopsis thaliana*, and is named *QUASIMODO1*. The *quasimodo1* (*qua1*) mutant is altered in a putative membrane-bound protein belonging to the CAZy family GT8, and has both reduced cell adhesion and a 25% decrease in GalA content, attributed to a lower amount of HG (Bouton *et al.*, 2002). Moreover, a 25% decrease in the HG GalAT synthase activity was observed in *qua1* plants (Orfila *et al.*, 2005).

2.2.4.2 *Rhamnogalacturonan I*

RG I is a pectic polysaccharide with a backbone repeat unit composed of the disaccharide 1,4-α-D-GalA-1,2-α-L-Rha (Figure 2.2). Galacturonic acid residues of the RG I backbone are often *O*-acetylated at the C-2 or C-3 position (Komalavilas and Mort, 1989; Lerouge *et al.*, 1993; Rihouey *et al.*, 1995; Pauly and Scheller, 2000). To a lesser extent, C-3 can be substituted by a β-D-GlcA residue (Renard *et al.*, 1999). From 20 to 80% of the C-4 position, rhamnose residues are substituted with four different oligosaccharide chains, all composed of neutral sugars: (1) 1,5-α-L-arabinans branched at the C-2 or C-3 position of the arabinose residue; (2) 1,4-β-D-galactans; (3) type I arabinogalactans composed of a 1,4-β-D-galactan backbone with C-3 arabinosyl or short arabinan side chains; and (4) type II arabinogalactans where 1,3-β-D-galactosyl and 1,6-β-D-galactosyl linkages also occur in the galactan backbone (Brett and Waldron, 1990). Almost nothing is known about RG I biosynthesis and no GT involved in this process has been characterized. However, galactosyltransferase activities making galactan RG I side chains have been measured in flax (*Linum usitatissimum*) and soybean (*Glycine max*) (Peugnet *et al.*, 2001; Konishi *et al.*, 2004).

Arabidopsis mutants, *rhm2* and *mum4*, for an NDP-L-rhamnose synthase have been characterized (Usadel *et al.*, 2004; Western *et al.*, 2004). The NDP-L-rhamnose synthase supplies a required activated sugar for RG I biosynthesis. *rhm2* and *mum4* are phenotypically characterized by a lack of mucilage biosynthesis and secretion around the seeds. Mucilage is principally composed of unsubstituted rhamnogalacturonan, while in these mutants the rhamnose content is decreased, leading to specific developmental defects of the seed coat.

Immunocytochemical studies with probes specific to the galactan (LM5) or arabinan (LM6) side chain (Jones *et al.*, 1997; Willats *et al.*, 1998) have been used to assess the heterogeneity of RG I side chain decorations in the roots, at the tissue as well as the cell level (Willats *et al.*, 2001). Interestingly, arabinan-substituted RG I was predominantly detected in the cell wall of proliferating meristem cells, whereas galactan-substituted RG I was specific to the elongation and differentiation zones of roots (Willats *et al.*, 1999; McCartney *et al.*, 2003). The relationship between the particular functions of those cells and their RG I composition are still unknown.

2.2.4.3 *Rhamnogalacturonan II*

RG II is a low abundance, highly complex pectic polysaccharide of the primary cell wall (O'Neill *et al.*, 1990; Whitcombe *et al.*, 1995) that occurs mainly (>95%) as a dimer (Kobayashi *et al.*, 1996; O'Neill *et al.*, 1996). Its biosynthesis requires at least 12 different glycosyl residues, distributed among four different side chains (named A, B, C and D) that are attached on the C-2 or C-3 position of a 1,4-α-D-GalA backbone. RG II fine structure has been challenging to establish and knowledge about its detailed structure has evolved over the past 10 years (Whitcombe *et al.*, 1995; Vidal *et al.*, 2000). While there is now consensus about the monosaccharide composition of the side chains, the arrangement of the side chains onto the galacturonan backbone is still speculative (O'Neill *et al.*, 2004) (Figure 2.3). The

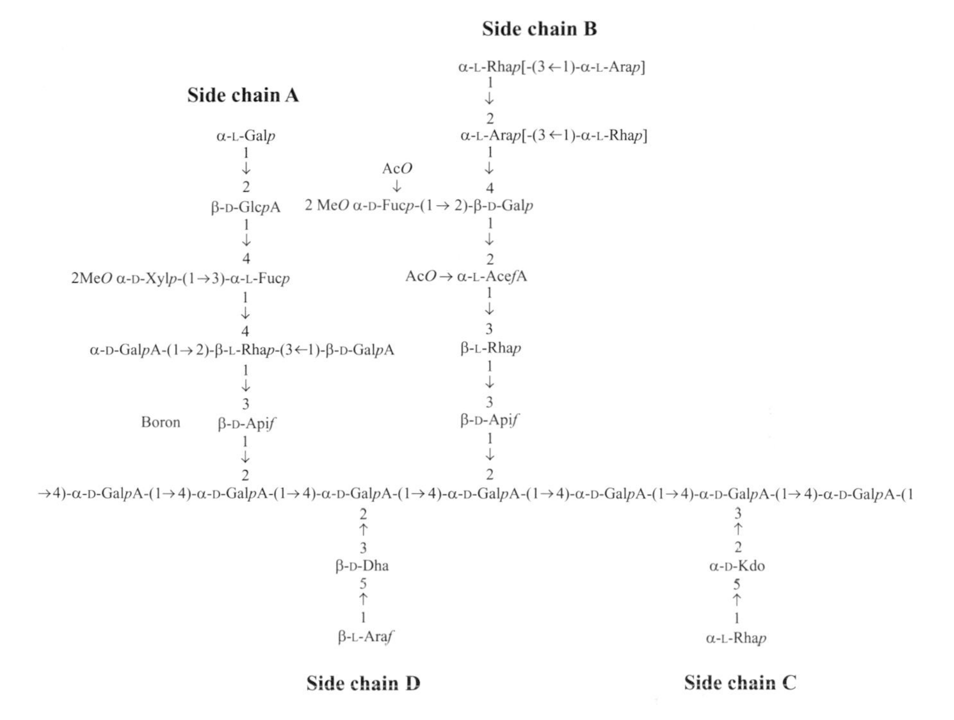

Figure 2.3 Composition and structure of RG II, adapted from O'Neill *et al.* (2004). Structure of the four side chains is detailed and a boron atom is symbolized next to the apiosyl residue (side chain A) involved in RG II dimerization. Rhamnosyl and arabinosyl residues represented in brackets in the side chain B are absent in some RG II, for example, in Arabidopsis. The 2-O-Me α-L-Fuc*p* (side chain B) residue is absent in *mur1* Arabidopsis mutant, and the (α-Gal*p*-(1→2)-β-D-Glc*p*A-1→) glycosyl chain (side chain A) is absent in *nolac-H18* tobacco mutant.

RG II main chains, designated A and B, are composed respectively of an octasaccharide and a nonasaccharide, whereas chains C and D are only a disaccharide of α-L-Rha-1,5-α-D-Kdo and β-L-Ara-1,5-β-D-(Dha). The apiosyl residue that links chain A to the GalA backbone is an essential residue, because it is involved in the homodimerization of RG II molecules by a borate ion (Figure 2.3). RG II structural complexity has been maintained during plant evolution, for example, RG II structure has been conserved among all vascular plants, from the pteridophytes to the angiosperms (Matsunaga *et al.*, 2004). While the reasons for this conservation are not well understood, it is clear that RG II plays an important role in wall function. *mur1* plants, Arabidopsis mutants deficient for GDP-fucose biosynthesis, have altered the RG II structure. Analysis of *mur1* plants confirms that a small alteration of the RG II structure leads to a strong decrease of the borate-dependent dimerization of RG II, and finally to growth and developmental defects (Ishii *et al.*, 2001; O'Neill *et al.*, 2001). Three putative GT activities involved in RG II biosynthesis have been reported to date. Iwai *et al.* (2002) have characterized the alteration of the RG II glucuronyltransferase (*NpGUT1*) in *nolac-H18* tobacco callus. Two putative

RG II xylosyltransferases not previously referenced in CAZy have been identified using an innovative bioinformatics approach (Egelund *et al.*, 2004). Interestingly, *NpGUT1* alteration in *nolac-H18* callus produces a structural modification of RG II, leading to almost a 50% decrease of the RG II dimerization, and finally to an intercellular attachment defect between tobacco cells, offering evidence for the RG II dimerization requirement in cellular adhesion.

2.2.5 Cell wall structural proteins

Primary cell walls are mainly composed of polysaccharides (90% dry weight); however, proteins are also present in the cell wall matrix, carrying out both structural and enzymatic functions. Cell wall structural proteins comprise a polypeptide backbone (accounting for 10% of the molecular mass) heavily substituted by branched carbohydrates. Structural proteins are generally divided into families because of their different polypeptide backbones, as well as their polysaccharide side chain characteristics. The main families are: (1) proteins whose polypeptide backbone is particularly rich in hydroxyproline (extensin or hydroxyproline-rich protein (HRP)), in proline (proline-rich protein (PRP)) or in glycine (glycine-rich protein (GRP)); and (2) proteins whose polypeptide backbone is heavily substituted by arabinose- and galactose-containing polysaccharides, called arabinogalactan proteins (AGPs). Extensins and AGPs are the most abundant structural proteins in the cell wall. They have certain common characteristics, as they both belong to the HRP family and therefore bear carbohydrate *O*-glycosylation of their hydroxyproline and serine (and sometimes threonine) residues, but they have different physicochemical properties, such as their solubility.

Extensin proteins contain a signal peptide responsible for their export into the cell wall, followed by a proline-rich region (Sommer-Knudsen *et al.*, 1998). These proline residues are modified by hydroxylation before the addition of sugars. Most of the genes encoding extensins are regulated during development, induced by wounding or elicitors during fungal infection (Jose and Puigdomenech, 1993). Although numerous genes encoding extensins have been cloned, few have been studied at the protein level (Jose-Estanyol and Puigdomenech, 2000). For example, a detailed analysis of an extensin from maize has revealed that *O*-glycosylation of hydroxyproline by short arabinan side chains accounts for 27% of the mass of the molecule (Kieliszewski and Lamport, 1994). This glycosylation is expected to stabilize the protein in the cell wall. Moreover, extensins are characterized by their ability to self-associate by isodityrosine bridges, leading to their reticulation and therefore the formation of insoluble polymers.

Arabinogalactan proteins are *O*-glycosylated at specific motifs by arabinose- and galactose-containing polysaccharides that constitute up to 95% of the mass of the molecule (Tan *et al.*, 2003; Shimizu *et al.*, 2005). Type II arabinogalactan proteins are characterized by a 1,3-β-D-galactose backbone with 1,6-β-D-galactose side chains bearing arabinosyl decoration (Showalter, 2001). Interestingly, it has been demonstrated that certain AGPs are present in the vicinity of the cell wall, attached to the plasma membrane via a glycosylphosphatidylinositol (GPI) anchor (Youl *et al.*, 1998). This observation, in the light of GPI anchor function in mammals, supports a potential role for AGPs in signaling, which may occur after a controlled release of the AGP by a phospholipase. In contrast to extensin, which bears structural function in the wall by interlocking cell wall polysaccharides and lignin polymers, AGPs are considered to be involved in cell–cell communication.

2.3 Cell wall architecture

Several different structural models have been put forward to explain how the various polysaccharide and protein components interact in the primary wall of plant cells (Keegstra *et al.*, 1973; McCann and Roberts, 1991; Carpita and Gibeaut, 1993; Somerville *et al.*, 2004). One common feature of most of these models is an emphasis on a network of cellulose and hemicellulose, usually xyloglucan, as the important feature that provides the structural integrity to resist the turgor of living plant cells (Plate 1A). In addition to this network, the matrix also contains significant levels of pectic polysaccharides and structural proteins (Plate 1B). In some models, these components are thought to form an independent network.

Because the cellulose and matrix components are synthesized at different locations, assembly is required after the various components arrive in the wall. While little is known about the molecular details of these assembly events, xyloglucan endotransglucosylase/hydrolase (XTH) enzymes play an important role in assembling the cellulose/xyloglucan network. More details on the activity and possible roles of these enzymes are described below.

2.4 Primary cell wall expansion and regulation

2.4.1 Cellulose deposition and orientation

Plant cell turgidity results from the pressure exerted by the vacuole upon the cell wall. This turgor pressure is used by plants for cell expansion and to stand erect. Without cell walls, plant cells would be unable to withstand turgor pressure. An ordered network of cellulose microfibrils embedded in matrix polysaccharides acts as a strong and flexible sheath to resist and harness turgor forces generated in plant cells. Plants use turgor-powered directional cell expansion to control form during development and in response to various environmental stimuli. In elongating cells, the pattern of cellulose microfibril orientation influences the directionality of cell elongation. Because cellulose microfibrils have high tensile strength, cells typically elongate perpendicular to the orientation of cellulose microfibrils in the cell wall. The orientation of cellulose microfibrils, both during and after deposition, is thus an important determinant of plant growth patterns. Actin and tubulin elements of the plant cytoskeleton both seem to influence the deposition patterns of cellulose microfibrils. However, the mechanisms and factors controlling the patterns of deposition and orientation of cellulose microfibrils are incompletely understood (Lloyd and Chan, 2004; Bannigan and Baskin, 2005; Smith and Oppenheimer, 2005).

Recent studies of several Arabidopsis mutants have linked actin organization with cell wall organization. Several *fragile fiber* (*fra*) mutants are characterized by decreased stem strength caused by defects in cell wall deposition. Secondary cell walls in fiber cells and vessel elements of the *fra4/rhd3* mutant, containing a lesion in a putative GTP-binding protein, have decreased thickness compared to their wild-type counterparts. The fine actin arrays observed in wild-type plants were replaced with thicker cables in the *fra4* mutant (Hu *et al.*, 2003). The *fra3* and *fra7* mutants also displayed defects in actin organization, resulting from lesions in a type II inositol polyphosphate 5-phosphatase, and a suppressor of actin domain-containing phosphoinositide phosphatase, respectively (Zhong *et al.*, 2004, 2005). Because actin arrays are thought to act as conduits for vesicular traffic, the cell wall defects observed in mutants affecting actin organization may result from aberrant delivery of critical components of the cell wall biosynthetic machinery (Bannigan and Baskin, 2005).

An interaction between cellulose microfibrils and microtubules has long been inferred because the two often display co-linearity (Smith and Oppenheimer, 2005). In addition, treatment of plants with certain microtubule-disrupting agents (e.g. oryzalin and colchicine) produces similar phenotypes as treatments with inhibitors of cellulose biosynthesis (e.g. isoxaben or 2,6-dichlorobenzonitrile (DCB)) (Burk and Ye, 2002; Himmelspach *et al.*, 2003). Multiple models have been advanced to explain the relationship and interactions between microtubules and cellulose microfibrils. The simplest model is the co-alignment hypothesis, which proposes that cellulose synthase complexes interact with cortical microtubules and that this interaction guides the deposition of cellulose microfibrils. While certain mutant plants (e.g. *bot1/fra2*; Bichet *et al.*, 2001; Burk and Ye, 2002) display defects in both microtubule and cellulose microfibril patterning, there are several lines of evidence indicating that the co-alignment hypothesis does not fully describe the interaction between these components. First, in tip growing cells such as root tips and pollen tubes, microtubules and cellulose microfibrils are not aligned. Studies of the *mor1* and *fra1* mutants illustrate some additional shortcomings of the co-alignment hypothesis. *mor1* is a temperature-sensitive mutant that displays shortened and disordered cortical microtubules when treated at the non-permissive temperature (Whittington *et al.*, 2001). Contrary to the prediction of the co-alignment hypothesis, cellulose deposition and microfibril arrangement are unaffected in *mor1* mutants treated at the non-permissive temperature, indicating that an ordered microtubule scaffold is not necessary for ordered microfibril alignment and deposition (Sugimoto *et al.*, 2003). Further studies of the *mor1* mutant have shown that normal cellulose deposition is restored after treatment with the cellulose synthesis inhibitor DCB in the absence of ordered microtubules (Himmelspach *et al.*, 2003). The phenotype of the *fra1* mutant also seems at odds with the co-alignment hypothesis. The *fra1* mutant has disorganized cellulose microfibrils despite having wild-type patterns of cortical microtubules (Zhong *et al.*, 2002).

More recent models describing the relationship between microtubules and cellulose microfibrils attempt to reconcile the inconsistencies between the phenotypes of mutant plants and the co-alignment hypothesis. The templated incorporation model predicts that plasma membrane integral membrane proteins bind to both the microtubules and cellulose microfibrils. These membrane proteins could serve as a scaffold to guide microfibril deposition in the absence of organized microtubules (Baskin, 2001). An alternative model is the microfibril length regulation model, which proposes that microtubule disruptions result in shortened cellulose microfibrils. The decreased length of the resulting cellulose microfibrils prevents them from being adequately cross-linked by matrix polysaccharides resulting in radial expansion in mutants with shortened microfibrils (Wasteneys, 2004). There is much left to learn about the nature and regulation of the interactions between the cytoskeleton and cellulose microfibrils. As additional factors are discovered, new models to explain the relationship between these elements will most certainly emerge.

2.4.2 Hemicelluloses and their reorganization

2.4.2.1 Expansins

Expansins were initially identified in extracts of cucumber (*Cucumis sativus*) seedling cell walls as proteins that have the ability to induce pH-dependent cell wall extension when applied exogenously to an array of dicots and monocots (McQueen-Mason *et al.*, 1992). Expansins are also involved in a variety of developmental processes, such as embryo development and seed germination (Chen *et al.*, 2001); root hair initiation and growth (Cho and Cosgrove, 2002); differential elongation of gravi-responding roots (Zhang and Hasenstein, 2000); leaf initiation, growth and abscission (see Chapter 6) (Fleming *et al.*, 1997; Cho and Cosgrove, 2000; Pien *et al.*, 2001); stem elongation as an adaptive response to submergence, wounding and gibberellic acid (Lee and Kende, 2001); floral opening and senescence (Gookin *et al.*, 2003); pollen tube invasion of the maternal material (Cosgrove *et al.*, 1997); and fruit ripening (see Chapter 8) (Brummell *et al.*, 1999).

Many expansin and expansin-like genes have been identified in diverse plants (for reviews, see Lee *et al.*, 2001, 2003; Cosgrove *et al.*, 2002; and http://www.bio.psu.edu/expansins/). Indeed, the size and complexity of the expansin superfamily necessitated the development of a nomenclature for expansin genes and proteins (Kende *et al.*, 2004). The EXPANSIN A family (EXPA, formerly α-expansin family) includes expansins that are phylogenetically related to the cucumber expansins originally discovered by McQueen-Mason *et al.* (1992), while the EXPANSIN B family (EXPB, formerly the β-expansin family) includes expansins phylogenetically related to the group-1 pollen allergens found in grasses. Members of the EXPANSIN-LIKE A (EXLA, formerly the Expansin-like family) and EXPANSIN-LIKE B (EXLB, formerly the Expansin-related family) families have a significant divergence in sequence with respect to EXPA and EXPB, and have not been shown to have expansin activity (Kende *et al.*, 2004). All expansin and expansin-like proteins contain an N-terminal endoplasmic reticulum/Golgi signal peptide, a putative N-terminal catalytic domain (domain I) and a C-terminal putative binding domain (domain II) (Cosgrove, 2000). Domain I is distantly related to the catalytic domain of hydrolases grouped in the CAZy family GH45 (Cosgrove, 1997); however, expansins appear to have neither hydrolytic activity

(McQueen-Mason and Cosgrove, 1995) nor transglycosylation activity (McQueen-Mason *et al.*, 1993). Domain II is distantly related to group-2 grass pollen allergens that have unknown functions (Cosgrove, 2000). While the precise mechanism of expansin activity is unknown, the proteins are thought to increase cell wall extensibility by disrupting the hydrogen bonds between cellulose and hemicelluloses (McQueen-Mason *et al.*, 1992; McQueen-Mason and Cosgrove, 1995).

2.4.2.2 Xyloglucan endotransglucosylase/hydrolases

The XTH family is composed of enzymes that have xyloglucan endotransglucosylase (XET) and/or xyloglucan endohydrolase (XEH) activities (for a detailed review of the nomenclature, see Rose *et al.*, 2002). XET activity hydrolyzes a molecule of xyloglucan into two fragments, and then grafts one of the fragments onto a different molecule of xyloglucan, if present (Baydoun and Fry, 1989; McDougall and Fry, 1991; Nishitani and Tominaga, 1991, 1992). Some members of the XTH family were shown to have XEH activity, namely, the ability to hydrolyze xyloglucan when no suitable acceptor xyloglucan molecules were present (Fanutti *et al.*, 1993; Maclachlan and Brady, 1994; Matsumoto *et al.*, 1997; Tabuchi *et al.*, 1997; Schröder *et al.*, 1998); however, other XTHs lacked this activity (Nishitani and Tominaga, 1992).

XTH genes are encoded by large multigene families and are classified as members of the CAZy family GH16 (Henrissat and Davies, 2000). XTH proteins typically contain an N-terminal signal peptide that targets the protein for secretion into the apoplast (Yokoyama and Nishitani, 2001) and a signature amino acid motif DEIDFEFLG, which is predicted to function as the catalytic site for both hydrolase and transferase activities (Okazawa *et al.*, 1993; Campbell and Braam, 1998). *XTH* genes have been found in an array of dicotyledonous and monocotyledonous plants, and they often have either tissue- or organ-specific expression patterns (Rose *et al.*, 2002; Strohmeier *et al.*, 2004; Yokoyama *et al.*, 2004).

There are 33 *XTH* genes in Arabidopsis, in which xyloglucan comprises over 20% of the primary cell wall carbohydrates (Rose *et al.*, 2002). Paradoxically, there are 29 *XTH* genes in rice and 22 in barley, yet xyloglucan constitutes only 4–8% of the total carbohydrate content of primary cell walls of these species (Strohmeier *et al.*, 2004; Yokoyama *et al.*, 2004). While xyloglucan is the only known substrate for all XTHs characterized to date (Rose *et al.*, 2002), none of the rice or barley *XTH* genes have been shown to encode proteins that have either XET or XEH activities (Strohmeier *et al.*, 2004; Yokoyama *et al.*, 2004). One intriguing possibility is that some of the rice and barley *XTH*s encode proteins that are involved in the transglycosylation of hemicelluloses other than xyloglucan (Strohmeier *et al.*, 2004; Yokoyama *et al.*, 2004). Indeed, there is one report of mannan transglycosylase activity from tomato cell wall extracts (Schröder *et al.*, 2004). However, it is unclear if the mannan transglycosylase is a member of the XTH family or is a member of a new family of transglycosylases.

In vivo XTHs are thought to have significant role in the assembly and restructuring of the cellulose–xyloglucan network during cell expansion (Rose *et al.*, 2002;

Yokoyama *et al.*, 2004). Theoretically, cell wall extensibility would be increased by XTH hydrolysis of the cross-linking xyloglucan, which would allow the distance between the cellulose microfibrils to increase due to the hydrostatic push of the protoplasm. As the cell wall expands, the XET activity of XTH would be involved in the restoration of the cellulose–xyloglucan network (Baydoun and Fry, 1989; Smith and Fry, 1991; Fry *et al.*, 1992; Rose *et al.*, 2002; Vissenberg *et al.*, 2005). There is a strong positive correlation between the growth rate of the cell and XET activity (see Rose *et al.*, 2002; Vissenberg *et al.*, 2005 for reviews), but conclusive evidence for this model is still lacking (Yokoyama *et al.*, 2004). While both XTHs and expansins are thought to be involved in cell wall extensibility, the exact role of each in the process remains to be determined.

2.4.3 Pectins involved in cell wall structure and intercellular adhesion

Pectins are essential macromolecules for plant cells, which occur at the junction zone of two cells, providing cellular adhesion in a structure shared between two adjacent cells called the middle lamella. Mainly composed of anionic monomers, pectic polymers provide a hydrated matrix around the cellulose/hemicellulose network, required not only at the cell level to enable enzymatic remodeling of the cell wall, but also at the plant level to provide the apoplast, a major pathway for water in the plant. Moreover, pectic polysaccharides are now considered important for their structural support to the cell wall. For example, compensation of cellulose deficiency in an isoxaben-habituated cell culture, by pectic polysaccharides, argues for their structural load-bearing ability (Manfield *et al.*, 2004). In cell wall organization models, pectins have been described as forming a network independent from the cellulose/hemicellulose network (Carpita and Gibeaut, 1993). However, the independence of the pectin network in the cell wall has been challenged by evidence of covalent linkages of the pectin with xyloglucan, a hemicellulosic polymer, in pea and rose (*Rosa* sp. '*Paul's Scarlet*') cell culture (Thompson and Fry, 2000; Cumming *et al.*, 2005).

All three major pectic domains are essential for plant cell wall expansion. First, xylogalacturonan, a discrete substituted HG, has been shown to be involved in plant cell separation, providing insight on cellular adhesion/separation at the molecular level (Willats *et al.*, 2004). Methylesterified domains of HG constitute another essential domain of pectin, which is a substrate for specific pectin methylesterases in the cell wall. Demethylesterification by these pectin methylesterases results in the release of strongly anionic molecules that can either prevent aggregation because of charge repulsion, or be specifically cross-linked by calcium bridges in a structure known as an 'egg-box', providing a regulation mechanism for cell expansion (Micheli, 2001).

One example of the involvement of HG in cell wall cohesion is shown by a study of the *colorless non-ripening* (*cnr*) tomato mutant. In the *cnr* mutant, an altered cell-to-cell contact phenotype was reported, due to decreased calcium-binding capacities of HG and RG I arabinan side chains (Orfila *et al.*, 2001, 2002). Another study pointed out that RG I arabinan side chains play a role in the control of pectin fluidity, by preventing HG calcium-dependent cross bridges (Jones *et al.*, 2003). These results indicate that both RG I and HG are required for cell wall cohesion, and that the RG I chain branching seems to determine the regulation of the calcium-dependent cohesion of the wall. RG I chain branch specificities have been correlated to not only the cellular adhesion process but also the cell expansion process. Indeed, Willats *et al.* (1999) has shown, using carrot cell cultures, that arabinan is present in proliferating cells in contrast to galactan in elongating cells. More evidence for RG I involvement in cell–cell cohesion is provided by the *nolac-H14* tobacco callus mutant, deficient in the arabinan branching of the RG I arabinan chain, which results in a cellular adhesion defect (Iwai *et al.*, 2001). Modification of RG I arabinan chains also occurs during ripening (see Chapter 8) and has been proposed to play a part in cell wall stiffness and adhesion, by preventing enzymatic HG hydrolysis (Pena and Carpita, 2004). Third, RG II molecules self assemble and are found in *muro*, almost exclusively in a borate cross-linked dimer structure, which is required for plant growth and development (O'Neill *et al.*, 2001); more specifically, to determine cell wall porosity and ensure intercellular adhesion (Iwai *et al.*, 2002). These three examples point out the role of pectic polysaccharides in the cell cohesion process. Other important biological activities related to pectin structure and organization are discussed in the literature, and without a doubt many more are waiting to be understood (Ridley *et al.*, 2001; Willats *et al.*, 2001).

2.5 Concluding remarks

As summarized above, and in more detail in other reviews, considerable progress has been made in defining the structure of various cell wall components. However, more progress is needed in defining the organization of these components within the wall matrix. Studies on the biosynthesis and assembly of the cell wall components are still in their early stages and only a few of the genes and proteins required for wall biosynthesis have been identified and characterized in detail. Finally, the regulation of wall synthesis and assembly is an important topic that needs more attention in order to understand the molecular details of plant growth and differentiation.

Acknowledgements

We thank Ms. Katherine Krive for excellent editorial assistance. Cell wall research in the laboratory of K.K. is supported in part by a grant from the Plant Genome Program at the National Science Foundation and in part by a grant from the US Department of Energy.

References

AGI (2000) Analysis of the genome sequence of the flowering plant *Arabidopsis thaliana*. *Nature* **408**, 796–815.

Akita, K., Ishimizu, T., Tsukamoto, T., Ando, T. and Hase, S. (2002) Successive glycosyltransfer activity and enzymatic characterization of pectic polygalacturonate 4-α-galacturunosyltransferase solubilized from pollen tubes of *Petunia axillaris* using pyridylaminated oligogalacturonates as substrates. *Plant Physiology* **130**, 374–379.

Arioli, T., Peng, L., Betzner, A.S., Burn, J., Wittke, W., Herth, W., Camilleri, C., Hofte, H., Plazinski, J., Birch, R., Cork, A., Glover, J., Redmond, J. and Williamson, R.E. (1998) Molecular analysis of cellulose biosynthesis in *Arabidopsis*. *Science* **279**, 717–720.

Bacic, A., Harris, P.J. and Stone, B.A. (1988) Structure and function of plant cell walls. In: *The Biochemistry of Plants: A Comprehensive Treatise* (ed. Preiss, J.), pp. 297–371. Academic Press, Inc., New York, NY.

Bailey, R.W. and Hassid, W.Z. (1966) Xylan synthesis from uridine-diphosphate-D-xylose by particulate preparations from immature corncobs. *Proceedings of the National Academy of Sciences of the United States of America* **56**, 1586–1593.

Bannigan, A. and Baskin, T.I. (2005) Directional cell expansion – turning toward actin. *Current Opinion in Plant Biology* **8**, 619–624.

Baskin, T.I. (2001) On the alignment of cellulose microfibrils by cortical microtubules: a review and a model. *Protoplasma* **215**, 150–171.

Bauer, W.D., Talmadge, K.W., Keegstra, K. and Albersheim, P. (1973) The structure of plant cell walls II. The hemicellulose of the walls of suspension-cultured sycamore cells. *Plant Physiology* **51**, 174–187.

Baydoun, E.A. and Fry, S.C. (1989) *In vivo* degradation and extracellular polymer-binding of xyloglucan nonasaccharide, a natural anti-auxin. *Journal of Plant Physiology* **134**, 453–459.

Baydoun, E.A., Waldron, K.W. and Brett, C.T. (1989) The interaction of xylosyltransferase and glucuronyltransferase involved in glucuronoxylan synthesis in pea (*Pisum sativum*) epicotyls. *Biochemistry Journal* **257**, 853–858.

Ben-Arie, R., Ordin, L. and Kindinger, J.I. (1973) A cell-free xylan synthesizing enzyme system from *Avena sativa*. *Plant and Cell Physiology* **14**, 427–434.

Bichet, A., Desnos, T., Turner, S.R., Grandjean, O. and Hofte, H. (2001) *BOTERO1* is required for normal orientation of cortical microtubules and anisotropic cell expansion in *Arabidopsis*. *Plant Journal* **25**, 137–148.

Bolwell, G.P. (1986) Microsomal arabinosylation of polysaccharide and elicitor-induced carbohydrate-binding glycoprotein in French bean. *Phytochemistry* **25**, 1807–1813.

Bolwell, G.P. and Northcote, D.H. (1983) Arabinan synthase and xylan synthase activities of *Phaseolus vulgaris*: subcellular localization and possible mechanism of action. *Biochemical Journal* **210**, 497–507.

Bonin, C.P., Potter, I., Vanzin, G.F. and Reiter, W.D. (1997) The *MUR1* gene of *Arabidopsis thaliana* encodes an isoform of GDP-D-mannose-4,6-dehydratase, catalyzing the first step in the de novo synthesis of GDP-L-fucose. *Proceedings of the National Academy of Sciences of the United States of America* **94**, 2085–2090.

Bouton, S., Leboeuf, E., Mouille, G., Leydecker, M.T., Talbotec, J., Granier, F., Lahaye, M., Hofte, H. and Truong, H.N. (2002) *QUASIMODO1* encodes a putative membrane-bound glycosyltransferase required for normal pectin synthesis and cell adhesion in *Arabidopsis*. *Plant Cell* **14**, 2577–2590.

Brett, C. and Waldron, K. (1990) *Physiology and Biochemistry of Plant Cell Walls*. Unwin Hyman, London.

Brown, D.M., Zeef, L.A.H., Ellis, J. and Turner, S.R. (2005) Identification of novel genes in Arabidopsis involved in secondary cell wall formation using expression profiling and reverse genetics. *Plant Cell* **17**, 2281–2295.

Brummell, D.A., Harpster, M.H., Civello, P.M., Palys, J.M., Bennett, A.B. and Dunsmuir, P. (1999) Modification of expansin protein abundance in tomato fruit alters softening and cell wall polymer metabolism during ripening. *Plant Cell* **11**, 2203–2216.

Buckeridge, M.S., Vergara, C.E. and Carpita, N.C. (1999) The mechanism of synthesis of a mixed-linkage (1,3),(1,4)β-D-glucan in maize. Evidence for multiple sites of glucosyl transfer in the synthase complex. *Plant Physiology* **120**, 1105–1116.

Buckeridge, M.S., Vergara, C.E. and Carpita, N.C. (2001) Insight into multi-site mechanisms of glycosyl transfer in (1,4)-ß-glycans provided by the cereal mixed-linkage (1,3),(1,4)-ß-glucan synthase. *Phytochemistry* **57**, 1045–1053.

Burk, D.H. and Ye, Z.-H. (2002) Alteration of oriented deposition of cellulose microfibrils by mutation of a katanin-like microtubule-severing protein. *Plant Cell* **14**, 2145–2160.

Campbell, P. and Braam, J. (1998) Co- and/or post-translational modifications are critical for TCH4 XET activity. *The Plant Journal for Cell and Molecular Biology* **15**, 553–561.

Carpita, N.C. (1996) Structure and biogenesis of the cell walls of grasses. *Annual Review of Plant Physiology and Plant Molecular Biology* **47**, 445–476.

Carpita, N.C. and Gibeaut, D.M. (1993) Structural models of primary cell walls in flowering plants: consistency of molecular structure with the physical properties of the walls during growth. *Plant Journal* **3**, 1–30.

Carpita, N.C. and McCann, M.C. (2000) The cell wall. In: *Biochemistry & Molecular Biology of Plants* (eds Buchanan, B. Gruissem, W. and Jones, R.), pp. 52–108. American Society of Plant Physiologists, Rockville.

Chen, F., Dahal, P. and Bradford, K.J. (2001) Two tomato expansin genes show divergent expression and localization in embryos during seed development and germination. *Plant Physiology* **127**, 928–936.

Cho, H.T. and Cosgrove, D.J. (2000) Altered expression of expansin modulates leaf growth and pedicel abscission in *Arabidopsis thaliana. Proceedings of the National Academy of Sciences of the United States of America* **97**, 9783–9788.

Cho, H.T. and Cosgrove, D.J. (2002) Regulation of root hair initiation and expansin gene expression in Arabidopsis. *Plant Cell* **14**, 3237–3253.

Colombani, A., Djerbi, S., Bessueille, L., Blomqvist, K., Ohlsson, A., Berglund, T., Teeri, T.T. and Bulone, V. (2004) *In vitro* synthesis of $(1\rightarrow3)$-β-D-glucan (callose) and cellulose by detergent extracts of membranes from cell suspension cultures of hybrid aspen. *Cellulose* **11**, 313–327.

Cosgrove, D.J. (1997) Relaxation in a high-stress environment: the molecular bases of extensible cell walls and cell enlargement. *Plant Cell* **9**, 1031–1041.

Cosgrove, D.J. (2000) Loosening of plant cell walls by expansins. *Nature* **407**, 321–326.

Cosgrove, D.J., Bedinger, P. and Durachko, D.M. (1997) Group I allergens of grass pollen as cell wall-loosening agents. *Proceedings of the National Academy of Sciences of the United States of America* **94**, 6559–6564.

Cosgrove, D.J., Li, L.C., Cho, H.-T., Hoffmann-Benning, S., Moore, R.C. and Blecker, D. (2002) The growing world of expansins. *Plant Cell Physiology* **43**, 1436–1444.

Coutinho, P.M., Deleury, E., Davies, G.J. and Henrissat, B. (2003) An evolving hierarchical family classification for glycosyltransferases. *Journal of Molecular Biology* **328**, 307–317.

Cumming, C.M., Rizkallah, H.D., McKendrick, K.A., Abdel-Massih, R.M., Baydoun, E.A. and Brett, C.T. (2005) Biosynthesis and cell-wall deposition of a pectin-xyloglucan complex in pea. *Planta* **222**, 546–555.

Dhugga, K.S., Barreiro, R., Whitten, B., Hazebroek, J., Randhawa, G., Dolan, M., Kinney, A., Tomes, D., Nichols, S. and Anderson, P. (2004) Guar seed beta-mannan synthase is a member of the cellulose synthase super gene family. *Science* **16**, 363–366.

Doblin, M.S., Kurek, I., Jacob-Wilk, D. and Delmer, D.P. (2002) Cellulose biosynthesis in plants: from genes to rosettes. *Plant & Cell Physiology* **43**, 1407–1420.

Doblin, M.S., Vergara, C.E., Read, S., Newbigin, E. and Bacic, A. (2003) Plant cell wall biosynthesis: making the bricks. In: *The Plant Cell Wall* (ed. Rose, J.K.C.), pp. 183–222. Blackwell Publishing Ltd., Oxford.

Doong, R.L. and Mohnen, D. (1998) Solubilization and characterization of a galacturonosyltransferase that synthesizes the pectic polysaccharide homogalacturonan. *Plant Journal* **13**, 363–374.

Ebringerová, A. and Heinze, T. (2000) Xylan and xylan derivatives – biopolymers with valuable properties, 1. Naturally occuring xylans structures, isolation procedures and properties. *Macromolecular Rapid Communications* **21**, 542–556.

Edwards, M.E., Choo, T.-S., Dickson, C.A., Scott, C., Gidley, M.J. and Reid, J.S.G. (2004) The seeds of lotus japonicus lines transformed with sense, antisense, and sense/antisense galactomannan galactosyltransferase constructs have structurally altered galactomannans in their endosperm cell walls. *Plant Physiology* **134**, 1153–1162.

Edwards, M.E., Dickson, C.A., Chengappa, S., Sidebottom, C., Gidley, M.J. and Reid, J.S. (1999) Molecular characterisation of a membrane-bound galactosyltransferase of plant cell wall matrix polysaccharide biosynthesis. *Plant Journal* **19**, 691–697.

Edwards, M.E., Marshall, E., Gidley, M.J. and Reid, J.S.G. (2002) Transfer specificity of detergent-solubilized fenugreek galactomannan galactosyltransferase. *Plant Physiology* **129**, 1391–1397.

Egelund, J., Skjot, M., Geshi, N., Ulvskov, P. and Petersen, B.L. (2004) A complementary bioinformatics approach to identify potential plant cell wall glycosyltransferase-encoding genes. *Plant Physiology* **136**, 2609–2620.

Fagard, M., Desnos, T., Desprez, T., Goubet, F., Refregier, G., Mouille, G., McCann, M., Rayon, C., Vernhettes, S. and Hofte, H. (2000) *PROCUSTE1* encodes a cellulose synthase required for normal cell elongation specifically in roots and dark-grown hypocotyls of *Arabidopsis*. *Plant Cell* **12**, 2409–2424.

Faik, A., Bar-Peled, M., DeRocher, A.E., Zeng, W., Perrin, R.M., Wilkerson, C.G., Raikhel, N.V. and Keegstra, K. (2000) Biochemical characterization and molecular cloning of an α-1,2-fucosyltransferase that catalyzes the last step of cell wall xyloglucan biosynthesis in pea. *Journal of Biological Chemistry* **275**, 15082–15089.

Faik, A., Price, N.J., Raikhel, N.V. and Keegstra, K. (2002) An *Arabidopsis* gene encoding an α-xylosyltransferase involved in xyloglucan biosynthesis. *Proceedings of the National Academy of Sciences of the United States of America* **99**, 7797–7802.

Fanutti, C., Gidley, M.J. and Reid, J.S. (1993) Action of a pure xyloglucan endo-transglycosylase (formerly called xyloglucan-specific endo (1→4)-β-D-glucanase) from the cotyledons of germinated nasturtium seeds. *Plant Journal* **3**, 691–700.

Fleming, A.J., McQueen-Mason, S., Mandel, T. and Kuhlemeier, C. (1997) Induction of leaf primordia by the cell wall protein expansin. *Science* **276**, 1415–1418.

Fry, S.C. (1989) The structure and functions of xyloglucan. *Journal of Experimental Botany* **40**, 1–11.

Fry, S.C., Smith, R.C., Renwick, K.F., Martin, D.J., Hodge, S.K. and Matthews, K.J. (1992) Xyloglucan endotransglycosylase, a new wall-loosening enzyme activity from plants. *The Biochemical Journal* **282**(Pt 3), 821–828.

Fry, S.C., York, W.S., Albersheim, P., Darvill, A., Hayashi, T., Joseleau, J.P., Kato, P., Kato, Y., Lorences, E.P., Maclachlan, G., McNeil, M., Mort, A., Reid, J.G., Seitz, H.U., Selvendran, R.R., Voragen, A.G. and White, A.R. (1993) An unambigous nomenclature for xyloglucan-derived oligosaccharides. *Physiologia Plantarum* **89**, 1–3.

Gibeaut, D.M. and Carpita, N. (1993) Synthesis of (1–3), (1–4)-ß-D-glucan in the Golgi apparatus of maize coleoptiles. *Proceedings of the National Academy of Sciences of the United States of America* **90**, 3850–3854.

Gillmor, C.S., Lukowitz, W., Brininstool, G., Sedbrook, J.C., Hamann, T., Poindexter, P. and Somerville, C. (2005) Glycosylphosphatidylinositol-anchored proteins are required for cell wall synthesis and morphogenesis in Arabidopsis. *Plant Cell* **17**, 1128–1140.

Gookin, T.E., Hunter, D.A. and Reid, M.S. (2003) Temporal analysis of alpha and beta-expansin expression during floral opening and senescence. *Plant Science* **164**, 769–781.

Hamann, T., Osborne, E., Youngs, H., Misson, J., Nussaume, L. and Somerville, C. (2004) Global expression analysis of *CESA* and *CSL* genes in Arabidopsis. *Cellulose* **11**, 279–286.

Hart, D.A. and Kindel, P.K. (1970) Isolation and partial characterization of apiogalacturonans from the cell wall of Lemna minor. *Biochemical Journal* **116**, 569–579.

Hatfield, R.D. and Nevins, D.J. (1987) Hydrolytic activity and substrate specificity of an endoglucanase from *Zea mays* seedling cell walls. *Plant Physiology* **83**, 203–207.

Hayashi, T. (1989) Xyloglucans in the primary cell wall. *Annual Review of Plant Physiology and Plant Molecular Biology* **40**, 139–168.

Hayashi, T. and Matsuda, K. (1981) Biosynthesis of xyloglucan in suspension-cultured soybean cells: synthesis of xyloglucan from UDP-glucose and UDP-xylose in the cell-free system. *Plant Cell Physiology* **22**, 517–523.

Hayashi, T., Ogawa, K. and Mitsuishi, Y. (1994a) Characterization of the adsorption of xyloglucan to cellulose. *Plant and Cell Physiology* **35**, 1199–1205.

Hayashi, T., Takeda, T., Ogawa, K. and Mitsuishi, Y. (1994b) Effects of the degree of polymerization on the binding of xyloglucans to cellulose. *Plant and Cell Physiology* **35**, 893–899.

Hazen, S.P., Scott-Craig, J.S. and Walton, J.D. (2002) Cellulose synthase-like genes of rice. *Plant Physiology* **128**, 336–340.

Henrissat, B. and Davies, G.J. (2000) Glycoside hydrolases and glycosyltransferases: families, modules, and implications for genomics. *Plant Physiology* **124**, 1515–1519.

Himmelspach, R., Williamson, R.E. and Wasteneys, G.O. (2003) Cellulose microfibril alignment recovers from DCB-induced disruption despite microtubule disorganization. *Plant Journal* **36**, 565–575.

Hoffman, M., Jia, Z., Pena, M.J., Cash, M., Harper, A., Blackburn 2nd, A.R., Darvill, A. and York, W.S. (2005) Structural analysis of xyloglucans in the primary cell walls of plants in the subclass *Asteridae*. *Carbohydrate Research* **340**, 1826–1840.

Hu, Y., Zhong, R., Morrison, W.H. and Ye, Z.-H. (2003) The Arabidopsis *RHD3* gene is required for cell wall biosynthesis and actin organization. *Planta* **217**, 912–921.

Huber, D.J. and Nevins, D.J. (1981) Partial purification of endo- and exo-β-glucanase enzymes from *Zea mays* L. seedlings and their involvement in cell wall autohydrolysis. *Planta* **151**, 206–214.

Ishii, T. (1997) O-acetylated oligosaccharides from pectins of potato tuber cell walls. *Plant Physiology* **113**, 1265–1272.

Ishii, T., Matsunaga, T. and Hayashi, N. (2001) Formation of rhamnogalacturonan II-borate dimer in pectin determines cell wall thickness of pumpkin tissue. *Plant Physiology* **126**, 1698–1705.

Iwai, H., Ishii, T. and Satoh, S. (2001) Absence of arabinan in the side chains of the pectic polysaccharides strongly associated with cell walls of *Nicotiana plumbaginifolia* non-organogenic callus with loosely attached constituent cells. *Planta* **213**, 907–915.

Iwai, H., Masaoka, N., Ishii, T. and Satoh, S. (2002) A pectin glucuronyltransferase gene is essential for intercellular attachment in the plant meristem. *Proceedings of the National Academy of Sciences of the United States of America* **99**, 16319–16324.

Jacobs, A.K., Lipka, V., Burton, R.A., Panstruga, R., Strizhov, N., Schulze-Lefert, P. and Fincher, G.B. (2003) An Arabidopsis callose synthase, GSL5, is required for wound and papillary callose formation. *Plant Cell* **15**, 2503–2513.

Jones, L., Milne, J.L., Ashford, D. and McQueen-Mason, S.J. (2003) Cell wall arabinan is essential for guard cell function. *Proceedings of the National Academy of Sciences of the United States of America* **100**, 11783–11788.

Jones, L., Seymour, G.B. and Knox, J.P. (1997) Localization of pectic galactan in tomato cell walls using a monoclonal antibody specific to $(1\rightarrow4)$-β-D-Galactan. *Plant Physiology* **113**, 1405–1412.

Jose, M. and Puigdomenech, P. (1993) Structure and expression of genes coding for structural proteins of the plant cell wall. *New Phytologist* **125**, 259–282.

Jose-Estanyol, M. and Puigdomenech, P. (2000) Plant cell wall glycoproteins and their genes. *Plant Physiology and Biochemistry* **38**, 97–108.

Keegstra, K. and Raikhel, N. (2001) Plant glycosyltransferases. *Current Opinion in Plant Biology* **4**, 219–224.

Keegstra, K., Talmadge, K.W., Bauer, W.D. and Albersheim, P. (1973) The structure of plant cell walls. III. A model of the walls of suspension-cultured sycamore cells based on the interconnections of the macromolecular components. *Plant Physiology* **51**, 188–196.

Kende, H., Bradford, K., Brummell, D., Cho, H.-T., Cosgrove, D., Fleming, A., Gehring, C., Lee, Y., Mcqueen-mason, S., Rose, J. and Voesenek, L. (2004) Nomenclature for members of the expansin superfamily of genes and proteins. *Plant Molecular Biology* **55**, 311–314.

Kiefer, L.L., York, W.S., Darvill, A.G. and Albersheim, P. (1989) Xyloglucan isolated from suspension-cultured sycamore cell walls is O-acetylated. *Phytochemistry* **28**, 2105–2107.

Kieliszewski, M.J. and Lamport, D.T.A. (1994) Extensin: repetitive motifs, functional sites, post-translational codes, and phylogeny. *Plant Journal* **5**, 157–172.

Kobayashi, M., Matoh, T. and Azuma, J. (1996) Two chains of rhamnogalacturonan II are cross-linked by borate-diol ester bonds in higher plant cell walls. *Plant Physiology* **110**, 1017–1020.

Komalavilas, P. and Mort, A.J. (1989) The acetylation of O-3 of galacturonic acid in the rhamnose-rich portion of pectins. *Carbohydrate Research* **189**, 261–272.

Konishi, T., Mitome, T., Hatsushika, H., Haque, M.A., Kotake, T. and Tsumuraya, Y. (2004) Biosynthesis of pectic galactan by membrane-bound galactosyltransferase from soybean (Glycine max Merr) seedlings. *Planta* **218**, 833–842.

Kudlicka, K. and Brown, R.M. (1997) Cellulose and callose biosynthesis in higher plants I. Solubilization and separation of (1→3)- and (1→4)-β-glucan synthase activities from mung bean. *Plant Physiology* **115**, 643–656.

Kurek, I., Kawagoe, Y., Jacob-Wilk, D. and Delmer, D.P. (2002) Dimerization of cotton fiber cellulose synthase catalytic subunits occurs via oxidation of the zinc-binding domains. *Proceedings of the National Academy of Sciences of the United States of America* **99**, 11109–11114.

Kuroyama, H. and Tsumuraya, Y. (2001) A xylosyltransferase that synthesizes β-(1,4)-xylans in wheat (*Triticum aestivum* L.) seedlings. *Planta* **213**, 231–240.

Lai-Kee-Him, J., Chanzy, H., Muller, M., Putaux, J.-L., Imai, T. and Bulone, V. (2002) *In vitro versus in vivo* cellulose microfibrils from plant primary wall synthases: structural differences. *Journal of Biological Chemistry* **277**, 36931–36939.

Lane, D.R., Wiedemeier, A., Peng, J., Hofte, H., Vernhettes, S., Desprez, T., Hocart, C.H., Birch, R.J., Baskin, T.I., Burn, J.E., Arioli, T., Betzner, A.S. and Williamson, R.E. (2001) Temperature-sensitive alleles of *RSW2* link the KORRIGAN endo-1,4-β-glucanase to cellulose synthesis in Arabidopsis. *Plant Physiology* **126**, 278–288.

Le Goff, A., Renard, C.M.G.C., Bonnin, E. and Thibault, J.F. (2001) Extraction, purification and chemical characterisation of xylogalacturonans from pea hulls. *Carbohydrate Polymers* **45**, 325–334.

Lee, Y., Choi, D. and Kende, H. (2001) Expansins: ever-expanding numbers and functions. *Current Opinion in Plant Biology* **4**, 527–532.

Lee, Y. and Kende, H. (2001) Expression of beta-expansins is correlated with internodal elongation in deepwater rice. *Plant Physiology* **127**, 645–654.

Lerouge, P., O'Neill, M.A., Darvill, A.G. and Albersheim, P. (1993) Structural characterization of endo-glycanase-generated oligoglycosyl side chains of rhamnogalacturonan I. *Carbohydrate Research* **243**, 359–371.

Levy, S., Maclachlan, G. and Staehelin, L.A. (1997) Xyloglucan sidechains modulate binding to cellulose during in vitro binding assays as predicted by conformational dynamics simulations. *Plant Journal* **11**, 373–386.

Li, X., Cordero, I., Caplan, J., Molhoj, M. and Reiter, W.D. (2004) Molecular analysis of 10 coding regions from *Arabidopsis* that are homologous to the MUR3 xyloglucan galactosyltransferase. *Plant Physiology* **134**, 940–950.

Li, Y., Jones, L. and McQueen-Mason, S. (2003) Expansins and cell growth. *Current Opinion in Plant Biology* **6**, 603–610.

Liepman, A.H., Wilkerson, C.G. and Keegstra, K. (2005) Expression of cellulose synthase-like (*Csl*) genes in insect cells reveals that *CslA* family members encode mannan synthases. *Proceedings of the National Academy of Sciences of the United States of America* **102**, 2221–2226.

Lloyd, C. and Chan, J. (2004) Microtubules and the shape of plants to come. *Nature Reviews* **5**, 13–22.

Lukowitz, W., Nickle, T.C., Meinke, D.W., Last, R.W., Conklin, P.L. and Somerville, C.R. (2001) Arabidopsis *cyt1* mutants are deficient in a mannose-1-phosphate guanylyltransferase and point to a requirement of N-linked glycosylation for cellulose biosynthesis. *Proceedings of the National Academy of Sciences of the United States of America* **98**, 2262–2267.

Maclachlan, G. and Brady, C. (1994) Endo-1,4-beta-glucanase, xyloglucanase, and xyloglucan endo-transglycosylase activities versus potential substrates in ripening tomotoes. *Plant Physiology* **105**, 965–974.

Madson, M., Dunand, C., Li, X., Verma, R., Vanzin, G.F., Caplan, J., Shoue, D.A., Carpita, N.C. and Reiter, W.-D. (2003) The *MUR3* gene of *Arabidopsis* encodes a xyloglucan galactosyltransferase that is evolutionarily related to animal exostosins. *Plant Cell* **15**, 1662–1670.

Manfield, I.W., Orfila, C., McCartney, L., Harholt, J., Bernal, A.J., Scheller, H.V., Gilmartin, P.M., Mikkelsen, J.D., Paul Knox, J. and Willats, W.G. (2004) Novel cell wall architecture of isoxaben-habituated Arabidopsis suspension-cultured cells: global transcript profiling and cellular analysis. *Plant Journal* **40**, 260–275.

Matsunaga, T., Ishii, T., Matsumoto, S., Higuchi, M., Darvill, A., Albersheim, P. and O'Neill, M.A. (2004) Occurrence of the primary cell wall polysaccharide rhamnogalacturonan II in pteridophytes, lycophytes, and bryophytes. Implications for the evolution of vascular plants. *Plant Physiology* **134**, 339–351.

Matsumoto, T., Sakai, F. and Hayashi, T. (1997) A xyloglucan-specific endo-1,4-beta-glucanase isolated from auxin-treated pea stems. *Plant Physiology* **114**, 661–667.

McCann, M.C. and Roberts, K. (1991) Architecture of the primary cell wall. In: *The Cytoskeletal Basis of Plant Growth and Form* (ed. Lloyd C.W.), pp. 109–129. Academic Press, London, England.

McCartney, L., Steele-King, C.G., Jordan, E. and Knox, J.P. (2003) Cell wall pectic (1→4)-β-D-galactan marks the acceleration of cell elongation in the Arabidopsis seedling root meristem. *Plant Journal* **33**, 447–454.

McDougall, G.J. and Fry, S.C. (1991) Purification and analysis of growth-regulating xyloglucan-derived oligosaccharides by high-pressure liquid chromatography. *Carbohydrate Research* **219**, 123–132.

McQueen-Mason, S.J. and Cosgrove, D.J. (1995) Expansin mode of action on cell walls (analysis of wall hydrolysis, stress relaxation, and binding). *Plant Physiology* **107**, 87–100.

McQueen-Mason, S.J., Durachko, D.M. and Cosgrove, D.J. (1992) Two endogenous proteins that induce cell wall extension in plants. *Plant Cell* **4**, 1425–1433.

McQueen-Mason, S.J., Fry, S.C., Durachko, D.M. and Cosgrove, D.J.(1993) The relationship between xyloglucan endotransglycosylase and in-vitro cell wall extension in cucumber hypocotyls. *Planta* **190**, 327–331.

Micheli, F. (2001) Pectin methylesterases: cell wall enzymes with important roles in plant physiology. *Trends in Plant Science* **6**, 414–419.

Mohnen, D. (1999) Biosynthesis of pectins and galactomannans. In: *Carbohydrates and Their Derivatives Including Tannins, Cellulose, and Related Lignins* (ed. Pinto B.M.), pp. 497–527. Elsevier, New York.

Moore, P.J., Swords, K.M., Lynch, M.A. and Staehelin, L.A. (1991) Spatial organization of the assembly pathways of glycoproteins and complex polysaccharides in the Golgi apparatus of plants. *Journal of Cell Biology* **112**, 589–602.

Nairn, C.J. and Haselkorn, T. (2005) Three loblolly pine *CesA* genes expressed in developing xylem are orthologous to secondary wall *CesA* genes of angiosperms. *New Phytologist* **166**, 907–915.

Nishitani, K. and Tominaga, R. (1991) *In vitro* molecular weight increase in xyloglucans by an apoplastic enzyme preparation from epicotyls of *Vigna angularis*. *Physiologia Plantarum* **82**, 490–497.

Nishitani, K. and Tominaga, R. (1992) Endo-xyloglucan transferase, a novel class of glycosyltransferase that catalyzes transfer of a segment of xyloglucan molecule to another xyloglucan molecule. *Journal of Biological Chemistry* **267**, 21058–21064.

O'Neill, M.A., Albersheim, P. and Darvill, A. (1990) *The Pectic Polysaccharides of Primary Cell Walls*. London Academic Press, London, England.

O'Neill, M.A., Eberhard, S. Albersheim, P. and Darvill, A.G. (2001) Requirement of borate cross-linking of cell wall rhamnogalacturonan II for Arabidopsis growth. *Science* **294**, 846–849.

O'Neill, M.A., Ishii, T., Albersheim, P. and Darvill, A.G. (2004) Rhamnogalacturonan II: structure and function of a borate cross-linked cell wall pectic polysaccharide. *Annual Review of Plant Biology* **55**, 109–139.

O'Neill, M.A., Warrenfeltz, D., Kates, K., Pellerin, P., Doco, T., Darvill, A.G. and Albersheim, P. (1996) Rhamnogalacturonan-II, a pectic polysaccharide in the walls of growing plant cell, forms a dimer that is covalently cross-linked by a borate ester. *In vitro* conditions for the formation and hydrolysis of the dimer. *Journal of Biological Chemistry* **271**, 22923–22930.

O'Neill, M.A. and York, W.S. (2003) The composition and struture of plant primary cell walls. In: *The Plant Cell Wall* (ed. Rose, J.K.C.), pp. 1–54. CRC Press LCC, Boca Raton.

Okazawa, K., Sato, Y., Nakagawa, T., Asada, K., Kato, I., Tomita, E. and Nishitani, K. (1993) Molecular cloning and cDNA sequencing of endoxyloglucan transferase, a novel class of glycosyltransferase that mediates molecular grafting between matrix polysaccharides in plant cell walls. *Journal of Biological Chemistry* **268**, 25364–25368.

Orfila, C., Huisman, M.M., Willats, W.G., van Alebeek, G.J., Schols, H.A., Seymour, G.B. and Knox, J.P. (2002) Altered cell wall disassembly during ripening of *Cnr* tomato fruit: implications for cell adhesion and fruit softening. *Planta* **215**, 440–447.

Orfila, C., Seymour, G.B., Willats, W.G., Huxham, I.M., Jarvis, M.C., Dover, C.J., Thompson, A.J. and Knox, J.P. (2001) Altered middle lamella homogalacturonan and disrupted deposition of (1→5)-α-L-arabinan in the pericarp of *Cnr*, a ripening mutant of tomato. *Plant Physiology* **126**, 210–221.

Orfila, C., Sorensen, S.O., Harholt, J., Geshi, N., Crombie, H., Truong, H.N., Reid, J.S., Knox, J.P. and Scheller, H.V. (2005) QUASIMODO1 is expressed in vascular tissue of Arabidopsis thaliana inflorescence stems, and affects homogalacturonan and xylan biosynthesis. *Planta*, 1–10.

Pagant, S., Bichet, A., Sugimoto, K., Lerouxel, O., Desprez, T., McCann, M., Lerouge, P., Vernhettes, S. and Hofte, H. (2002) *KOBITO1* encodes a novel plasma membrane protein necessary for normal synthesis of cellulose during cell expansion in Arabidopsis. *Plant Cell* **14**, 2001–2013.

Pauly, M. and Scheller, H.V. (2000) O-acetylation of plant cell wall polysaccharides: identification and partial characterization of a rhamnogalacturonan *O*-acetyl-transferase from potato suspension-cultured cells. *Planta* **210**, 659–667.

Pear, J.R., Kawagoe, Y., Schrechengost, W.E., Delmer, D.P. and Stalker, D.M. (1996) Higher plants contain homologs of the bacterial *celA* genes encoding the catalytic subunit of cellulose synthase. *Proceedings of the National Academy of Sciences of the United States of America* **93**, 12637–12642.

Pena, M.J. and Carpita, N.C. (2004) Loss of highly branched arabinans and debranching of rhamnogalacturonan I accompany loss of firm texture and cell separation during prolonged storage of apple. *Plant Physiology* **135**, 1305–1313.

Perrin, R.M., DeRocher, A.E., Bar-Peled, M., Zeng, W., Norambuena, L., Orellana, A., Raikhel, N.V. and Keegstra, K. (1999) Xyloglucan fucosyltransferase, an enzyme involved in plant cell wall biosynthesis. *Science* **284**, 1976–1979.

Perrin, R.M., Jia, Z., Wagner, T.A., O'neal, M., Sarria, R., York, W.S., Raikhel, N.V. and Keegstra, K. (2003) Analysis of xyloglucan fucosylation in *Arabidopsis*. *Plant Physiology* **132**, 678–778.

Peugnet, I., Goubet, F., Bruyant-Vannier, M.P., Thoiron, B., Morvan, C., Schols, H.A. and Voragen, A.G. (2001) Solubilization of rhamnogalacturonan I galactosyltransfrases from membranes of a flax cell suspension. *Planta* **213**, 435–445.

Pien, S., Wyrzykowska, J., McQueen-Mason, S., Smart, C. and Fleming, A. (2001) From the cover: local expression of expansin induces the entire process of leaf development and modifies leaf shape. *Proceedings of the National Academy of Sciences of the United States of America* **98**, 11812–11817.

Porchia, A. and Scheller, H.V. (2000) Arabinoxylan biosynthesis: identification and partial characterization of β-1,4-xylosyltransferase from wheat. *Physiologia Plantarum* **110**, 350–356.

Porchia, A.C., Sorensen, S.O. and Scheller, H.V. (2002) Arabinoxylan biosynthesis in wheat: characterization of arabinosyltransferase activity in Golgi membranes. *Plant Physiology* **130**, 432–441.

Ray, P.M. (1980) Cooperative action of beta-glucan synthetase and UDP-xylose xylosyl transferase of Golgi membranes in the synthesis of xyloglucan-like polysaccharide. *Biochimica et Biophysica Acta* **629**, 431–444.

Ray, P.M., Shininger, T.L. and Ray, M.M. (1969) Isolation of ß-glucan synthetase particles from plant cells and identification with golgi membranes. *Proceedings of the National Academy of Sciences of the United States of America* **64**, 605–612.

Read, S. and Bacic, A. (2002) Prime time for cellulose. *Science* **295**, 59–60.

Reid, J.S.G., Edwards, M.E., Dickson, C.A., Scott, C. and Gidley, M.J. (2003) Tobacco transgenic lines that express fenugreek galactomannan galactosyltransferase constitutively have structurally altered galactomannans in their seed endosperm cell walls. *Plant Physiology* **131**, 1487–1495.

Reiter, W.D., Chapple, C. and Somerville, C. (1993) Altered growth and cell walls in a fucose-deficient mutant of *Arabidopsis*. *Science* **261**, 1032–1035.

Reiter, W.D., Chapple, C. and Somerville, C.R. (1997) Mutants of *Arabidopsis thaliana* with altered cell wall polysaccharide composition. *Plant Journal* **12**, 335–345.

Renard, C.M., Crepeau, M.J. and Thibault, J.F. (1999) Glucuronic acid directly linked to galacturonic acid in the rhamnogalacturonan backbone of beet pectins. *European Journal of Biochemistry* **266**, 566–574.

Renard, C.M., Lahaye, M., Mutter, M., Voragen, F.G. and Thibault, J.F. (1997) Isolation and structural characterisation of rhamnogalacturonan oligomers generated by controlled acid hydrolysis of sugar-beet pulp. *Carbohydrate Research* **305**, 271–280.

Richmond, T. (2000) Higher plant cellulose synthases. *Genome Biology* **1**, 3001.3001–3001.3006.

Richmond, T. and Somerville, C. (2000) The cellulose synthase superfamily. *Plant Physiology* **124**, 495–498.

Richmond, T. and Somerville, C. (2001) Integrative approaches to determining *Csl* function. *Plant Molecular Biology* **47**, 131–143.

Ridley, B.L., O'Neill, M.A. and Mohnen, D. (2001) Pectins: structure, biosynthesis, and oligogalacturonide-related signaling. *Phytochemistry* **57**, 929–967.

Rihouey, C., Morvan, B., Borissova, I., Jauneau, A., Demarty, M. and Jarvis, M.C. (1995) Structural features of CDTA-soluble pectins from flax hypocotyls. *Carbohydrate Polymers* **28**, 159–166.

Rose, J.K.C., Braam, J., Fry, S.C. and Nishitani, K. (2002) The XTH family of enzymes involved in xyloglucan endotransglucosylation and endohydrolysis: current perspectives and a new unifying nomenclature. *Plant Cell Physiology* **43**, 1421–1435.

Sarria, R., Wagner, T.A., O'Neill, M.A., Faik, A., Wilkerson, C.G., Keegstra, K. and Raikhel, N.V. (2001) Characterization of a family of *Arabidopsis* genes related to xyloglucan fucosyltransferase1. *Plant Physiology* **127**, 1595–1606.

Saxena, I.M. and Brown, R.M. (2005) Cellulose biosynthesis: current views and evoloving concepts. *Annals of Botany* **96**, 9–21.

Scheible, W.-R. and Pauly, M. (2004) Glycosyltransferases and cell wall biosynthesis: novel players and insights. *Current Opinion in Plant Biology* **7**, 285–295.

Scheller, H.V., Doong, R.L., Ridley, B.L. and Mohnen, D. (1999) Pectin biosynthesis: a solubilized a1,4-galacturonosyltransferase from tobacco catalyzes the transfer of galacturonic acid from UDP-galacturonic acid onto the non-reducing end of homogalacturonan. *Planta* **207**, 512.

Schols, H.A., Vierhuis, E., Bakx, E.J. and Voragen, A.G. (1995) Different populations of pectic hairy regions occur in apple cell walls. *Carbohydrate Research* **275**, 343–360.

Schröder, R., Atkinson, R.G., Langenkamper, G. and Redgwell, R.J. (1998) Biochemical and molecular characterisation of xyloglucan endotransglycosylase from ripe kiwifruit. *Planta* **204**, 242–251.

Schröder, R., Wegrzyn, T.F., Bolitho, K.M. and Redgwell, R.J. (2004) Mannan transglycosylase: a novel enzyme activity in cell walls of higher plants. *Planta* **219**, 590–600.

Schultz, C.J., Rumsewicz, M.P., Johnson, K.L., Jones, B.J., Gaspar, Y.M. and Bacic, A. (2002) Using genomic resources to guide research directions. The arabinogalactan protein gene family as a test case. *Plant Physiology* **129**, 1448–1463.

Shimizu, M., Igasaki, T., Yamada, M., Yuasa, K., Hasegawa, J., Kato, T., Tsukagoshi, H., Nakamura, K., Fukuda, H. and Matsuoka, K. (2005) Experimental determination of proline hydroxylation and hydroxyproline arabinogalactosylation motifs in secretory proteins. *Plant Journal* **42**, 877–889.

Showalter, A.M. (2001) Arabinogalactan-proteins: structure, expression and function. *Cellular and Molecular Life Sciences* **58**, 1399–1417.

Sims, I.M., Munro, S.L., Currie, G., Craik, D. and Bacic, A. (1996) Structural characterisation of xyloglucan secreted by suspension-cultured cells of *Nicotiana plumbaginifolia*. *Carbohydrate Research* **293**, 147–172.

Smith, R.C. and Fry, S.C. (1991) Endotransglycosylation of xyloglucans in plant cell suspension cultures. *Biochemical Journal* **279**(Part 2), 529–535.

Smith, L.G. and Oppenheimer, D.G. (2005) Spatial control of cell expansion by the plant cytoskeleton. *Annual Review of Cell and Developmental Biology* **21**, 271–295.

Somerville, C., Bauer, S., Brininstool, G., Facette, M., Hamann, T., Milne, J., Osborne, E., Paredez, A., Perrson, S., Raab, T., Vorwerk, S. and Youngs, H. (2004) Toward a systems approach to understanding plant cell walls. *Science* **306**, 2206–2211.

Sommer-Knudsen, J., Bacic, A. and Clarke, A.E. (1998) Hydroxyproline-rich plant glycoproteins. *Phytochemistry* **4**, 483–497.

Staehelin, L.A. and Moore, I. (1995) The plant Golgi apparatus: structure, functional organization and trafficking mechanisms. *Annual Review of Plant Physiology* **46**, 261–288.

Staudte, R., Woodward, J., Fincher, G.B. and Stone, B.A. (1983) Water-soluble (1,3)-(1,4)-β-D-glucans from barley (*Hordeum vulgare*) endosperm. III. Distribution of cellotriosyl and cellotetraosyl residues. *Carbohydrate Polymers* **3**, 299–312.

Sterling, J.D., Quigley, H.F., Orellana, A. and Mohnen, D. (2001) The catalytic site of the pectin biosynthetic enzyme alpha-1,4-galacturonosyltransferase is located in the lumen of the Golgi. *Plant Physiology* **127**, 360–371.

Stone, B.A. and Clarke, A.E. (1992) *Chemistry and Biology of (1→3)-β-Glucans*. La Trobe University Press, Victoria.

Strohmeier, M., Hrmova, M., Fischer, M., Harvey, A.J., Fincher, G.B. and Pleiss, J. (2004) Molecular modeling of family GH16 glycoside hydrolases: potential roles for xyloglucan transglucosylases/hydrolases in cell wall modification in the poaceae. *Protein Science* **13**, 3200–3213.

Sugimoto, K., Himmelspach, R., Williamson, R.E. and Wasteneys, G.O. (2003) Mutation or drug-dependent microtubule disruption causes radial swelling without altering parallel cellulose microfibril deposition in Arabidopsis root cells. *Plant Cell* **15**, 1414–1429.

Szyjanowicz, P.M.J., McKinnon, I., Taylor, N.G., Gardiner, J., Jarvis, M.C. and Turner, S.R. (2004) The *irregular xylem 2* mutant is an allele of *korrigan* that affects the secondary cell wall of *Arabidopsis thaliana*. *Plant Journal* **37**, 730–740.

Tabuchi, A., Kamisaka, S. and Hoson, T. (1997) Purification of xyloglucan hydrolase/endotransferase from cell walls of azuki bean epicotyls. *Plant and Cell Physiology* **38**, 653–658.

Tan, L., Leykam, J.F. and Kieliszewski, M.J. (2003) Glycosylation motifs that direct arabinogalactan addition to arabinogalactan-proteins. *Plant Physiology* **132**, 1362–1369.

Tanaka, K., Murata, K., Yamazaki, M., Onosato, K., Miyao, A. and Hirochika, H. (2003) Three distinct rice cellulose synthase catalytic subunit genes required for cellulose synthesis in the secondary wall. *Plant Physiology* **133**, 73–83.

Taylor, N.G., Gardiner, J.C., Whiteman, R. and Turner, S.R. (2004) Cellulose synthesis in the Arabidopsis secondary cell wall. *Cellulose* **11**, 329–338.

Taylor, N.G., Scheible, W.-R., Cutler, S., Somerville, C.R. and Turner, S.R. (1999) The *irregular xylem3* locus of Arabidopsis encodes a cellulose synthase required for secondary cell wall synthesis. *Plant Cell* **11**, 769–779.

Thibault, J.-F., Renard, C.M.G.C., Axelos, M.A.V., Roger, P. and Crepeau, M.-J. (1993) Studies of the length of homogalacturonic regions in pectins by acid hydrolysis. *Carbohydrate Research* **238**, 271.

Thompson, J.E. and Fry, S.C. (2000) Evidence for covalent linkage between xyloglucan and acidic pectins in suspension-cultured rose cells. *Planta* **211**, 275–286.

Urbanowicz, B.R., Rayon, C. and Carpita, N.C. (2004) Topology of the maize mixed linkage (1→3),(1→4)-{beta}-D-glucan synthase at the Golgi membrane. *Plant Physiology* **134**, 758–768.

Usadel, B., Kuschinsky, A.M., Rosso, M.G., Eckermann, N. and Pauly, M. (2004) RHM2 is involved in mucilage pectin synthesis and is required for the development of the seed coat in Arabidopsis. *Plant Physiology* **134**, 286–295.

Valent, B.S. and Albersheim, P. (1974) The structure of plant cell walls. V. On the binding of xyloglucan to the cellulose fibers. *Plant Physiology* **98**, 369–379.

Vanzin, G.F., Madson, M., Carpita, N.C., Raikhel, N.V., Keegstra, K. and Reiter, W.D. (2002) The *mur2* mutant of Arabidopsis thaliana lacks fucosylated xyloglucan because of a lesion in fucosyltransferase AtFUT1. *Proceedings of the National Academy of Sciences of the United States of America* **99**, 3340–3345.

Verma, D.P.S. and Hong, Z. (2001) Plant callose synthase complexes. *Plant Molecular Biology* **47**, 693–701.

Vidal, S., Doco, T., Williams, P., Pellerin, P., York, W.S., O'Neill, M.A., Glushka, J., Darvill, A.G. and Albersheim, P. (2000) Structural characterization of the pectic polysaccharide rhamnogalacturonan II: evidence for the backbone location of the aceric acid-containing oligoglycosyl side chain. *Carbohydrate Research* **326**, 277–294.

Vincken, J.P., York, W.S., Beldman, G. and Voragen, A.G. (1997) Two general branching patterns of xyloglucan, XXXG and XXGG. *Plant Physiology* **114**, 9–13.

Vissenberg, K., Fry, S.C., Pauly, M., Hofte, H. and Verbelen, J.-P. (2005) XTH acts at the microfibril-matrix interface during cell elongation. *Journal of Experimental Botany* **56**, 673–683.

Vorwerk, S., Somerville, S. and Somerville, C. (2004) The role of plant cell wall polysaccharide composition in disease resistance. *Trends in Plant Science* **9**, 203–209.

Wasteneys, G.O. (2004) Progress in understanding the role of microtubules in plant cells. *Current Opinion in Plant Biology* **7**, 651–660.

Western, T.L., Young, D.S., Dean, G.H., Tan, W.L., Samuels, A.L. and Haughn, G.W. (2004) *MUCILAGE-MODIFIED4* encodes a putative pectin biosynthetic enzyme developmentally regulated by *APETALA2, TRANSPARENT TESTA GLABRA1*, and *GLABRA2* in the Arabidopsis seed coat. *Plant Physiology* **134**, 296–306.

Whitcombe, A.J., O'Neill, M.A., Steffan, W., Albersheim, P. and Darvill, A.G. (1995) Structural characterization of the pectic polysaccharide, rhamnogalacturonan-II. *Carbohydrate Research* **271**, 15–29.

White, A.R., Xin, Y. and Pezeshk, V. (1993) Xyloglucan glucosyltransferase in Golgi membranes from Pisum sativum (pea). *Biochemical Journal* **294** (Pt 1), 231–238.

Whittington, A.T., Vugrek, O., Wei, K.J., Hasenbein, N.G., Sugimoto, K., Rashbrooke, M.C. and Wasteneys, G.O. (2001) MOR1 is essential for organizing cortical microtubules in plants. *Nature* **411**, 610–613.

Willats, W.G., Marcus, S.E. and Knox, J.P. (1998) Generation of monoclonal antibody specific to (1→5)-alpha-L-arabinan. *Carbohydrate Research* **308**, 149–152.

Willats, W.G., McCartney, L., Mackie, W. and Knox, J.P. (2001) Pectin: cell biology and prospects for functional analysis. *Plant Molecular Biology* **47**, 9–27.

Willats, W.G., McCartney, L., Steele-King, C.G., Marcus, S.E., Mort, A., Huisman, M., van Alebeek, G.J., Schols, H.A., Voragen, A.G., Le Goff, A., Bonnin, E., Thibault, J.F. and Knox, J.P. (2004) A xylogalacturonan epitope is specifically associated with plant cell detachment. *Planta* **218**, 673–681.

Willats, W.G., Steele-King, C.G., Marcus, S.E. and Knox, J.P. (1999) Side chains of pectic polysaccharides are regulated in relation to cell proliferation and cell differentiation. *Plant Journal* **20**, 619–628.

Williamson, R.E., Burn, J.E., Birch, R., Baskin, T.I., Arioli, T., Betzner, A.S. and Cork, A. (2001) Morphology of *rsw1*, a cellulose-deficient mutant of *Arabidopsis thaliana*. *Protoplasma* **215**, 116–127.

Wood, P., Weisz, J. and Blackwell, B. (1994) Structural studies of (1,3)-(1,4)-B-D-glucans by ^{13}C-nuclear magnetic resonance spectroscopy and by rapid analysis of cellulose-like regions using high-performance anion-exchange chromatography of oligosaccharides released by lichenanase. *Cereal Chemistry* **71**, 301–307.

Yokoyama, R. and Nishitani, K. (2001) Endoxyloglucan transferase is localized both in the cell plate and in the secretary pathway destined for the apoplast in tobacco cells. *Plant and Cell Physiology* **42**, 292–300.

Yokoyama, R., Rose, J.K.C. and Nishitani, K. (2004) A surprising diversity and abundance of xyloglucan endotransglucosylase/hydrolases in rice. Classification and expression analysis. *Plant Physiology* **134**, 1088–1099.

York, W.S., Oates, J.E., van Halbeek, H., Darvill, A.G., Albersheim, P., Tiller, P.R. and Dell, A. (1988) Location of the O-acetyl substituents on a nonasaccharide repeating unit of sycamore extracellular xyloglucan. *Carbohydrate Research* **173**, 113–132.

Youl, J.J., Bacic, A. and Oxley, D. (1998) Arabinogalactan-proteins from *Nicotiana alata* and *Pyrus communis* contain glycosylphosphatidylinositol membrane anchors. *Proceedings of the National Academy of Sciences of the United States of America* **95**, 7921–7926.

Zhang, N. and Hasenstein, K.H. (2000) Distribution of expansins in graviresponding maize roots. *Plant Cell Physiology* **41**, 1305–1312.

Zhong, R., Burk, D.H., Morrison, W.H. and Ye, Z.-H. (2002) A kinesin-like protein is essential for oriented deposition of cellulose microfibrils and cell wall strength. *Plant Cell* **14**, 3101–3117.

Zhong, R., Burk, D.H., Morrison, W.H. and Ye, Z.-H. (2004) *Fragile fiber3*, an Arabidopsis gene encoding a type II inositol polyphosphate 5-phosphatase, is required for secondary wall synthesis and actin organization in fiber cells. *Plant Cell* **16**, 3242–3259.

Zhong, R., Burk, D.H., Nairn, C.J., Wood-Jones, A., Morrison, W.H. and Ye, Z.-H. (2005) Mutation of SAC1, an Arabidopsis SAC domain phosphoinositide phosphatase, causes alterations in cell morphogenesis, cell wall synthesis, and actin organization. *Plant Cell* **17**, 1449–1466.

3 Vascular cell differentiation

Hideo Kuriyama and Hiroo Fukuda

3.1 Tracheary Element (TE) differentiation as a model of cell–cell connection

Similar to higher plants, certain types of multicellular organisms consist largely of masses of sessile cells that adhere to each other through extracellular matrices. Such organization patterns are indisputably essential for the maintenance of organismal integrity. As described in other chapters of this volume, hydrolase-mediated cell separation often takes place in specific differentiated cells such as the endosperm or aleurone cells of germinating seeds, root cap cells, guard cells that form stomata, cells in abscission zones and specific flower organ cells (Roberts *et al.*, 2002). Artificial breakages of plant tissues (e.g. upon wounding, herbivory and pathogen attacks) also result in the separation of constituent cells and invoke elaborate, highly systematic responses in individual plants. Some of these breakages make it possible to alter the fate of cells by affecting intrinsic genetic programmes (Torrey *et al.*, 1971; Vorwerk *et al.*, 2004). TEs are highly specialized cells that arise from vascular tissue (procambial or cambial) cells under the strict control of genetic mechanisms (e.g. Kuriyama and Fukuda, 2001; Ye, 2002; Scarpella and Meijer, 2004). The functional form of TEs is a hollow tubular cell corpse that consists of cell walls with unique morphological features (Plate 4). This clearly indicates that cytodifferentiation processes of TEs include cell wall morphogenesis and programmed cell death (PCD) followed by protoplast digestion (e.g. Fukuda, 2004). These series of events mark differentiating cells particularly well. Thus, TE differentiation has long been regarded as a model system for the analysis of plant cell differentiation mechanisms (Torrey *et al.*, 1971; Aloni, 1987). In addition, the formation of continuous TE columns suggests that intercellular signals govern TE differentiation (Nelson and Dengler, 1997; Berleth *et al.*, 2000). Experimental findings on the artificial cues that promote TE differentiation accelerated the momentum for analyses of TE differentiation mechanisms. Simon (1908) and Freundlich (1909) found that the wounding of higher plant tissues led to the generation of new TEs near the wounded sites. Jacobs (1952) related this phenomenon to the action of the plant hormone auxin, which was transported basipetally and served to restore vessel (and vascular) continuity. Since then, various experimental systems that employ plant tissue cultures, organ explants, calli and isolated cells grown in appropriate media have been developed (Fukuda, 1992). In order to control more strictly the cellular environment, it is ideal to use homogeneous, isolated single cells. In doing so, each cell has equal access to components in the culture media. Kohlenbach and Schmidt (1975) and Fukuda and Komamine (1980) prepared single cells from *Zinnia elegans* L. leaves by crushing them with a pestle and mortar under appropriate conditions. The isolated cells became differentiated synchronously and at a high frequency into TEs with a combination of auxin and cytokinin. The proper hormonal control of a basal culture medium, which is otherwise optimized (e.g. with nutrients and vitamins) for maximum levels of TE differentiation, enables 40–45% of total isolated cells to transdifferentiate into TEs (Fukuda and Komamine, 1980). Using this system, molecular mechanisms of vascular and TE cells have been extensively studied (Fukuda, 2004).

3.2 Early processes induced by cell separation

It is easy to separate mesophyll cells mechanically from leaves of some plant species such as peanuts, asparagus, lettuce, tobacco, *Macleaya cordata* and *Z. elegans* (Fukuda and Komamine, 1985). Importantly, such cells do not split in interior cell parts: instead, adhering cells are separated just at their boundary. The successful isolation of single mesophyll cells by mechanical means depends on plant leaf mesophyll cell wall structures/properties, which vary among plant species. Growth conditions also influence leaf structures and wall biochemical properties. The first leaves of *Zinnia* seedlings grown in a light/dark cycle for 14 days have expanded intercellular spaces that allow us to separate mesophyll cells mechanically (Figure 3.1A). However, the first leaves of 7-day-old seedlings have less intercellular spaces and cannot be used for mechanical cell separation (Figure 3.1B). The cell wall properties of *Zinnia* mesophyll cells might also be suitable for the preparation of single cells. Separated cells, when cultured in the presence of plant hormones, proliferate and differentiate into specific types of cells such as TEs. In *Zinnia*, the separation of mesophyll cells was necessary for the strong induction of TE differentiation. This was because only a few TEs were formed from mesophyll cells in *Zinnia* leaf segments cultured with auxin and cytokinin (Church and Galston, 1989; H. Fukuda, unpublished). Protoplasts may also provide similar systems to those of single-cell cultures in the absence of cell walls. Kohlenbach and co-workers succeeded in inducing TE differentiation using *Z. elegans* leaf mesophyll (Kohlenbach and Schöpke, 1981) and cultured *Brassica napus* haploid stem embryo protoplasts (Kohlenbach *et al.*, 1982). However, the frequency of TE differentiation was much lower than that in *Zinnia* single-cell cultures. If these protoplast systems could be improved to produce many more TEs, molecular, biological or biotechnological analyses of single differentiating TE cells would become feasible using gene transfection experiments. Plant tissue cultures require cutting plant parts in order to incubate them under artificially determined conditions. Such a process invokes a wound response in the cut tissue explants. Stem wounding often induces ectopic TE differentiation around the lesions (Torrey *et al.*, 1971). In particular, when continuous vascular bundles or TE cell files are broken, TEs arise to regenerate the continuity of such tissues. Based on such findings, wounding signals have been implicated in the induction of TE differentiation. In *Zinnia* leaves, mechanical isolation

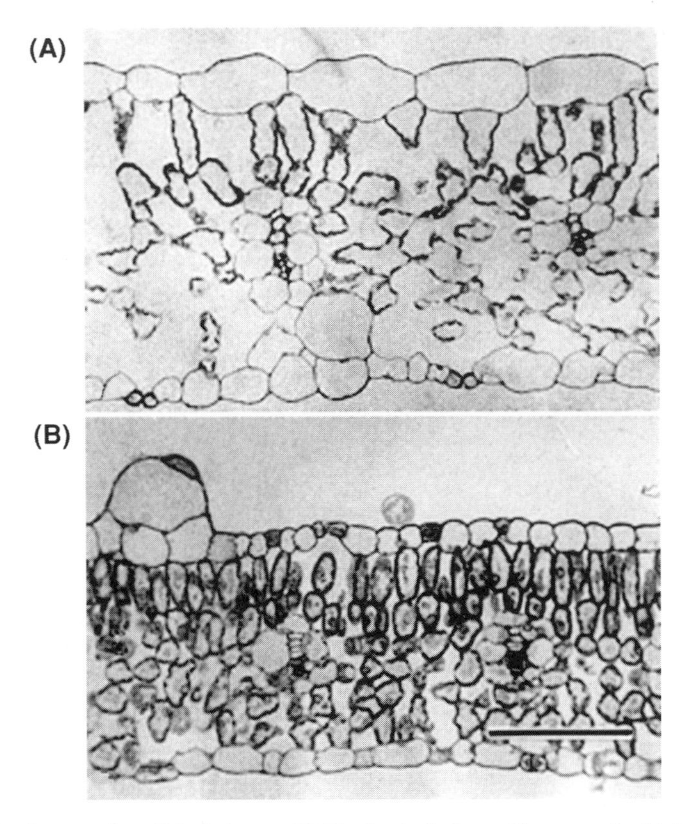

Figure 3.1 Cross section of *Zinnia* leaves. (A) The first leaf of a seedling grown for 14 days. (B) The first leaf of a seedling grown for 7 days. The bar indicates 50 μm.

procedures to prepare single mesophyll cells inevitably impose a severe wounding stress on the cells. Scanning electron microscopy showed a broken plasmodesmata-like structure at the longitudinal end of separated *Zinnia* palisade mesophyll cells (Burgess and Linstead, 1984). Although the incubation of detached *Zinnia* leaves in TE differentiation-inductive media does not induce TE differentiation, the peeling of leaf epidermal cells before such incubation invokes the transdifferentiation of both mesophyll and epidermal cells into TEs (Church and Galston, 1989). Nevertheless, plants can produce TE cells in the proper positions without any wounding stimulus. Thus, wounding signals may be necessary for the initiation of transdifferentiation into TEs. Indeed, ethylene, which can modulate wounding signals (O'Donnell *et al.*, 1996), promotes TE differentiation (reviewed in Fukuda, 1992; Kuriyama and Fukuda, 2001). The *Arabidopsis VEP1/AWI31* gene is wound-inducible, the down-regulation of which results in reduced xylem vessels in the leaves and stems. This suggests its involvement in the formation of normal vascular structures (Jun *et al.*, 2002). Therefore, both wounding and developmental cues may induce a gene(s) like *VEP1/A WI31* to initiate TE differentiation. Sulfonic pentapeptide phytosulfokine-α (PSK-α) promotes TE differentiation in *Zinnia* culture (Matsubayashi *et al.*, 1999).

PSK-α may regulate cell wound responses in early stages of TE differentiation (H. Motose, unpublished).

3.3 Factors that regulate TE cell differentiation

It is known that various factors are involved in the programme of TE differentiation. Because a recent review described such factors in detail (Kuriyama and Fukuda, 2001; Fukuda, 2004), in this chapter, we introduce only briefly the factors that regulate the continuous formation of TE cells.

3.3.1 Auxin

The plant hormone auxin plays a pivotal role in the regulation of meristematic cell differentiation into (pro)vascular or TE cells (e.g. Fukuda, 2004; Scarpella and Meijer, 2004). Recent studies revealed that the *Arabidopsis* auxin receptor *TIR1* gene product is a member of the Skp1, cullin and F-box (SCF) complex, E3 ubiquitin ligase, which binds ubiquitins to AUX/IAA transcription factors. This promotes the degradation of such proteins in response to increased auxin levels (Dharmasiri *et al.*, 2005; Kepinski and Leyser, 2005). Because AUX/IAA proteins interact with transcription activators such as auxin response factors (ARFs) to suppress their activity and abolish the initiation of auxin-inducible gene expression, the selective degradation of AUX/IAAs results in the activation of auxin-inducible gene expression (Hellmann and Estelle, 2002). Mutations that occur in *Arabidopsis* genes involved in the above auxin signalling machinery such as *AUXIN RESISTANT(AXR)6/AtCUL1* (Hellmann *et al.*, 2003), *BODENLOS(BDL)/IAA12* (Hamann *et al.*, 2002) and *MONOPTEROS(MP)/ARF5* (Hardtke and Berleth, 1998) cause severe defects to vascular bundle structures (Przemeck *et al.*, 1996; Hamann *et al.*, 1999; Hobbie *et al.*, 2000). Polar auxin transport is crucial for the vascular tissue development of plant organs (Berleth *et al.*, 2000). Basipetally flowing auxin induces and promotes vascular cell differentiation as it progresses through the plant (e.g. Sachs, 2000). Only a few cell types at specific sites such as the stipules of leaf primordia, the developing hydathodes of young leaves and the base of trichomes can synthesize auxin (Aloni, 2004). As such, auxin movement patterns are implicated in the patterning of vascular tissues in plant organs (Berleth *et al.*, 2000). The auxin influx and efflux carrier genes *AUX1* (Swarup *et al.*, 2001) and *AtPIN1* (Gälweiler *et al.*, 1998) are localized in the plasma membrane of the cellular apical and basal ends of particular vascular tissue cells, respectively. Files of these cells correspond to routes of auxin transport. The GNOM protein has a role in the polar membrane traffic of the AtPIN1 protein and regulates polar auxin transport (Steinmann *et al.*, 1999), and the *PINOID* (*PID*) gene encodes a serine/threonine kinase that is expressed in xylem precursor and xylem parenchyma cells (Benjamins *et al.*, 2001). The *PID* gene product controls the polar distribution of the AtPIN1 auxin efflux carrier in such cells, thereby influencing polar auxin transport to organs (Friml *et al.*, 2004).

The *Arabidopsis VAN3* gene, which regulates another mechanism of auxin-related vascular development, encodes the ADP-ribosylation factor guanosine triphosphatase (GTPase)-activating protein (Koizumi *et al.*, 2005). *VAN3* expression is auxin inducible, self-dependent, *in planta*, and is detected ubiquitously in cotyledons and leaves at very young stages of organ development. At this stage, *VAN3* expression is widespread in the plant vasculature and then converges to the trichomes (Koizumi *et al.*, 2005). The VAN3 protein localizes in specific trans-Golgi network compartments and the plasma membrane, and it may mediate the vesicular transport of auxin signalling components (Koizumi *et al.*, 2005). Dynamin-related protein 1A (DRP1A) was found to interact with the VAN3 protein (Sawa *et al.*, 2005b). DRP1A is a microtubule-associated protein that interacts with the VAN3 protein in the microtubule-attached trans-Golgi network and functions in vesicle budding (Sawa *et al.*, 2005b). GUS-indicated *DPR1A* promoter activity is detected in both whole juvenile cotyledons and leaves. This includes trichomes and the vascular bundles in developed cotyledons and leaves (Sawa *et al.*, 2005b). The majority of these areas overlap with those of *VAN3* promoter activity. The *drp1a* mutant shows moderate defects in vascular continuity that is dramatically strengthened in combination with the *van3* or *gnom* mutant gene (Sawa *et al.*, 2005b). *DPR1A*-mediated vesicle trafficking is very important for auxin signal-regulated vascular differentiation.

3.3.2 Plant sterols

Plant sterol molecules have a role in the formation of normal, continuous vascular bundles and vessels. Several *Arabidopsis* plant lines carrying a mutation in brassinosteroid-synthesis-related genes have xylem cells in the vascular bundles that exhibit structural defects (Szekeres *et al.*, 1996; Choe *et al.*, 1999). In *Zinnia* culture, uniconazole or brassinazole, which are known to inhibit brassinosteroid biosynthesis, suppressed TE differentiation by preventing initiation into stage III during culture (Plate 4; Yamamoto *et al.*, 1997). The exogenous application of sitosterol or brassinolide reverses uniconazole-induced TE differentiation suppression (Yamamoto *et al.*, 2001). Exogenous brassinosteroids are dispensable in normal inductive culture. Therefore, endogenously synthesized brassinosteroids function in initiation into stage III of TE differentiation. This brassinosteroid function may be performed through the induction of *homeodomain leucine zipper* (*HD-ZIP*) *class III* genes that are involved in procambium and xylem differentiation (Ohashi-Ito and Fukuda, 2003, Ohashi-Ito *et al.*, 2005). Thus, brassinosteroids, which are synthesized endogenously by cultured cells, are essential for TE differentiation. This is consistent with the data obtained from *Arabidopsis* mutants defective in brassinolide synthesis.

Arabidopsis hydra1 (Topping *et al.*, 1997; Souter *et al.*, 2002), *fackel/hydra2/ell1* (Mayer *et al.*, 1991; Jang *et al.*, 2000; Schrick *et al.*, 2000; Souter *et al.*, 2002), *sterol methyltransferase1* (*smt1* or *orc*; Scheres *et al.*, 1996; Diener *et al.*, 2000; Schrick *et al.*, 2002; Willemsen *et al.*, 2003) and *cotyledon vascular pattern 1* (*smt2*; Carland *et al.*, 1999, 2002) mutants affect (post)embryonic development and cotyledon vascular patterning. All of these mutations occur in enzyme genes required for sterol biosynthesis. Nevertheless, exogenous brassinosteroids could not reverse such mutant phenotypes (Schrick *et al.*, 2000; Carland *et al.*, 2002), suggesting that unknown sterols are required for a distinct signal transduction pathway to form a continuous vascular network. Some of these sterols appear to promote the establishment of cell polarity to influence asymmetrical tissue cell patterning. Interactions between sterols and the sterol/lipid binding domains of *HD-ZIP III* homeobox genes are also possible mechanisms that can account for the roles of sterols (Ponting and Aravind, 1999).

3.3.3 Xylogen

Xylogen is a novel xylogenesis-inducing factor that regulates local promotive effects on TE differentiation in vascular tissues (Motose *et al.*, 2004). Of particular interest, the xylogen protein sequence contains a characteristic motif of the one-cell-wall structural protein, i.e. an arabinogalactan protein (AGP; see below). The core structure of *Zinnia* xylogen is a small 16-kD protein encoded by the *ZeXYP1* gene. ZeXYP1 proteins consist of four distinct domain sequences: a signal peptide, the AGP domain, a non-specific lipid transfer protein (nsLTP) domain and a glycosyl phosphatidylinositol (GPI) anchor domain. ZeXYP1 undergoes extensive modification from the binding of sizable carbohydrates to form a 'hard-to-diffuse' local signalling factor. As a result, the molecular weight of mature *Zinnia* xylogen is much larger than that of the core protein itself. Expression of the *ZeXYP1* mRNA and protein occurs in procambium and immature xylem cells. Immunohistochemical analyses of the anti-ZeXYP1 antibody points to ZeXYP1 localization in the cell walls of only one side tip of the cell wall. There are two closely related *Arabidopsis ZeXYP1* homologues: *AtXYP1* and *AtXYP2*. Double knockout plants of these two homologues exhibit disconnected vessels or a lack of lateral vessel loops in the cotyledons and leaves. Synthetic xylogen in 'biosynthetic factory' tobacco BY-2 cells binds to stigmasterol and, less strongly, to brassicasterol with a high specificity. However, it does not bind to other lipidic molecules generated via brassinosteroid synthesis pathway such as campesterol and brassinolide. The effect of xylogen may depend on either the lipid-based signalling of cell polarity-related pathways or other unidentified pathways.

3.4 Effects of tissue organization on cell differentiation

Genetic analyses are unravelling the mechanisms of stem (Dengler, 2001; Fletcher, 2002), root (Paquette and Benfey, 2001) and abaxial/adaxial tissue pattern formation (Bowman *et al.*, 2002; Canales *et al.*, 2005). Various transcription factors and cell signalling machinery genes (such as, receptor-like protein kinase genes; Clay and Nelson, 2002; Yin *et al.*, 2002) are differentially expressed around organs. Transcriptional regulator genes such as *HD-ZIP II* and *III* (Zhong and Ye, 1999; Scarpella *et al.*, 2000; Baima *et al.*, 2001; McConnell *et al.*, 2001; Ohashi-Ito *et al.*, 2005),

NAC (Kubo *et al.*, 2005; Mitsuda *et al.*; 2005), *knotted1-like homeobox* (*KNOX*; Lincoln *et al.*, 1994), *MYB* (Waites *et al.*, 1998; Bonke *et al.*, 2003), *YABBY* (Sawa *et al.*, 1999; Siegfried *et al.*, 1999) and *KANADI* (Emery *et al.*, 2003) genes, as well as micro RNAs (Kim *et al.*, 2005; Williams *et al.*, 2005) and RNA interference (RNAi) machinery genes (Bohmert *et al.*, 1998), which control the determination of radial patterns in stem vascular tissues, participate in the regulation of TE and other types of vascular cell differentiation (Ye, 2002; Dinneny and Yanofsky, 2004). Single or overlapped expression patterns of these genes determine the fate of specific types of cell differentiation (reviewed in Fukuda, 2004). Based on such characteristics, the structure of wall-partitioned immobilized cell clusters and the actions of local signalling factors gives rise to appropriate amounts of specialized cells at appropriate positions in vascular tissues. In other words, they restrict the emergence of such specialized cells at specific tissue sites. Once tissue cells are released, similar to isolated mesophyll cells in *Zinnia* culture, mechanisms for the positional restriction of cell specification would be compromised. In such a situation, the exogenously supplied excess (local) signalling factors can promote the production of much greater amounts of specific types of cells than that normally occurs in tissues (Fukuda and Komamine, 1980). As a result, the frequency of specific cell differentiation increases, as can be seen by the high efficiency of TE differentiation in *Zinnia* xylogenic culture supplemented with many inductive factors (Matsubayashi *et al.*, 1999; Motose *et al.*, 2004). Ultimately, cell-wall-mediated plant cell adhesion helps construct the highly ordered organ and tissue structures that consist of various cell types. Artificial cell separation can alter the balance of specific cell population levels, which is applicable to the massive production of specific cells.

3.5 Cell wall components characteristic of TE and/or vascular cells

The cell walls of TEs consist of primary and secondary walls. TE primary cell wall components include pectin, hemicellulose, cellulose and additional structural and enzymatic proteins (e.g. Fukuda, 1992; Milioni *et al.*, 2001; Reiter, 2002). This is essentially the same structure as that of general plant cells. Detailed compositions, properties and biosynthetic regulation of general plant primary walls are specified in Chapter 2 of this volume. Thus, we focus only on the primary walls of TEs in this section to describe their additional specific characters. In contrast, the secondary walls are unique to TEs and several other specific types of cells such as sclerenchyma, xylem, phloem and interfascicular fibres. In addition to general cell wall materials, there are distinct components in secondary walls such as lignin. Explanations of the characteristic properties and biosynthetic regulation of secondary walls are also discussed.

3.5.1 Cellulose

Cellulose is the major matrix polymer in the primary and secondary cell walls of TE cells. Upon TE morphogenesis, the synthesis of a large amount of cellulose microfibrils occurs to construct the characteristic spiral or reticulate secondary walls (Plate 5). This patterning of nascent cellulose microfibril bundles is controlled by the orientation of cortical microtubules. The dynamic rearrangement of the microtubules, the orientation of which changes from longitudinal to transverse along the cellular axis, becomes evident in developing TEs (Falconer and Seagull, 1985a,b). Rosette-shaped cellulose synthase complexes move along the microtubule-based 'guard rail', thereby arranging the orientations of nascent microfibrils in parallel to microtubules (Roberts *et al.*, 2004; Wasteneys, 2004). However, pharmacological analyses strongly suggest that the orientations of developing TE microtubules are regulated by actin filaments, which also show dramatic rearrangement just before TE secondary wall development (Kobayashi *et al.*, 1988). As a result, TE actin filaments determine the orientations of TE microtubules, which in turn determine the orientation of the cellulose microfibrils in TE secondary walls (Fukuda, 1992). In *Arabidopsis*, there are ten known cellulose synthase subunit genes that are extensively characterized in terms of molecular genetics (Scheible and Pauly, 2004). Single mutations of *IRREGULAR XYLEM1* (*IRX1*), *IRX3* and *IRX5*, which correspond to *AtCESA8*, *AtCESA7* and *AtCESA4*, respectively, result in TE collapse in inflorescence stems (Turner and Somerville, 1997; Taylor *et al.*, 1999, 2000, 2003). This cytological phenotype of TEs is due to reduced cellulose synthesis that weakens the mechanical strength of secondary walls. Such walls cannot tolerate the negative pressure generated by the transpirational stream. All of these genes encode subunits of a cellulose synthase complex that interact with one another to form a rosette-like higher molecular structure (Gardiner *et al.*, 2003; Taylor *et al.*, 2003). The expression patterns of these genes are almost identical, and a lack of only one of these component molecules precludes the formation of the complex (Gardiner *et al.*, 2003; Taylor *et al.*, 2003). Although other cellulose synthase subunit genes also exist in *Arabidopsis*, their function is either in primary cell wall synthesis or unknown (Desprez *et al.*, 2002; Scheible and Pauly, 2004).

In the case of cotton, oxidative conditions enhance the activity of cellulose synthases (Kurek *et al.*, 2002). Cotton CESA homologues include a zinc finger motif from which subunit molecules can bind to one another (Kurek *et al.*, 2002). Oxidative conditions help their assembly via their intramolecular disulfide bonds, which reside in the zinc finger motifs (Kurek *et al.*, 2002). In developing *Zinnia* TEs, the levels of reactive oxygen species (ROS) such as hydrogen peroxide (H_2O_2) increase (Ogawa *et al.*, 1997; Ros Barceló, 1998, 2005). Such ROS may have a role in promoting the formation of the cellulose synthase complexes involved in the production of TE secondary walls. Inhibitors of cellular H_2O_2 production can reduce the thickening of the secondary walls of cultured *Zinnia* cells (Karlsson *et al.*, 2005).

3.5.2 Hemicellulose

In contrast to cellulose, hemicellulosic materials are synthesized in the endoplasmic reticulum (ER) and Golgi apparatus. From here, vesicular transport systems carry the polymeric carbohydrates that form the major components of the cell wall.

Wheat germ agglutinin is a lectin that can specifically bind to *N*-acetylglucosamine carbohydrate residues. This lectin binds to a hemicellulosic material(s) in TE secondary walls (Hogetsu, 1990; Plate 5). Whereas monoclonal antibodies that recognize polysaccharide epitopes of fucosylated xyloglucan (CCRC-M1) can only slightly label TE primary walls (Stacey *et al.*, 1995), those of unsubstituted or low-substituted xylan (LM10) and arabinoxylan (LM11) give rise to highly specific immunocytochemical signals in xylem secondary walls (McCartney *et al.*, 2005). β-1,4-mannosyl residue-containing hemicellulosic polysaccharides are also localized in TE secondary walls, as well as in epidermal, xylem parenchyma, and xylary fibre cell walls (Handford *et al.*, 2003). Glycosyltransferases (GTs) may play critical roles in the synthesis of hemicellulosic materials in TE secondary walls (Zhong and Ye, 2003). The isolation and sequencing of massive *Zinnia* expression sequence tags (ESTs) have identified many clones that encode fragment sequences of GT homologue genes (Demura *et al.*, 2002; Milioni *et al.*, 2002). Functional analyses of corresponding genes of such ESTs will reveal the regulatory mechanisms of secondary wall hemicellulose synthesis.

Recently, the *Arabidopsis* GT gene *FRAGILE FIBER8* (*FRA8*) was functionally characterized (Zhong *et al.*, 2005). *FRA8*-encoded GT belongs to the gene family GT47 and is structurally similar to the GTs of animal exostosins that catalyse the synthesis of an extracellular material (heparan sulphate). Analyses of the *fra8* mutant plant revealed that this gene was involved in the thickening of xylem TEs and interfascicular fibre secondary walls. The FRA8 protein resides in the Golgi apparatus of developing xylem and maturing interfascicular fibre cells, and putatively possesses a glucuronyltransferase activity that enables the synthesis of glucuronoxylan, a secondary wall hemicellulose material.

3.5.3 Pectin

Pectin is another major plant cell wall matrix polymer. Homogalacturonan (HG), rhamnogalacturonan I (RG I), and rhamnogalacturonan II (RG II) are representative pectic polysaccharides (O'Neill *et al.*, 2004). The distribution of pectic materials is examined using specific antibodies that are raised against particular pectic polysaccharides. In xylem TE cells, such materials exist in the primary cell wall, which has been confirmed using at least four different antibodies against different epitopes of pectic molecules. Monoclonal antibodies that are raised against the de-esterified HG epitope JIM5 (Stacey *et al.*, 1995; Ryser, 2003), and against the partially methyl-esterified HG epitope JIM7 (Ryser, 2003), bind to TE primary cell walls. JIM5 antibodies stained TE primary walls much more heavily than they did adjacent parenchyma cells (Ryser, 2003). The LM5 antibody recognizes an epitope of (1$\rightarrow$4)-β-D-galactan and materials in TE primary walls, particularly those located just beneath thickening secondary walls (Guillemin *et al.*, 2005). In contrast, the LM6 antibody binds to (1$\rightarrow$5)-α-L-arabinan and provides almost even immunocytochemical signals to TE primary walls (Guillemin *et al.*, 2005). Electron microscopic observations of cultured *Zinnia* TE cell walls subjected to PATAg staining suggested the 'transient' presence of pectic materials in the secondary walls

(Nakashima *et al.*, 2004). However, the regulatory mechanisms of TE pectin synthesis remain unknown. Rather, decreases in pectin synthesis during TE differentiation have been reported (Bolwell and Northcote, 1981; Bolwell *et al.*, 1985). Although recent findings strongly suggest that several GTs can synthesize or modify pectic polysaccharides (Bouton *et al.*, 2002), neither activity of this enzyme nor the regulatory mechanisms of cognate gene expression in TEs have been studied to date.

3.5.4 Lignin

Lignin is a heterogeneous amorphous macromolecule in which substrate (*p*-hydroxyphenyl, guaiacyl and syringyl) units are polymerized by oxidative coupling reactions. Monomeric lignin substrates are synthesized via the phenylpropanoid pathway. Inhibition of lignin precursor synthesis by α-aminooxi-β-phenylpropionic (AOPP) acid or 2-aminoindan-2-phosphonic (AIP) acid results in the production of TEs with unlignified secondary walls (Sato *et al.*, 1993). The first substrates for the phenylpropanoid pathway are aromatic amino acids such as phenylalanine and tyrosine, which are produced through another pathway of secondary metabolic reactions (called the shikimate pathway). These aromatic amino acids are essential components for both lignin production and anthocyanin and protein biosynthesis. Thus, the occurrence of shikimate pathways may not necessarily be associated with vascular or TE cells. Lignification is a fairly unique process in the construction of TE secondary walls and sclerenchyma, xylem and phloem fibres. The specific regulation of lignin synthesis depends on the control of phenylpropanoid pathways. Thus far, considerable effort has been expended towards understanding the genetics and molecular biology of lignin production (Anterola and Lewis, 2002). The down-regulation or mutation of the genes that encode phenylpropanoid pathway enzymes alters the lignin amount/composition in the plant and often generates characteristic phenotypes in whole plants and/or TE secondary walls. In particular, the down-regulation and mutation of *phenylalanine ammonia-lyase* (*PAL*), p-*coumarate 3-hydroxylase* (*C3H*), *4-coumarate-CoA ligase* (*4-CL*), *caffeic acid 3-O-methyltransferase* (*COMT*), *cinnamoyl CoA reductase* (*CCR*) and *cinnamyl alcohol dehydrogenase* (*CAD*; Sibout *et al.*, 2005) genes result in the collapse of xylem TEs. This is probably due to the abortion of lignification-mediated mechanical strength conferred to the secondary walls. In contrast, *cinnamate-4-hydroxylase* (*C4H*), *caffeoyl CoA 3-O-methyltransferase* (*CCOMT*) and probably *ferulate-5-hydroxylase* (*F5H*) down-regulation/mutation do not, or do only slightly, affect the integrity of vascular, xylem, or TE cells, although lignin contents are significantly affected (reduced). Instead, some of these genetic manipulations produce coloured (pink-red or red-brown) xylem tissues that account for the modifications in monomeric lignin precursors (Anterola and Lewis, 2002). Peroxidases (Ros Barceló, 1998), and possibly laccases (Liu *et al.*, 1994), catalyse the final step of lignin polymerization. The evolution of H_2O_2 occurs in developing TEs and may have a role in peroxidase-mediated lignin deposition (Ros Barceló, 1998). In contrast, laccases and polyphenol oxidases possibly promote lignin formation without H_2O_2 (Liu *et al.*, 1994). As was the case in *Arabidopsis CAD-C* and *CAD-D* (Sibout *et al.*, 2005), gene redundancy may prevent the expression of complete loss-of-function phenotypes in *C4H*, *CCOMT* and *F5H*. This may also be the case for peroxidases and laccases, which often form multigene families in many plant species (Demura *et al.*, 2002; Veitch, 2004).

The regulation of expression of these enzyme genes is important not only for lignin synthesis but also for the construction of robust whole secondary walls (Anterola and Lewis, 2002). Genetic manipulations, including overexpression, of these genes affect the composition of lignins (Anterola and Lewis, 2002). Moreover, there are many reports indicating that the production and composition of other cell wall materials (cellulose and hemicellulose) are also affected (Boudet *et al.*, 2003). These findings indicate the occurrence of a complex crosstalk regulation of cell wall material synthesis.

Certain *MYB* class transcription factors control the co-ordinate expression of phenylpropanoid pathway enzyme genes. Such *MYB* genes have been reported in *Antirrhinum* (Tamagnone *et al.*, 1998), *Arabidopsis* (Borevitz *et al.*, 2000), pine (Patzlaff *et al.*, 2003), hybrid aspen (Karpinska *et al.*, 2004) and *Eucalyptus* (Goicoechea *et al.*, 2005). These *MYB* genes can regulate the expression of a set of phenylpropanoid pathway enzyme genes. When they are heterologously overexpressed or repressed, the lignin content and composition are significantly altered in transgenic plants (Goicoechea *et al.*, 2005). Promoter regions of the enzyme genes often contain characteristic *cis* elements to which the MYB factors can bind. Some of these elements promote expression of these genes (Goicoechea *et al.*, 2005), while others confer tissue specificity on their expression patterns (Hatton *et al.*, 1995).

Some phenylpropanoid pathway enzyme genes are expressed not only in TE cells but also in surrounding xylem parenchyma cells in the vascular bundles (Smith *et al.*, 1994). Xylem parenchyma cells also produce monomeric lignin precursors and probably supply them to nearby developing TEs (Savidge, 1989). The lignification of TE secondary walls appears to proceed even after TE PCD occurs (Ros Barceló, 2005). These supplied substances and hydrolysis-resistant peroxidase/laccases may be responsible for the post-mortem development of secondary cell walls, which is consistent with TE secondary wall formation in *Zinnia* culture (Fukuda and Komamine, 1982; Hosokawa *et al.*, 2001; Tokunaga *et al.*, 2005). This is a novel regulatory mechanism for the collective cell wall construction of specific cell types.

3.5.5 Cell wall component proteins

Plant cell walls contain distinct classes of structural proteins. These consist of glycine-rich protein (GRP), AGP, hydroxyproline-rich protein (HPRP), expansins, and other proteins that contain repetitive sequences and/or carbohydrates (e.g. Showalter, 2001). These cell wall component proteins are considered to confer additional properties to cell walls.

GRP genes have been isolated from various plant species such as the French bean (Keller *et al.*, 1989). An antibody raised against French bean GRP1.8 produced immunogold signals in the primary cell walls of vascular cells (Keller *et al.*, 1989). In particular, it recognizes the filamentous structure of the primary wall of protoxylem TEs, which seems to connect secondary wall hoops (Ryser, 2003). Although GRP1.8 associates itself with the pectin RG I, it forms a distinct structural network that may support TE wall structures against tension forces. Their occurrence is inevitable as the surrounding living tissue cells expand (Ryser *et al.*, 2004). Wall materials containing GRP1.8 are resistant to detergent (sodium dodecyl sulfate (SDS)) treatments combined with various hydrolytic activities such as those of proteinase K, pectinase and cellulase (Ryser *et al.*, 2004). The GRP1.8-containing network persists even after TEs died and autolytic activities were released into the cell walls (Ryser *et al.*, 2004). Interestingly, although the GRP1.8 protein is produced in xylem parenchyma cells, it is secreted into TE cell walls, where it functions (Ryser and Keller, 1992).

The addition of Yariv reagent suppresses TE differentiation (Motose *et al.*, 2004), which suggests the involvement of AGPs in the progression of TE differentiation. In addition to xylogen *ZeXYP1* and its *Arabidopsis* homologues *AtXYP1* and *2*, the *TRACHEARY ELEMENT DIFFERENTIATION3* (*TED3*) protein, which can be classified into the AGP class, may also be a candidate for such members (Milioni *et al.*, 2001). The *TED3* protein sequence includes an Asn-Gly-Tyr repeat motif that is characteristic of some AGPs (Demura *et al.*, 1994). Although *TED3* expression patterns mark the progression of stage II in *Zinnia* culture (Fukuda, 1996) and are well associated with TE precursor cells in situ (Igarashi *et al.*, 1998), the biological functions of such proteins remain to be elucidated.

Extensins belong to a distinct class of hydroxy-proline-rich glycoproteins (HRGPs) and include characteristic motifs of Ser-$(Hyp)_4$ and Tyr-Lys-Tyr in a repetitive fashion. They have putative roles in reinforcing the mechanical strength of cell walls and in responses to wounding/infection. A loblolly pine extensin-like protein is localized in the secondary walls of developing/developed xylem cells (Bao *et al.*, 1992).

Expansins are involved in cell wall loosening during cell growth. In stems of *Zinnia* seedlings, some of these expansin-like genes (*ZeExp1* and *2*) (Im *et al.*, 2000) are expressed in developing xylem cells, especially in those adjacent to primary xylem TEs. In hybrid aspen, expansin family genes whose expression is associated with wood, xylem fibres or secondary growth have also been characterized (Gray-Mitsumune *et al.*, 2004). Expansins or expansin-like proteins have not been reported at present in TE secondary walls.

3.6 The degradation of TE primary cell walls and pore formation

While differentiating TEs actively synthesize secondary walls, the primary walls located at the interval of thickened reticulate and annular secondary walls undergo degradation during TE maturation (O'Brien, 1970). In contrast, primary wall regions that provide scaffolds for secondary wall thickenings do not seem to suffer from degradation. The degradation of specific primary wall regions proceeds during TE PCD-mediated protoplast degeneration, suggesting that hydrolases exuded from the disintegrating TE protoplasts are responsible for the degradation processes (Esau and Charvat, 1978). However, the possibility that primary wall hydrolysis is due to the active secretion of hydrolases by live TE cells cannot be ruled out because O'Brien (1981) observed primary wall degradation in developing immature TE cells. A compatible explanation for these two varying results is that there is no absolute order between primary wall degradation and TE protoplast disruption. Rather, the progression of such processes may depend on species-, organ-, or tissue-specific situations in TE development.

Stacey *et al.* (1995) measured the levels of cell wall carbohydrates in *Zinnia* xylogenic cultures using specific antibodies. They detected several cell wall components in their culture media. Such substances may result from the secretion of polysaccharides and/or glycoproteins into the culture media from differentiating cells. While the HG epitope and fucosylated epitope of xyloglucan (which are recognized by monoclonal JM7 and CCRC-M1 antibodies, respectively) appear in both inductive and non-inductive culture media, an AGP epitope (which is recognized by JIM13) becomes far more abundant in media of inductive cultures than in non-inductive media. The HG epitope continues to remain in the culture media but exhibits significant turnover following cycles of continuous synthesis, secretion and degradation. The timing of AGP epitope appearance in the media coincides with the initiation of TE morphogenesis for a substantial percentage of cultured cells. Interestingly, the CCRC-M1-reactive fucosylated xyloglucan, which is released into media just before the beginning of secondary wall thickening, almost completely disappears for some time when TE morphogenesis proceeds. It then reappears in the media after developing TE cells undergo maturation. The medium of the developing TE-containing culture can block the binding of CCRC-Ml to the synthetic epitope. Heat, cold and protease treatments of such media compromised or reduced their ability to block the binding. This result strongly suggests that a specific hydrolytic activity for some epitope-containing carbohydrates occurs in the inductive medium. Developing TE cells are likely to synthesize and secrete the hydrolytic enzymes involved in the degradation of specific cell wall materials.

Ohdaira *et al.* (2002) addressed this problem by first analysing degrading TE cell walls at the biochemical level. They prepared cell wall specimens from cultured *Zinnia* cells, subjected them to a direct enzyme activity assay and monitored cell wall material hydrolysis in which both enzymes and substrates were derived from the same specimens. Cell wall material hydrolysis was monitored by the quantification of sugars released into a medium. According to their data, cells in inductive D-medium commence drastic cell wall material hydrolysis 48 h after culture initiation. In contrast, cells in non-inductive control media exhibit only low levels of cell wall degradation throughout the time course in culture. Hydrolytic reactions proceed faster at pH 5.5–6.5 than at other medium pH conditions. Sugars of various molecular weights were detected by gel matrix-based fractioning. Among such released sugars, uronic acids (glucuronic and galacturonic acids) are most abundant as monomeric components, and gas chromatographic analyses indicated that arabinose, galactose, glucose, rhamnose, xylose, fucose and mannose are included in cell wall hydrolytes. Results of linkage analyses indicated that sugars that formed (1→4)β-D-galactan, (1→5)α-L-arabinan, xyloglucan and (1→4)β-D-xylan are released into the medium. Therefore, drastic pectin and hemicellulose degradation occur upon TE differentiation. Alcian blue staining, which visualizes cell wall pectins in situ through specific reactions at pH 2.5, confirmed the reduction of pectic materials in TE cell walls. The previously observed primary cell wall degradation of TEs (O'Brien, 1981) was an example of pectic material (RG and HG) degradation.

Recently, Ryser (2003) made closer observations of primary cell wall degradation in soya bean hypocotyl protoxylem TEs. The author employed several monoclonal antibodies against plant cell wall polysaccharides as tools for monitoring degradation. One antibody, CCRC-M1, recognizes a fucosyl epitope of xyloglucan, and to a lesser extent, RG I. In immunoelectron microscopic images, the levels of CCRC-M1 binding to TE primary cell wall are scarce, whereas those in xylem parenchyma cell cytoplasm and walls are abundant. The middle lamella between these two different cell types showed slight levels of immunogold signals. TE primary cell wall xyloglucans disappear with the occurrence of TE PCD. These observations are consistent with the results of biochemical analyses of TE cell wall autolysis (Ohdaira *et al.*, 2002), which demonstrated active wall material mobilization and sugar release from xyloglucan. The degradation of xyloglucan may facilitate the passive extension of TEs (McCann and Roberts, 1991; Carpita and Gibeaut, 1993). In addition, the disappearance of HG at junctions between the primary and secondary cell walls is also observed with the JIM5 antibody (Ryser, 2003).

Characteristics of hydrolytic enzymes that are likely to be involved in such TE cell wall degradation have been reported. Cultured *Zinnia* cells specifically express a polygalacturonase isoenzyme (ZePG1) under a TE differentiation-inductive condition (Nakashima *et al.*, 2004). The ZePG1 protein appears at relatively early stages during the course of induction and continues to exist beyond the final stage of TE differentiation. *ZePG1* mRNA localizes in TEs and phloem tissues on *Zinnia* stem sections (Nakashima *et al.*, 2004). Immunoelectron microscopic analyses revealed that ZePG1 resides in TE primary and secondary cell walls and the Golgi bodies and vesicles at the subcellular level (Nakashima *et al.*, 2004). Some non-TE cells also contain ZePG1 in their Golgi bodies, vesicles and extending primary cell walls (Nakashima *et al.*, 2004). These results suggest that ZePG1 can exert its activity not only on the primary walls but also on the secondary wall regions of TEs.

A *Zinnia* pectate lyase gene, *Zepel*, has also been molecularly and enzymatically characterized (Domingo *et al.*, 1998). In cultured *Zinnia* cells, this gene exhibits auxin-induced expression at probably stage II, when active cell elongation occurs. *Zepel* transcripts are localized in vascular bundles and young leaf primordia in situ. Although biochemical characteristics of this Zepel pectin lyase suggest that its activity remains sub-optimal under physiological conditions, it is still possible that it functions in loosening cell walls to allow cell expansion by degrading pectic materials. Of course, the possibility that this enzyme is involved in the degradation of TE primary wall pectin cannot be ruled out. However, cultured *Zinnia* cells that seem to express this gene undergo tip growth under differentiation-inducible conditions (Roberts and Uhnak, 1998). Thus, it is particularly interesting to examine subcellular localization patterns of the *Zepel* gene product to explore a visible evidence for the detailed molecular basis of cellular tip growth.

Sequences similar to pectin methyl esterases, pectin acetyl esterases, polygalacturonases (PGs), and glucosyl hydrolase family enzymes such as β-galactosidases, α-L-arabinofuranosidases, and endo(1$\rightarrow$4)-β-glucanases are present in the *Zinnia* EST collection (Milioni *et al.*, 2001). Corresponding genes of these sequences are expressed in the middle to late stages of TE differentiation in *Zinnia* cell culture time courses.

Similarly, an expression analysis using a *Zinnia* EST-derived DNA microarray system identified cDNA fragments, the corresponding genes of which encode characteristic TE-associated glycosyl hydrolases (Demura *et al.*, 2002). Such a group of enzymes includes β-xylosidase, arabinoxylan arabinofuranohydrolase, β-galactosidase, β-glucosidase, β-xylanase, α-mannosidase and PG. Some *Arabidopsis* gene homologues that encode such glycosyl hydrolases are suggested to work in the secondary walls of xylem and other tissues (Goujon *et al.*, 2003).

Of these, a xylanase gene was further characterized. The promoter region of *Atxylanase-3* (*Atxyn3*) was fused with yellow fluorescent protein, attached with nuclear localizing signals, and introduced into *Arabidopsis* plants to examine tissue-specific expression patterns. The *Atxyn-3* promoter exhibited TE-associated activity in almost all tissues examined (Sawa *et al.*, 2005a). Using this transformant line as a background, *Arabidopsis* mutants showing ectopic *pAtxyn3* activities were isolated. Such mutants (*ate1*, *2* and *3*) contain ectopic TE cells at the site of ectopic expression of the reporter protein. Secondary wall thickening, lignification and cell death are also apparent in such cells. Even in these mutants, the expression of *Atxyn-3* is tightly associated with the formation of TEs, suggesting that the degradation of primary cell walls is (at least in part) systematically regulated by a genetic programme of TE differentiation.

Xyloglucan endotransglucosylases/hydrolases (XTHs) form a cluster of multigene families. One *Arabidopsis* XTH isoform, *AtXTH27*, exhibits vascular/TE-associated expression patterns in its rosette leaves (Matsui *et al.*, 2005). Both allelic T-DNA knockout plants with the *AtXTH27* gene showed smaller TE cells and fewer tertiary veins in the leaves than those of wild-type plants (Matsui *et al.*, 2005). Thus, *AtXTH27* is involved in the hydrolysis of TE primary walls by loosening cell walls so that TEs can elongate co-ordinately with the development of surrounding cells, even after TE PCD occurs.

Important roles of TE cell wall degradation are to alter wall properties to enable them to tolerate the high pressure of the transpiration water stream, and to confer mechanical strength to both vascular tissues and those of the whole plant body (Nakashima *et al.*, 2004). As noted above, lignin is a major component of mature TE cell walls that can serve in the construction of waterproof and robust walls. Enzymes like ZePG1 may work in the pectic wall material degradation of TEs for the substitution of hydrophilic (secondary) wall material for lignin-based water-resistant (hydrophobic) substances.

In vascular tissues, the formation of continuous vessel cell files, which are comprised of tandem arrays of single TEs, is necessary for efficient water transport. At the junctions of neighbouring single TE cells, pores appear that connect the interior spaces of these cells. Such characteristic structures result from the modification of TE cell walls. Partial cell wall degradation accounts for the formation of such porous structures. Differentiated TEs in the *Zinnia* suspension culture also form the cell wall pore structure. Nakashima *et al.* (2000) carried out scanning electron microscopic observations of cultured *Zinnia* TE cells and found that such structures develop only on one side of each cell. These pore structures arise from the primary cell wall of TEs. It is likely that hydrolytic enzymes of primary cell wall materials are involved in pore formation.

Why does such an asymmetrical structure occur in a single cell? This may be a consequence of the expression or maintenance of cellular polarity. Even in a single mesophyll cell, structures along longitudinal and transversal axes differ in general. The direction of the spiral of the secondary cell wall may also depend on cellular polarity. Several specific vascular cell-related biomolecules exhibit asymmetrical distribution patterns within a cell. The auxin influx and efflux carrier proteins, AUX1 (Swarup *et al.*, 2001) and PIN1 (Gälweiler *et al.*, 1998), respectively, are localized on the plasma membrane at either one side end of longitudinal cell axes. A putative hemicellulosic molecule, which can be recognized by immunofluorescence microscopy with the CN8 antibody, is detected on cultured *Zinnia* cell walls in a polarized manner (Shinohara *et al.*, 2000). Moreover, a local signalling factor for TE differentiation induction, xylogen, also resides at only the upper part of stem xylem-differentiating TE cells (Motose *et al.*, 2004). A system of polar vesicle transport may be involved in such asymmetric material deposition patterns on the cell wall or membrane proteins on the plasmalemma (Xu and Scheres, 2005). These molecules will provide precious clues for further detailed studies on the identification of plant cell polarity.

3.7 Co-regulation of cell wall degradation and PCD

Cell wall degradation sometimes accompanies PCD during plant development. PCD plays a critical role in successful plant development and defensive actions (Pennell and Lamb, 1997). In vascular plants, TEs, root cap cells, senescent organ cells and reproductive organ cells commit developmental PCD to aid with efficient water/mineral transport, nutrition, nutrient recycling and reproduction at the organismal level (Pennell and Lamb, 1997; Kuriyama and Fukuda, 2002). In addition, plant (mainly leaf) cells undergo PCD to fight infection following invasion by various pathogens (Pennell and Lamb, 1997). There are several examples that associate or implicate cell death with cell wall degradation, e.g. aerenchyma formation in rice root cortex cells (Evans, 2004), death of the stem cortex, pith, pericycle cells (van Doorn and Woltering, 2005) and root cap cells (Wang *et al.*, 1996), characteristic leaf shape development in lace plants (Gunawardena *et al.*, 2004), collapse of the endosperm leading to nutrient mobilization (van Doorn and Woltering, 2005) and TE cell death (Ohdaira *et al.*, 2002). Even in a hypersensitive response, which can serve as a powerful plant defense mechanism against invasive pathogens (Pennell and Lamb, 1997), traces of cell wall modifications can be assumed (Schenk *et al.*, 2000). Although cell wall degradation manifests during organ abscission, the question as to whether cells in such places commit PCD has not been fully addressed.

In the development of root cap cells and endosperm cells (sometimes called aleurone cells in *Arabidopsis* seeds; Haughn and Chaudhury, 2005), cell separation occurs. Columella root cap cells perceive gravity with their characteristic large amyloplasts (Blancaflor, 2002). However, the cells lose such amyloplasts to become vacuolated, a process that occurs to the very ends of the root tips (Plate 6). At the surface of the root, such cells begin to separate from tissues (Barlow, 2002; and see Chapter 5). Lateral root cap cells also detach from roots at their surfaces (Barlow, 2002). Generally, dead cells can be observed on root surfaces (Wang *et al.*, 1996), i.e. cell death occurs concomitantly with, or even after, the initiation of cell separation. Some *Arabidopsis* aleurone cells (Haughn and Chaudhury, 2005) become vacuolated, spherical and isolated following germination, radicle protrusion and embryo hatching (Plate 6). The death of such cells then becomes apparent.

Plant PCD sometimes causes the disappearance of dead cells, such as those of the root cortex, stem pith, pericycle, tapetum, septum, stomium, three of the four nascent meiotic angiosperm megaspores, unpollinated ovaries, endosperm and embryonic suspensor cells (Pennell and Lamb, 1997; van Doorn and Woltering, 2005). Sculpturing leaf shapes in lace plant development also involves PCD (Gunawardena *et al.*, 2004). At first, expanding lace plant leaves are filled with cells. However, many perforations become conspicuous as the leaves further develop. These perforations arise from mesophyll cells in non-vascular tissue areas. The disappearance of these cells postulates the complete cell wall digestion of dead or dying cells. Active cell wall hydrolases must occur upon PCD in such cases. Mechanisms to integrate cell death and cell wall digestion should work in these systems.

How does cell wall degradation coincide with autolytic digestion in TEs? TE PCD mechanisms involve the synthesis and/or activation of various hydrolytic enzymes, changes in vacuolar membrane properties, the subsequent disintegration of the vacuolar membrane and the digestion of cellular contents (autolysis; e.g. Fukuda, 2000). Studies on such autolysis-related enzyme activities revealed that DNases, RNases and proteases work in degradative processes (Thelen and Northcote, 1989; Minami and Fukuda, 1995; Ye and Droste, 1996; Ye and Varner, 1996). The *ZEN1* gene encodes a Zn^{2+}-dependent nuclease (Aoyagi *et al.*, 1998), and the expression patterns of *ZEN1* mRNA and its protein product correlate well with the occurrence of TE PCD/autolysis (Aoyagi *et al.*, 1998; Ito and Fukuda, 2002). Transformation of cultured *Zinnia* cells with a construct for the expression of an antisense *ZEN1* cRNA culminates in the repression of nuclear DNA digestion in dead TE cells (Ito and Fukuda, 2002). Thus, ZEN1 is required for the destruction of dead TE nuclei.

ZRNase I mRNA also exhibits transient expression patterns with a peak of expression level at the timing of semisynchronous TE morphogenesis (Ye and Droste, 1996). The *ZRNase I* protein contains an ER-retention signal (HDEL; Ye and Droste, 1996), implying that a protein transport mechanism of differentiating TEs shares some common features with one of those observed in *Ricinus communis* endosperm

(Schmid *et al.*, 1999), *Vigna mungo* endosperm (Toyooka *et al.*, 2000) and *Arabidopsis* cotyledon epidermal cells (Hayashi *et al.*, 2001). In these cells, ER-retention signal-containing hydrolases (proteases) migrate via ER-derived organelles.

p48h-17 and *ZCP4* are genes of a distinct class of cysteine proteases, the sequences of which are highly homologous to those of papain (Ye and Varner, 1996). In culture, these genes exhibit TE-specific expression patterns. Protease activities of their putative products become the highest under acid pH conditions, suggesting that they degrade proteinaceous molecules in TEs following TE PCD (Minami and Fukuda, 1995; Beers and Freeman, 1997).

Many of these hydrolytic enzymes are considered to accumulate in TE vacuoles (Fukuda, 2000), the disruption of which plays a pivotal role in the execution of TE PCD (Fukuda, 1992; Jones and Dangl, 1996). However, vacuolar disruption itself seems to be sufficient to cause an immediate, catastrophic and irreversible damage to cell viability (Groover *et al.*, 1997). Thus, the TE vacuole needs to be, at least in part, functional for some time after the onset of TE morphogenesis (i.e. secondary wall thickening; Obara *et al.*, 2001; Kuriyama and Fukuda, 2002). The earliest sign of vacuolar malfunction during TE development is a change in tonoplast semipermeability, which appears almost concurrently with the secondary walls (Kuriyama, 1999). Fluorescein, a fluorescent probe, cannot reside in TE vacuoles. Two mechanisms account for such a vacuolar change. The first is that fluorescein cannot enter TE vacuoles by moving across the tonoplasts. The second is that TE vacuoles cannot retain the dye. Intriguingly, fluorescein transport into the vacuoles of non-TE cells can also be inhibited by the administration of the lipophilic drug, probenecid. The probenecid treatment of cultured *Zinnia* cells results in the accelerated disruption of TE vacuoles, leading to TE cell death. However, the vacuoles of non-TE cells do not collapse as quickly with this treatment: it takes a long time to kill non-TE cells with probenecid. Probenecid is an inhibitor of ABC transporter function (Wright and Oparka, 1994). Generally, the transport of xenobiotic molecules such as fluorescein is mediated by certain classes of ABC transporters (Wright and Oparka, 1994). In light of these findings, some ABC transporters in the tonoplasts relate TE vacuole disruption to changes in semipermeability. TE vacuole disruption depends on the protein synthesis of differentiating TEs (Kuriyama, 1999).

The genes of TE-associated cell wall material hydrolases, secondary wall synthetic enzymes (such as a putative lignin-polymerizing peroxidase, *ZPO-C*; Sato *et al.*, 2006) and vacuolar hydrolases begin to be expressed in parallel (Demura *et al.*, 2002). Fukuda and co-workers consider that the expression of genes of these enzymes is commonly regulated by endogenous brassinosteroids (Yamamoto *et al.*, 2001). In the *Zinnia* cell culture system, the brassinosteroid-mediated expression of these genes provides the molecular basis for the tight coupling of secondary wall thickening, primary wall hydrolysis and the autolysis of TEs (Fukuda, 2000; Demura *et al.*, 2002).

The analysis of developmental roles of a cell wall degrading enzyme suggests the existence of another type of mechanism that couples cell death with cell separation. Orozco-Cárdenas and Ryan (2003) showed that the antisense suppression of a PG β-subunit gene led to a disorder in the regulation of PG activity in tomato plants.

Based on their results, the PG β-subunit is required for the maintenance of proper levels of PG activity. Resultant excessively high PG activity causes significant abscission in developing flowers, invokes severe wounding responses in unwounded tissues, promotes ROS production and induces cell death in the leaves and other tissues of this transformant. It is currently unknown as to whether this enhanced PG activation exerts similar effects on other cell death cases. However, it cannot be ruled out that some cell wall hydrolases can execute cell death directly or via the control of a specific genetic programme. *Zinnia* PG or other cell wall hydrolases might also assist in TE cell death.

In addition, leaves suffer from extensive lesion formation in *AtXTH27* knockout plants (Matsui *et al.*, 2005). This appears to be the case opposite to that of tomato plants with enhanced PG activity (Orozco-Cárdenas and Ryan, 2003). Because comprehensive tissue-specific activity assays of cell wall degrading enzymes have yet to be performed, the activity really responsible for lesion formation in knockout plants remains obscure. However, if the missing (or reduced) total XTH activity directly causes lesion formation, cell death may be induced not only by excess degradation of cell wall materials (as was the case in the above hyperactive PG; Orozco-Cárdenas and Ryan, 2003) but also by a lack of proper hydrolysis-mediated cell wall expansion during tissue development. Cell wall hydrolases might exert dual effects on cell death regulation.

3.8 Conclusion

Although the effects have not been conspicuously highlighted, processes of vascular/TE cell differentiation associate or implicate cell separation-related events. Genes of wall-degrading enzymes that appear to be isotypes of those involved in organ abscission, dehiscence and fruit ripening are expressed concomitantly with the timing of TE morphogenesis and PCD. Some are considered to work in the development of the proper architecture of vessel-integrating leaves (e.g. Matsui *et al.*, 2005). Others may be involved in alterations of TE cell wall properties (e.g. Nakashima *et al.*, 2004). Future analyses will focus on understanding the functional details and regulatory mechanisms of TE-associated cell wall hydrolysing enzyme genes. Understanding molecular mechanisms, such as how mechanical cell wall breakage mediates cellular wounding responses, and how such responses are involved in vascular/TE cell differentiation, is also very important. Additionally, it has been shown that some wall-related molecules (such as oligosaccharides) can induce specific cellular responses such as defense reactions (Pilling and Höfte, 2003; Vorwerk *et al.*, 2004). Thus, it is possible that there are other novel cell wall components that act as signalling molecules to regulate fine aspects of vascular tissue development such as TE differentiation. To explore such wall-derived signalling substances, media of *Zinnia* xylogenic culture may be particularly useful. To date, at least two different molecules have been successfully isolated from this culture medium (Roberts *et al.*, 1997; Motose *et al.*, 2004). Such analytical methods will reveal novel aspects of plant cell wall functions in vascular/TE cell development.

References

Aloni, R. (1987) Differentiation of vascular tissues. *Annual Review of Plant Physiology* **38**, 179–204.

Aloni, R. (2004) The induction of vascular tissue by auxin. In: *Plant Hormones: Biosynthesis, Signal Transduction, Action!* (ed. Davies, P.J.), pp. 471–492. Kluwer Academic Publishers, Dordrecht, Boston, London.

Anterola, A.M. and Lewis, N.G. (2002) Trends in lignin modification: a comprehensive analysis of the effects of genetic manipulations/mutations on lignification and vascular integrity. *Phytochemistry* **61** (3), 221–294.

Aoyagi, S., Sugiyama, M. and Fukuda, H. (1998) *BEN1* and *ZEN1* cDNAs encoding S1-type DNases that are associated with programmed cell death in plants. *FEBS Letters* **429** (2), 134–138.

Asami, T., Min, Y.K., Nagata, N., Yamagishi, K., Takatsuto, S., Fujioka, S., Murofushi, N., Yamaguchi, I. and Yoshida, S. (2000) Characterization of brassinazole, a triazole-type brassinosteroid biosynthesis inhibitor. *Plant Physiology* **123** (1), 93–100.

Baima, S., Possenti, M., Matteucci, A., Wisman, E., Altamura, M.M., Ruberti, I. and Morelli, G. (2001) The Arabidopsis ATHB-8 HD-Zip protein acts as a differentiation-promoting transcription factor of the vascular meristems. *Plant Physiology* **126** (2), 643–655.

Bao, W., O'Malley, D.M. and Sederoff, R.R. (1992) Wood contains a cell-wall structural protein. *Proceedings of the National Academy of Sciences of the United States of America* **89** (14), 6604–6608.

Barlow, P.W. (2002) The root cap: cell dynamics, cell differentiation and cap function. *Journal of Plant Growth Regulation* **21** (4), 261–286.

Beers, E.P. and Freeman, T.B. (1997) Proteinase activity during tracheary element differentiation in zinnia mesophyll cultures. *Plant Physiology* **113** (3), 873–880.

Benjamins, R., Quint, A., Weijers, D., Hooykaas, P. and Offringa, R. (2001) The PINOID protein kinase regulates organ development in *Arabidopsis* by enhancing polar auxin transport. *Development* **128** (20), 4057–4067.

Berleth, T., Mattsson, J. and Hardtke, C.S. (2000) Vascular continuity and auxin signals. *Trends in Plant Sciences* **5** (9), 387–393.

Blancaflor, E.B. (2002) The cytoskeleton and gravitropism in higher plants. *Journal of Plant Growth Regulation* **21** (2), 120–136.

Bohmert, K., Camus, I., Bellini, C., Bouchez, D., Caboche, M. and Benning, C. (1998) *AGO1* defines a novel locus of *Arabidopsis* controlling leaf development. *EMBO Journal* **17** (1), 170–180.

Bolwell, G.P., Dalessandro, G. and Northcote, D.H. (1985) Decrease of polygalacturonic acid synthase during xylem differentiation in sycamore. *Phytochemistry* **24** (4), 699–702.

Bolwell, G.P. and Northcote, D.H. (1981) Control of hemicellulose and pectin synthesis during differentiation of vascular tissue in bean (*Phaseolus vulgaris*) callus and in bean hypocotyls. *Planta* **152** (3), 225–233.

Bonke, M., Thitamadee, S., Mähönen, A.P., Hauser, M.T. and Helariutta, Y. (2003) APL regulates vascular tissue identity in *Arabidopsis*. *Nature* **426** (6963), 181–186.

Borevitz, J.O., Xia, Y., Blount, J., Dixon, R.A. and Lamb, C. (2000) Activation tagging identifies a conserved MYB regulator of phenylpropanoid biosynthesis. *Plant Cell* **12** (12), 2383–2394.

Boudet, A.M., Kajita, S., Grima-Pettenati, J. and Goffner, D. (2003) Lignins and lignocellulosics: a better control of synthesis for new and improved uses. *Trends in Plant Science* **8** (12), 576–581.

Bouton, S., Leboeuf, E., Mouille, G., Leydecker, M.T., Talbotec, J., Granier, F., Lahaye, M., Höfte, H. and Truong, H.N. (2002) *QUASIMODO1* encodes a putative membrane-bound glycosyltransferase required for normal pectin synthesis and cell adhesion in Arabidopsis. *Plant Cell* **14** (10), 2577–2590.

Bowman, J.L., Eshed, Y. and Baum, S.F. (2002) Establishment of polarity in angiosperm lateral organs. *Trends in Genetics* **18** (3), 134–141.

Burgess, J. and Linstead, P. (1984) *In-vitro* tracheary element formation: structural studies and the effect of tri-iodobenzonic acid. *Planta* **160** (6), 481–489.

Canales, C., Grigg, S. and Tsiantis, M. (2005) The formation and patterning of leaves: recent advances. *Planta* **221** (6), 752–756.

Carland, F.M., Berg, B.L., FitzGerald, J.N., Jianamornphongs, S., Nelson, T. and Keith, B. (1999) Genetic regulation of vascular tissue patterning in Arabidopsis. *Plant Cell* **11** (11), 2123–2137.

Carland, F.M., Fujioka, S., Takatsuto, S., Yoshida, S. and Nelson, T. (2002) The identification of *CVP1* reveals a role for sterols in vascular patterning. *Plant Cell* **14** (9), 2045–2058.

Carpita, N.C. and Gibeaut, D.M. (1993) Structural models of primary cell walls in flowering plants: consistency of molecular structure with the physical properties of the walls during growth. *Plant Journal* **3** (1), 1–30.

Choe, S., Noguchi, T., Fujioka, S., Takatsuto, S., Tissier, C.P., Gregory, B.D., Ross, A.S., Tanaka, A., Yoshida, S., Taxl, F.E. and Feldmann, K.A. (1999) The *Arabidopsis dwf7/ste1* mutant is defective in the Δ^7 sterol C-5 desaturation step leading to brassinosteroid biosynthesis. *Plant Cell* **11** (2), 207–221.

Church, D.L. and Galston, A.W. (1989) Hormonal induction of vascular differentiation in cultured *Zinnia* leaf-disks. *Plant and Cell Physiology* **30** (1), 73–78.

Clay, N.K. and Nelson, T. (2002) VH1, a provascular cell-specific receptor kinase that influences leaf cell patterns in *Arabidopsis*. *Plant Cell* **14** (11), 2707–2722.

Demura, T. and Fukuda, H. (1994) Novel vascular cell-specific genes whose expression is regulated temporally and spatially during vascular system development. *Plant Cell* **6** (7), 967–981.

Demura, T., Tashiro, G., Horiguchi, G., Kishimoto, N., Kubo, M., Matsuoka, N., Minami, A., Nagata-Hiwatashi, M., Nakalmura, K., Okamura, Y., Sassa, N., Suzuki, S., Yazaki, J., Kikuchi, S. and Fukuda, H. (2002) Visualization by comprehensive microarray analysis of gene expression programs during transdifferentiation of mesophyll cells into xylem cells. *Proceedings of the National Academy of Sciences of the United States of America* **99** (24), 15794–15799.

Dengler, N. (2001) Regulation of vascular development. *Journal of Plant Growth Regulation* **20** (1), 1–13.

Desprez, T., Vernhettes, S., Fagard, M., Refrégier, G., Desnos, T., Aletti, E., Py, N., Pelletier, S., and Höfte, H. (2002) Resistance against herbicide isoxaben and cellulose deficiency caused by distinct mutations in same cellulose synthase isoform CESA6. *Plant Physiology* **128** (2), 482–490.

Dharmasiri, N., Dharmasiri, S. and Estelle, M. (2005) The F-box protein TIR1 is an auxin receptor. *Nature* **435** (7041), 441–445.

Diener, A.C., Li, H., Zhou, W., Whoriskey, W.J., Nes, W.D. and Fink, G.R. (2000) *STEROL METHYLTRANSFERASE 1* controls the level of cholesterol in plants. *Plant Cell* **12** (6), 853–870.

Dinneny, J.R. and Yanofsky, M.F. (2004) Vascular patterning: xylem or phloem? *Current Biology* **14** (3), R112–R114.

Domingo, C., Roberts, K., Stacey, N., Connerton, I., Ruíz-Teran, F. and McCann, M. (1998) A pectate lyase from *Zinnia elegans* is auxin inducible. *Plant Journal* **13** (1), 17–28.

Emery, J.F., Floyd, S.K., Alvarez, J., Eshed, Y., Hawker, N.P., Izhaki, A., Baum, S.F. and Bowman, J.L. (2003) Radial patterning of Arabidopsis shoots by class III HD-ZIP and KANADI genes. *Current Biology* **13** (20), 1768–1774.

Esau, K. and Charvat, I. (1978) Vessel member differentiation in bean (*Phaseolus vulgaris* L.). *Annals of Botany* **42** (179), 665–677.

Evans, D.E. (2004) Aerenchyma formation. *New Phytologist* **161** (1), 35–49.

Falconer, M.M. and Seagull, R.W. (1985a) Immunofluorescent and calcofluor white staining of developing tracheary elements in *Zinnia elegans* L. suspension cultures. *Protoplasma* **125** (3), 190–198.

Falconer, M.M. and Seagull, R.W. (1985b) Xylogenesis in tissue culture: taxol effects on microtubule reorientation and lateral association in differentiating cells. *Protoplasma* **128** (2–3), 157–166.

Fletcher, J.C. (2002) Coordination of cell proliferation and cell fate decisions in the angiosperm shoot apical meristem. *Bioessays* **24** (1), 27–37.

Freundlich, H.F., (1909) Entwicklung und Regeneration von Gefässbündeln in Blattgebilden. *Jahrbücher für wissenschaftliche Botanik* **46** (2), 137–206.

Friml, J., Yang, X., Michniewicz, M., Weijers, D., Quint, A., Tietz, O., Benjamins, R., Ouwerkerk, P.B., Ljung, K., Sandberg, G., Hooykaas, P.J., Palme, K. and Offringa, R. (2004) A PINOID-dependent binary switch in apical-basal PIN polar targeting directs auxin efflux. *Science* **306** (5697), 862–865.

Fukuda, H. (1992) Tracheary element formation as a model system of cell differentiation. *International Review of Cytology – a Survey of Cell Biology* **136**, 289–332.

Fukuda, H. (1996) Xylogenesis: initiation, progression and cell death. *Annual Review in Plant Physiology and Plant Molecular Biology* **47**, 299–325.

Fukuda, H. (2000) Programmed cell death of tracheary elements as a paradigm in plants. *Plant Molecular Biology* **44** (3), 245–253.

Fukuda, H. (2004) Signals that control plant vascular cell differentiation. *Nature Reviews Molecular Cell Biology* **5** (5), 379–391.

Fukuda, H. and Komamine, A. (1980) Establishment of an experimental system for the tracheary element differentiation from single cells isolated from the mesophyll of *Zinnia elegans*. *Plant Physiology* **65** (1), 57–60.

Fukuda, H. and Komamine, A. (1982) Lignin synthesis and its related enzymes as markers of tracheary-element differentiation in single cells isolated from the mesophyll of *Zinnia elegans*. *Planta* **155** (5), 423–430.

Fukuda, H. and Komamine, A. (1985) Cytodifferentiation. In: *Cell Culture and Somatic Cell Genetics in Plants* , Vol. 2 (ed. Vasil, I.K.), pp. 149–212. Academic Press, Orlando, FL.

Gälweiler, L., Guan, C., Müller, A., Wisman, E., Mendgen, K., Yephremov, A. and Palme, K. (1998) Regulation of polar auxin transport by *AtPIN1* in *Arabidopsis* vascular tissue. *Science* **282** (5397), 2226–2230.

Gardiner, J.C., Taylor, N.G. and Turner, S.R. (2003) Control of cellulose synthase complex localization in developing xylem. *Plant Cell* **15** (8), 1740–1748.

Goicoechea, M., Lacombe, E., Legay, S., Mihaljevic, S., Rech, P., Jauneau, A., Lapierre, C., Pollet, B., Verhaegen, D., Chaubet Gigot, N. and Grima-Pettenati, J. (2005) *EgMYB2*, a new transcriptional activator from *Eucalyptus* xylem, regulates secondary cell wall formation and lignin biosynthesis. *Plant Journal* **43** (4), 553–567.

Goujon, T., Minic, Z., El Amrani, A., Lerouxel, O., Aletti, E., Lapierre, C., Joseleau, J.P. and Jouanin, L. (2003) *AtBXL1*, a novel higher plant (*Arabidopsis thaliana*) putative beta-xylosidase gene, is involved in secondary cell wall metabolism and plant development. *Plant Journal* **33** (4), 677–690.

Gray-Mitsumune, M., Mellerowicz, E.J., Abe, H., Schrader, J., Winzell, A., Sterky, F., Blomqvist, K., McQueen-Mason, S., Teeri, T.T. and Sundberg, B. (2004) Expansins abundant in secondary xylem belong to subgroup A of the α-expansin gene family. *Plant Physiology* **135** (3), 1552–1564.

Groover, A., DeWitt, N., Heidel, A. and Jones, A. (1997) Programmed cell death of plant tracheary elements differentiating *in vitro*. *Protoplasma* **196** (3–4), 197–211.

Guillemin, F., Guillon, F., Bonnin, E., Devaux, M.F., Chevalier, T., Knox, J.P., Liners, F. and Thibault, J.F. (2005) Distribution of pectic epitopes in cell walls of the sugar beet root. *Planta* **222** (2), 355–371.

Gunawardena, A.H., Greenwood, J.S. and Dengler, N.G. (2004) Programmed cell death remodels lace plant leaf shape during development. *Plant Cell* **16** (1), 60–73.

Hamann, T., Benkova, E., Bäurle, I., Kientz, M. and Jürgens, G. (2002) The *Arabidopsis BODENLOS* gene encodes an auxin response protein inhibiting MONOPTEROS-mediated embryo patterning. *Genes and Development* **16** (13), 1610–1615.

Hamann, T., Mayer, U. and Jürgens, G. (1999) The auxin-insensitive *bodenlos* mutation affects primary root formation and apical-basal patterning in the *Arabidopsis* embryo. *Development* **126** (7), 1387–1395.

Handford, M.G., Baldwin, T.C., Goubet, F., Prime, T.A., Miles, J., Yu, X. and Dupree, P. (2003) Localisation and characterisation of cell wall mannan polysaccharides in *Arabidopsis thaliana*. *Planta* **218** (1), 27–36.

Hardtke, C.S. and Berleth, T. (1998) The *Arabidopsis* gene *MONOPTEROS* encodes a transcription factor mediating embryo axis formation and vascular development. *EMBO Journal* **17** (5), 1405–1411.

Hatton, D., Sablowski, R., Yung, M.H., Smith, C., Schuch, W. and Bevan, M.(1995) Two classes of *cis* sequences contribute to tissue-specific expression of a *PAL2* promoter in transgenic tobacco. *Plant Journal* **7** (6), 859–876.

Haughn, G. and Chaudhury, A. (2005) Genetic analysis of seed coat development in *Arabidopsis*. *Trends in Plant Science* **10** (10), 472–477.

Hayashi, Y., Yamada, K., Shimada, T., Matsushima, R., Nishizawa, N.K., Nishimura, M. and Hara-Nishimura, I. (2001) A proteinase-sorting body that prepares for cell death or stresses in the epidermal cells of Arabidopsis. *Plant and Cell Physiology* **42** (9), 894–899.

Hellmann, H. and Estelle, M. (2002) Plant development: regulation by protein degradation. *Science* **297** (5582), 793–797.

Hellmann, H., Hobbie, L., Chapman A., Dharmasiri, S., Dharmasiri, N., del Pozo, C., Reinhardt, D. and Estelle, M. (2003) Arabidopsis *AXR6* encodes CUL1 implicating SCF E3 ligases in auxin regulation of embryogenesis. *EMBO Journal* **22** (13), 3314–3325.

Hobbie, L., McGovern, M., Hurwitz, L.R., Pierro, A., Liu, N.Y., Bandyopadhyay, A. and Estelle, M. (2000) The *axr6* mutants of *Arabidopsis thaliana* define a gene involved in auxin response and early development. *Development* **127** (1), 23–32.

Hogetsu, T. (1990) Detection of hemicelluloses specific to the cell-wall of tracheary elements and phloem cells by fluorescein-conjugated lectins. *Protoplasma* **156** (1–2), 67–73.

Hosokawa, M., Suzuki, S., Umezawa, T. and Sato, Y. (2001) Progress of lignification mediated by intercellular transportation of monolignols during tracheary element differentiation of isolated *Zinnia* mesophyll cells. *Plant and Cell Physiology* **42** (9), 959–968.

Igarashi, M., Demura, T. and Fukuda, H. (1998) Expression of the *Zinnia TED3* promoter in developing tracheary elements of transgenic *Arabidopsis*. *Plant Molecular Biology* **36** (6), 917–927.

Im, K.H., Cosgrove, D.J. and Jones, A.M. (2000) Subcellular localization of expansin mRNA in xylem cells. *Plant Physiology* **123** (2), 463–470.

Ito, J. and Fukuda, H. (2002) ZEN1 is a key enzyme in degradation of nuclear DNA during programmed cell death of tracheary elements. *Plant Cell* **14** (12), 3201–3211.

Jacobs, W.P. (1952) The role of auxin in differentiation of xylem around a wound. *American Journal of Botany* **39** (5), 301–309.

Jang, J.C., Fujioka, S., Tasaka, M., Seto, H., Takatsuto, S., Ishii, A., Aida, M., Yoshida, S. and Sheen, J. (2000) A critical role of sterols in embryonic patterning and meristem programming revealed by the *fackel* mutants of *Arabidopsis thaliana*. *Genes and Development* **14** (12), 1485–1497.

Jones, A.M. and Dangl, J.L. (1996) Logjam at the Styx: programmed cell death in plants. *Trends in Plant Science* **1** (4), 114–119.

Jun, J.H., Ha, C.M. and Nam, H.G. (2002) Involvement of the *VEP1* gene in vascular strand development in *Arabidopsis thaliana*. *Plant and Cell Physiology* **43** (3), 323–330.

Karlsson, M., Melzer, M., Prokhorenko, I., Johansson, T. and Wingsle, G. (2005) Hydrogen peroxide and expression of hipI-superoxide dismutase are associated with the development of secondary cell walls in *Zinnia elegans*. *Journal of Experimental Botany* **56** (418), 2085–2093.

Karpinska, B., Karlsson, M., Srivastava, M., Stenberg, A., Schrader, J., Sterky, F., Bhalerao, R. and Wingsle, G. (2004) MYB transcription factors are differentially expressed and regulated during secondary vascular tissue development in hybrid aspen. *Plant Molecular Biology* **56** (2), 255–270.

Keller, B., Schmid, J. and Lamb, C.J. (1989) Vascular expression of a bean cell wall glycine-rich protein-β-glucuronidase gene fusion in transgenic tobacco. *EMBO Journal* **8** (5), 1309–1314.

Kepinski, S. and Leyser, O. (2005) The Arabidopsis F-box protein TIR1 is an auxin receptor. *Nature* **435** (7041), 446–451.

Kim, J., Jung, J.H., Reyes, J.L., Kim, Y.S., Kim, S.Y., Chung, K.S., Kim, J.A., Lee, M., Lee, Y., Narry Kim, V., Chua, N.H. and Park, C.M. (2005) microRNA-directed cleavage of *ATHB15* mRNA regulates vascular development in Arabidopsis inflorescence stems. *Plant Journal* **42** (1), 84–94.

Kobayashi, H., Fukuda, H. and Shibaoka, H. (1988) Interrelation between the spatial disposition of actin-filaments and microtubules during the differentiation of tracheary elements in cultured *Zinnia* cells. *Protoplasma* **143** (1), 29–37.

Kohlenbach, H.W., Korber, M. and Li, L. (1982) Cytodifferentiation of protoplasts isolated from a stem embryo system of *Brassica napus* to tracheary elements. *Zeitschrift für Pflanzenphysiologie* **107** (4), 367–371.

Kohlenbach, H.W. and Schmidt, B. (1975) Cytodifferentiation in mode of a direct transformation of isolated mesophyll-cells to tracheids. *Zeitschrift für Pflanzenphysiologie* **75** (4), 369–374.

Kohlenbach, H.W. and Schöpke, C. (1981) Cytodifferentiation to tracheary elements from isolated mesophyll protoplasts of *Zinnia elegans*. *Naturwissenschaften* **68** (11), 576–577.

Koizumi, K., Naramoto, S., Sawa, S., Yahara, N., Ueda, T., Nakano, A., Sugiyama, M. and Fukuda, H. (2005) VAN3 ARF-GAP-mediated vesicle transport is involved in leaf vascular network formation. *Development* **132** (7), 1699–1711.

Kubo, M., Udagawa, M., Nishikubo, N., Horiguchi, G., Yamaguchi, M., Ito, J., Mimura, T., Fukuda, H. and Demura, T. (2005) Transcription switches for protoxylem and metaxylem vessel formation. *Genes and Development* **19** (16), 1855–1860.

Kurek, I., Kawagoe, Y., Jacob-Wilk, D., Doblin, M. and Delmer, D. (2002) Dimerization of cotton fiber cellulose synthase catalytic subunits occurs via oxidation of the zinc-binding domains. *Proceedings of the National Academy of Sciences of the United State of America* **99** (17), 11109–11114.

Kuriyama, H. (1999) Loss of tonoplast integrity programmed in tracheary element differentiation. *Plant Physiology* **121** (3), 763–774.

Kuriyama, H. and Fukuda, H. (2001) Regulation of tracheary element differentiation. *Journal of Plant Growth Regulation* **20** (1), 35–51.

Kuriyama, H. and Fukuda, H. (2002) Developmental programmed cell death in plants. *Current Opinion in Plant Biology* **5** (6), 568–573.

Lincoln, C., Long, J., Yamaguchi, J., Serikawa, K. and Hake, S. (1994) A *knotted1*-like homeobox gene in Arabidopsis is expressed in the vegetative meristem and dramatically alters leaf morphology when overexpressed in transgenic plants. *Plant Cell* **6** (12), 1859–1876.

Liu, L., Dean, J.F.D., Friedman, W.E. and Eriksson, K.E.L. (1994) A laccase-like phenoloxidase is correlated with lignin biosynthesis in *Zinnia elegans* stem tissues. *Plant Journal* **6** (2), 213–224.

Matsubayashi, Y., Takagi, L., Omura, N., Morita, A. and Sakagami, Y. (1999) The endogenous sulfated pentapeptide phytosulfokine-α stimulates tracheary element differentiation of isolated mesophyll cells of zinnia. *Plant Physiology* **120** (4), 1043–1048.

Matsui, A., Yokoyama, R., Seki, M., Ito, T., Shinozaki, K., Takahashi, T., Komeda, Y. and Nishitani, K. (2005) AtXTH27 plays an essential role in cell wall modification during the *Development* of tracheary elements. *Plant Journal* **42** (4), 525–534.

Mayer, U., Torres-Ruiz, R.A., Berleth, T., Miséra, S. and Jürgens, G. (1991) Mutations affecting body organization in the *Arabidopsis* embryo. *Nature* **353** (6343), 402–407.

McCann, M.C., and Roberts, K. (1991) Architecture of the primary cell wall. In: *The Cytoskeletal Basis of Plant Growth and Form* (ed. Lloyd, C.W.), pp. 109–129. Academic Press, London.

McCartney, L., Marcus, S.E. and Knox, J.P. (2005) Monoclonal antibodies to plant cell wall xylans and arabinoxylans. *Journal of Histochemistry and Cytochemistry* **53** (4), 543–546.

McConnell, J.R., Emery, J., Eshed, Y., Bao, N., Bowman, J. and Barton M.K. (2001) Role of *PHABULOSA* and *PHAVOLUTA* in determining radial patterning in shoots. *Nature* **411** (6838), 709–713.

Milioni, D., Sado, P.E., Stacey, N.J., Domingo, C., Roberts, K. and McCann, M.C. (2001) Differential expression of cell-wall-related genes during the formation of tracheary elements in the *Zinnia* mesophyll cell system. *Plant Molecular Biology* **47** (1–2), 221–238.

Milioni, D., Sado, P.E., Stacey, N.J., Roberts, K. and McCann, M.C. (2002) Early gene expression associated with the commitment and differentiation of a plant tracheary element is revealed by cDNA-amplified fragment length polymorphism analysis. *Plant Cell* **14** (11), 2813–2824.

Minami, A. and Fukuda, H. (1995). Transient and specific expression of a cysteine endopeptidase associated with autolysis during differentiation of *Zinnia* mesophyll cells into tracheary elements. *Plant and Cell Physiology* **36** (8), 1599–1606.

Mitsuda, N., Seki, M., Shinozaki, K. and Ohme-Takagi, M. (2005) The NAC transcription factors NST1 and NST2 of Arabidopsis regulate secondary wall thickenings and are required for anther dehiscence. *Plant Cell* **17** (11), 2993–3006.

Motose, H., Sugiyama, M. and Fukuda, H. (2004) A proteoglycan mediates inductive interaction during plant vascular *Development*. *Nature* **429** (6994), 873–878.

Nakashima, J., Endo, S. and Fukuda, H. (2004) Immunocytochemical localization of polygalacturonase during tracheary element differentiation in *Zinnia elegans*. *Planta* **218** (5), 729–739.

Nakashima, J., Takabe, K., Fujita, M. and Fukuda H. (2000) Autolysis during in vitro tracheary element differentiation: formation and location of the perforation. *Plant and Cell Physiology* **41** (11), 1267–1271.

Nelson, T. and Dengler, N. (1997) Leaf vascular pattern formation. *Plant Cell* **9** (7), 1121–1135.

Obara, K., Kuriyama, H. and Fukuda, H. (2001) Direct evidence of active and rapid nuclear degradation triggered by vacuole rupture during programmed cell death in zinnia. *Plant Physiology* **125** (2), 615–626.

O'Brien, T.P. (1970) Further observations on hydrolysis of cell wall in xylem. *Protoplasma* **69** (1), 1–14.

O'Brien T.P. (1981) The primary xylem. In: *Xylem Cell Development* (ed.Barnett, J.R.), pp. 14–46. Castle House Publications, Tunbridge Wells, UK.

O'Donnell, P.J., Calvert, C., Atzorn, R., Wasternack, C., Leyser, H.M.O. and Bowles, D.J. (1996) Ethylene as a signal mediating the wound response of tomato plants. *Science* **274** (5294), 1914–1917.

Ogawa, K., Kanematsu, S. and Asada, K. (1997) Generation of superoxide anion and localization of CuZn-superoxide dismutase in the vascular tissue of spinach hypocotyls: their association with lignification. *Plant and Cell Physiology* **38** (10), 1118–1126.

Ohashi-Ito, K. and Fukuda, H. (2003) HD-zip III homeobox genes that include a novel member, *ZeHB-13* (*Zinnia*)/*ATHB-15* (*Arabidopsis*), are involved in procambium and xylem cell differentiation. *Plant and Cell Physiology* **44** (12), 1350–1358.

Ohashi-Ito, K., Kubo, M., Demura, T. and Fukuda H. (2005) Class III homeodomain leucine-zipper proteins regulate xylem cell differentiation. *Plant and Cell Physiology* **46** (10), 1646–1656.

Ohdaira, Y., Kakegawa, K., Amino, S., Sugiyama, M. and Fukuda, H. (2002) Activity of cell-wall degradation associated with differentiation of isolated mesophyll cells of *Zinnia elegans* into tracheary elements. *Planta* **215**(2), 177–184.

O'Neill, M.A., Ishii, T., Albersheim, P. and Darvill, A.G. (2004) Rhamnogalacturonan II: structure and function of a borate cross-linked cell wall pectic polysaccharide. *Annual Review of Plant Biology* **55**, 109–139.

Orozco-Cárdenas, M.L. and Ryan, C.A. (2003) Polygalacturonase β-subunit antisense gene expression in tomato plants leads to a progressive enhanced wound response and necrosis in leaves and abscission of developing flowers. *Plant Physiology* **133** (2), 693–701.

Paquette, A.J. and Benfey, P.N. (2001) Axis formation and polarity in plants. *Current Opinion in Genetics and Development* **11** (4), 405–409.

Patzlaff, A., Newman, L.J., Dubos, C., Whetten, R.W., Smith, C., McInnis, S., Bevan, M.W., Sederoff, R.R. and Campbell, M.M. (2003) Characterisation of *Pt* MYB1, an R2R3-MYB from pine xylem. *Plant Molecular Biology* **53** (4), 597–608.

Pennell, R.I. and Lamb, C. (1997) Programmed cell death in plants. *Plant Cell* **9** (7), 1157–1168.

Pilling, E. and Höfte, H. (2003) Feedback from the wall. *Current Opinion in Plant Biology* **6** (6), 611–616.

Ponting, C.P. and Aravind, L. (1999) START: a lipid-binding domain in StAR, HD-ZIP and signalling proteins. *Trends in Biochemical Sciences* **24** (4), 130–132.

Przemeck, G.K.H., Mattsson, J., Hardtke, C.S., Sung, Z.R. and Berleth, T. (1996) Studies on the role of the *Arabidopsis* gene *MONOPTEROS* in vascular development and plant cell axialisation. *Planta* **200** (2), 229–237.

Reiter, W.D. (2002) Biosynthesis and properties of the plant cell wall. *Current Opinion in Plant Biology* **5** (6), 536–542.

Roberts, A.W., Donovan, S.G. and Haigler, C.H. (1997) A secreted factor induces cell expansion and formation of metaxylem-like tracheary elements in xylogenic suspension cultures of zinnia. *Plant Physiology* **115** (2), 683–692.

Roberts, A.W., Frost, A.O., Roberts, E.M. and Haigler, C.H. (2004) Roles of microtubules and cellulose microfibril assembly in the localization of secondary-cell-wall deposition in developing tracheary elements. *Protoplasma* **224** (3–4), 217–229.

Roberts, A.W. and Uhnak, K.S. (1998) Tip growth in xylogenic suspension cultures of *Zinnia elegans* L.: implications for the relationship between cell shape and secondary-cell-wall pattern in tracheary elements. *Protoplasma* **204** (1–2), 103–113.

Roberts, J.A., Elliott, K.A. and González-Carranza, Z.H. (2002) Abscission, dehiscence, and other cell separation processes. *Annual Review of Plant Biology* **53**, 131–158.

RosBarceló, A. (1998) The generation of H_2O_2 in the xylem of *Zinnia elegans* is mediated by an NADPH-oxidase-like enzyme. *Planta* **207** (2), 207–216.

RosBarceló, A. (2005) Xylem parenchyma cells deliver the H_2O_2 necessary for lignification in differentiating xylem vessels. *Planta* **220** (5), 747–756.

Ryser, U. (2003) Protoxylem: the deposition of a network containing glycine-rich cell wall proteins starts in the cell corners in close association with the pectins of the middle lamella. *Planta* **216** (5), 854–864.

Ryser, U. and Keller, B. (1992) Ultrastructural localization of a bean glycine-rich protein in unlignified primary walls of protoxylem cells. *Plant Cell* **4** (7), 773–783.

Ryser, U., Schorderet, M., Guyot, R. and Keller, B. (2004) A new structural element containing glycine-rich proteins and rhamnogalacturonan I in the protoxylem of seed plants. *Journal of Cell Science* **117** (7), 1179–1190.

Sachs, T. (2000) Integrating cellular and organismic aspects of vascular differentiation. *Plant and Cell Physiology* **41** (6), 649–656.

Sato, Y., Demura, T., Yamawaki, K., Inoue, Y., Sato, S., Sugiyama, M. and Fukuda, H. (2006) Isolation and characterization of a novel peroxidase gene *ZPO-C* whose expression and function are closely associated with lignification during tracheary element differentiation, *Plant and Cell Physiology* **47** (4), 493–503.

Sato, Y., Sugiyama, M., Górecki, R.J., Fukuda, H. and Komamine, A. (1993) Interrelationship between lignin deposition and the activities of peroxidase isoenzymes in differentiating tracheary elements of *Zinnia*. *Planta* **189** (4), 584–589.

Savidge, R.A. (1989) Coniferin, a biochemical indicator of commitment to tracheid differentiation in conifers. *Canadian Journal of Botany* **67** (9), 2663–2668.

Sawa, S., Demura, T., Horiguchi, G., Kubo, M. and Fukuda, H. (2005a) The *ATE* genes are responsible for repression of transdifferentiation into xylem cells in Arabidopsis. *Plant Physiology* **137** (1), 141–148.

Sawa, S., Koizumi, K., Naramoto, S., Demura, T., Ueda, T., Nakano, A. and Fukuda, H. (2005b) *DRP1A* is responsible for vascular continuity synergistically working with *VAN3* in Arabidopsis. *Plant Physiology* **138** (2), 819–826.

Sawa, S., Watanabe, K., Goto, K., Liu, Y.G., Shibata, D., Kanaya, E., Morita, E.H. and Okada, K. (1999) *FILAMENTOUS FLOWER*, a meristem and organ identity gene of *Arabidopsis*, encodes a protein with a zinc finger and HMG-related domains. *Genes and Development* **13** (9), 1079–1088.

Scarpella, E. and Meijer, A.H. (2004) Pattern formation in the vascular system of monocot and dicot plant species. *New Phytologist* **164** (2), 209–242.

Scarpella, E., Rueb, S., Boot, K.J.M., Hoge, J.H.C. and Meijer, A.H. (2000) A role for the rice homeobox gene *Oshox1* in provascular cell fate commitment. *Development* **127** (17), 3655–3669.

Scheible, W.R. and Pauly, M.L. (2004) Glycosyltransferases and cell wall biosynthesis: novel players and insights. *Current Opinion in Plant Biology* **7** (3), 285–295.

Schenk, P.M., Kazan, K., Wilson, I., Anderson, J.P., Richmond, T., Somerville, S.C. and Manners, J.M. (2000) Coordinated plant defense responses in *Arabidopsis* revealed by microarray analysis. *Proceedings of the National Academy of Sciences of the United States of America* **97** (21), 11655–11660.

Scheres, B., McKhann, H.I. and van den Berg, C. (1996) Roots redefined: anatomical and genetic analysis of root development. *Plant Physiology* **111** (4), 959–964.

Schmid, M., Simpson, D. and Gietl, C. (1999) Programmed cell death in castor bean endosperm is associated with the accumulation and release of a cysteine endopeptidase from ricinosomes.

Proceedings of the National Academy of Sciences of the United States of America **96** (24), 14159–14164.

Schrick, K., Mayer, U., Horrichs, A., Kuhnt, C., Bellini, C., Dangl, J., Schmidt, J. and Jürgens, G. (2000) FACKEL is a sterol C-14 reductase required for organized cell division and expansion in *Arabidopsis* embryogenesis. *Genes and Development* **14** (12), 1471–1484.

Schrick, K., Mayer, U., Martin, G., Bellini, C., Kuhnt, C., Schmidt, J. and Jürgens, G. (2002) Interactions between sterol biosynthesis genes in embryonic development of *Arabidopsis*. *Plant Journal* **31** (1), 61–73.

Shinohara, N., Demura, T. and Fukuda, H. (2000) Isolation of a vascular cell wall-specific monoclonal antibody recognizing a cell polarity by using a phage display subtraction method. *Proceedings of the National Academy of Sciences of the United States of America* **97** (6), 2585–2590.

Showalter, A.M. (2001) Introduction: plant cell wall proteins. *Cellular and Molecular Life Sciences* **58** (10), 1361–1362.

Sibout, R., Eudes, A., Mouille, G., Pollet, B., Lapierre, C., Jouanin, L. and Séguin, A. (2005) *CINNAMYL ALCOHOL DEHYDROGENASE-C* and *-D* are the primary genes involved in lignin biosynthesis in the floral stem of Arabidopsis. *Plant Cell* **17** (7), 2059–2076.

Siegfried, K.R., Eshed, Y., Baum, S.F., Otsuga, D., Drews, G.N. and Bowman, J.L. (1999) Members of the YABBY gene family specify abaxial cell fate in *Arabidopsis*. *Development* **126** (18), 4117–4128.

Simon, S., (1908) Experimentelle Untersuchungen über die Entstehung von Gefässverbindungen. *Berichte der Deutschen Botanischen Gesellschaft* **26**, 364–396.

Smith, C.G., Rodgers, M.W., Zimmerlin, A., Ferdinando, D. and Bolwell, G.P. (1994) Tissue and subcellular immunolocalization of enzymes of lignin synthesis in differentiating and wounded hypocotyl tissue of French bean (*Phaseolus vulgaris* L.). *Planta* **192** (2), 155–164.

Souter, M., Topping, J., Pullen, M., Friml, J., Palme, K., Hackett, R., Grierson, D. and Lindsey, K. (2002) *hydra* mutants of Arabidopsis are defective in sterol profiles and auxin and ethylene signaling. *Plant Cell* **14** (5), 1017–1031.

Stacey, N.J., Roberts, K., Carpita, N.C., Wells, B., and McCann, M.C. (1995) Dynamic changes in cell surface molecules are very early events in the differentiation of mesophyll cells from *Zinnia elegans* into tracheary elements. *Plant Journal* **8** (6), 891–906.

Steinmann, T., Geldner, N., Grebe, M., Mangold, S., Jackson, C.L., Paris, S., Gälweiler, L., Palme, K. and Jürgens, G. (1999) Coordinated polar localisation of auxin efflux carrier PIN1 by GNOM ARF GEF. *Science* **286** (5438), 316–318.

Swarup, R., Friml, J., Marchant, A., Ljung, K., Sandberg, G., Palme, K. and Bennett, M. (2001) Localization of the auxin permease AUX1 suggests two functionally distinct hormone transport pathways operate in the *Arabidopsis* root apex. *Genes and Development* **15** (20), 2648–2653.

Szekeres, M., Németh, K., Koncz-Kálmán, Z., Mathur, J., Kauschmann, A., Altmann, T., Rédei, G.P., Nagy, F., Schell, J. and Koncz, C. (1996) Brassinosteroids rescue the deficiency of CYP90 a cytochrome P450 controlling cell elongation and de-etiolation in *Arabidopsis*. *Cell* **85** (2), 171–182.

Tamagnone, L., Merida, A., Parr, A., Mackay, S., Culianez-Macia, F.A., Roberts, K. and Martin, C. (1998) The AmMYB308 and AmMYB330 transcription factors from antirrhinum regulate phenylpropanoid and lignin biosynthesis in transgenic tobacco. *Plant Cell* **10** (2), 135–154.

Taylor, N.G., Howells, R.M., Huttly, A.K., Vickers, K. and Turner, S.R. (2003) Interactions among three distinct CesA proteins essential for cellulose synthesis. *Proceedings of the National Academy of Sciences of the United States of America* **100** (3), 1450–1455.

Taylor, N.G., Laurie, S. and Turner, S.R. (2000) Multiple cellulose synthase catalytic subunits are required for cellulose synthesis in Arabidopsis. *Plant Cell* **12** (12), 2529–2540.

Taylor, N.G., Scheible, W.R., Cutler, S., Somerville, C.R. and Turner, S.R. (1999) The *irregular xylem3* locus of Arabidopsis encodes a cellulose synthase required for secondary cell wall synthesis. *Plant Cell* **11** (5), 769–780.

Thelen, M.P. and Northcote, D.H. (1989) Identification and purification of a nuclease from *Zinnia elegans* L.: a potential marker for xylogenesis. *Planta* **179** (2), 181–195.

Tokunaga, N., Sakakibara, N., Umezawa, T., Ito, Y., Fukuda, H. and Sato, Y. (2005) Involvement of extracellular dilignols in lignification during tracheary element differentiation of isolated *Zinnia* mesophyll cells. *Plant and Cell Physiology* **46** (1), 224–232.

Topping, J.F., May, V.J., Muskett, P.R. and Lindsey, K. (1997) Mutations in the *HYDRA1* gene of *Arabidopsis* perturb cell shape and disrupt embryonic and seedling morphogenesis. *Development* **124** (21), 4415–4424.

Torrey, J.G., Fosket, D.E. and Hepler, P.K. (1971) Xylem formation: a paradigm of cytodifferentiation in higher plants. *American Scientist* **59** (3), 338–352.

Toyooka, K., Okamoto, T. and Minamikawa, T. (2000) Mass transport of proform of a KDEL-tailed cysteine proteinase (SH-EP) to protein storage vacuoles by endoplasmic reticulum-derived vesicle is involved in protein mobilization in germinating seeds. *Journal of Cell Biology* **148** (3), 453–464.

Tumer, S.R. and Somerville, C.R. (1997) Collapsed xylem phenotype of Arabidopsis identifies mutants deficient in cellulose deposition in the secondary cell wall. *Plant Cell* **9** (5), 689–701.

van Doorn, W.G. and Woltering, E.J. (2005) Many ways to exit? Cell death categories in plants. *Trends in Plant Science* **10** (3), 117–122.

Veitch, N.C. (2004) Horseradish peroxidase: a modern view of a classic enzyme. *Phytochemistry* **65** (3), 249–259.

Vorwerk, S., Somerville, S. and Somerville, C. (2004) The role of plant cell wall polysaccharide composition in disease resistance. *Trends in Plant Science* **9** (4), 203–209.

Waites, R., Selvadurai, H.R.N., Oliver, I.R. and Hudson, A. (1998) The *PHANTASTICA* gene encodes a MYB transcription factor involved in growth and dorsoventrality of lateral organs in *Antirrhinum*. *Cell* **93** (5), 779–789.

Wang, H., Li, J., Bostock, R.M. and Gilchrist, D.G. (1996) Apoptosis: a functional paradigm for programmed plant cell death induced by a host-selective phytotoxin and invoked during development. *Plant Cell* **8** (3), 375–391.

Wasteneys, G.O. (2004) Progress in understanding the role of microtubules in plant cell s. *Current Opinion in Plant Biology* **7** (6), 651–660.

Willemsen, V., Friml, J., Grebe, M., van den Toorn, A., Palme, K. and Scheres, B. (2003) Cell polarity and PIN protein positioning in Arabidopsis require *STEROL METHYLTRANSFERASE1* function. *Plant Cell* **15** (3), 612–625.

Williams, L., Grigg, S.P., Xie, M., Christensen, S., and Fletcher, J.C. (2005) Regulation of *Arabidopsis* shoot apical meristem and lateral organ formation by microRNA *miR166g* and its *AtHD-ZIP* target genes. *Development* **132** (16), 3657–3668.

Wright, K.M. and Oparka, K.J. (1994) Physicochemical properties alone do not predict the movement and compartmentation of fluorescent xenobiotics. *Journal of Experimental Botany* **45** (270), 35–44.

Xu, J. and Scheres, B. (2005) Dissection of Arabidopsis ADP-RIBOSYLATION FACTOR 1 function in epidermal cell polarity. *Plant Cell* **17** (2), 525–536.

Yamamoto, R., Demura, T. and Fukuda, H (1997) Brassinosteroids induce entry into the final stage of tracheary element differentiation in cultured *Zinnia* cells. *Plant and Cell Physiology* **38** (8), 980–983.

Yamamoto, R., Fujioka, S., Demura, T., Takatsuto, S., Yoshida, S. and Fukuda, H.(2001) Brassinosteroid levels increase drastically prior to morphogenesis of tracheary elements. *Plant Physiology* **125** (2), 556–563.

Ye, Z.-H. (2002) Vascular tissue differentiation and pattern formation in plants. *Annual Review of Plant Physiology and Plant Molecular Biology* **53**, 183–202.

Ye, Z.-H. and Droste, D.L. (1996) Isolation and characterization of cDNAs encoding xylogenesis-associated and wounding-induced ribonucleases in *Zinnia elegans*. *Plant Molecular Biology* **30** (4), 697–709.

Ye, Z.-H. and Varner, J.E. (1996) Induction of cysteine and serine proteases during xylogenesis in *Zinnia elegans*. *Plant Molecular Biology* **30** (6), 1233–1246.

Yin, Y., Wu, D. and Chory, J. (2002) Plant receptor kinases: systemin receptor identified. *Proceedings of the National Academy of Sciences of the United States of America* **99** (14), 9090–9092.

Yoshida, S., Kuriyama, H. and Fukuda, H. (2005) Inhibition of transdifferentiation into tracheary elements by polar auxin transport inhibitors through intracellular auxin depletion. *Plant and Cell Physiology* **46** (12), 2019–2028.

Zhong, R., Peña, M.J., Zhou, G.K., Nairn, C.J., Wood-Jones, A., Richardson, E.A., Morrison III, W.H., Darvill, A.G., York, W.S. and Ye Z.-H. (2005) Arabidopsis *Fragile Fiber8*, which encodes a putative glucuronyltransferase, is essential for normal secondary wall synthesis. *Plant Cell* **17** (12), 3390–3408.

Zhong, R. and Ye, Z.-H. (1999) *IFL1*, a gene regulating interfascicular fiber differentiation in Arabidopsis encodes a homeodomain-leucine zipper protein. *Plant Cell* **11** (11), 2139–2152.

Zhong, R. and Ye, Z.-H. (2003) Unraveling the functions of glycosyltransferase family 47 in plants. *Trends in Plant Science* **8** (12), 565–568.

4 Cell adhesion, separation and guidance in compatible plant reproduction

Jean-Claude Mollet, Céline Faugeron and Henri Morvan

4.1 Introduction

Cell adhesion and separation are common mechanisms in many biological systems and an increasing number of studies have shown that cell adhesion molecules (CAMs) are not just sticky to maintain the cohesion of tissues and organs, but can act as signal receptors, eliciting changes in the cytoskeleton and gene expression (Harwood and Coates, 2004). Interestingly, despite the detection of immunologically related animal CAMs, true homologs are apparently not present in plants indicating that they have evolved distinct molecules, probably due to their diverse extracellular matrices (ECM) (Baluska *et al.*, 2003). Indeed, plants possess a structurally elaborated primary wall (see Chapter 1), composed mainly of polysaccharides (cellulose, hemicellulose and pectins) with some proteins and glycoproteins (Carpita and McCann, 2000), and cell adhesion (cell-cell and cell-matrix interactions) is widely observed throughout plant development. Cell separation is also a fundamental process implicated in fruit ripening (see Chapter 8), organ abscission (see Chapter 6) and pollen release (see Chapter 7).

Reproduction in most flowering plants requires a tight, spatial and temporal, regulation to allow the double fertilization of the female gametophyte by the two sperm cells, carried by the tip-growing pollen tubes (McCormick, 2004; Boavida *et al.*, 2005a). In several reviews, tip-localized pollen tube growth has been considered as a special case of cell movement (Lord, 2000) and the guidance mechanism, regulating this process, has been compared to neuron outgrowth (Palanivelu and Preuss, 2000). Indeed, as the male gametophyte comes in contact with the female tissues, cell–cell adhesion and interaction, production of short- and long-range signals by the pistil and perception of the signal cues are required to guide the pollen tubes to their final target, the female gametophyte (Wheeler *et al.*, 2001; Kim *et al.*, 2004; Sanchez *et al.*, 2004). In the past 10 years, considerable progress has been made in our understanding of plant reproduction mechanisms and our current knowledge on self-incompatibility and intracellular machinery implicated in tip growing plant cells has been recently reviewed (Wheeler *et al.*, 2001; Hiscock and McInnis, 2003; Holdaway-Clarke and Hepler, 2003; Drøbak *et al.*, 2004; Feijó *et al.*, 2004; Gu *et al.*, 2004). Here, we aim to cover, throughout pollen formation in the anther and its development in the pistil, specific events focused on cell separation, cell adhesion and cell guidance required for the proper delivery of the sperm cells to the ovule.

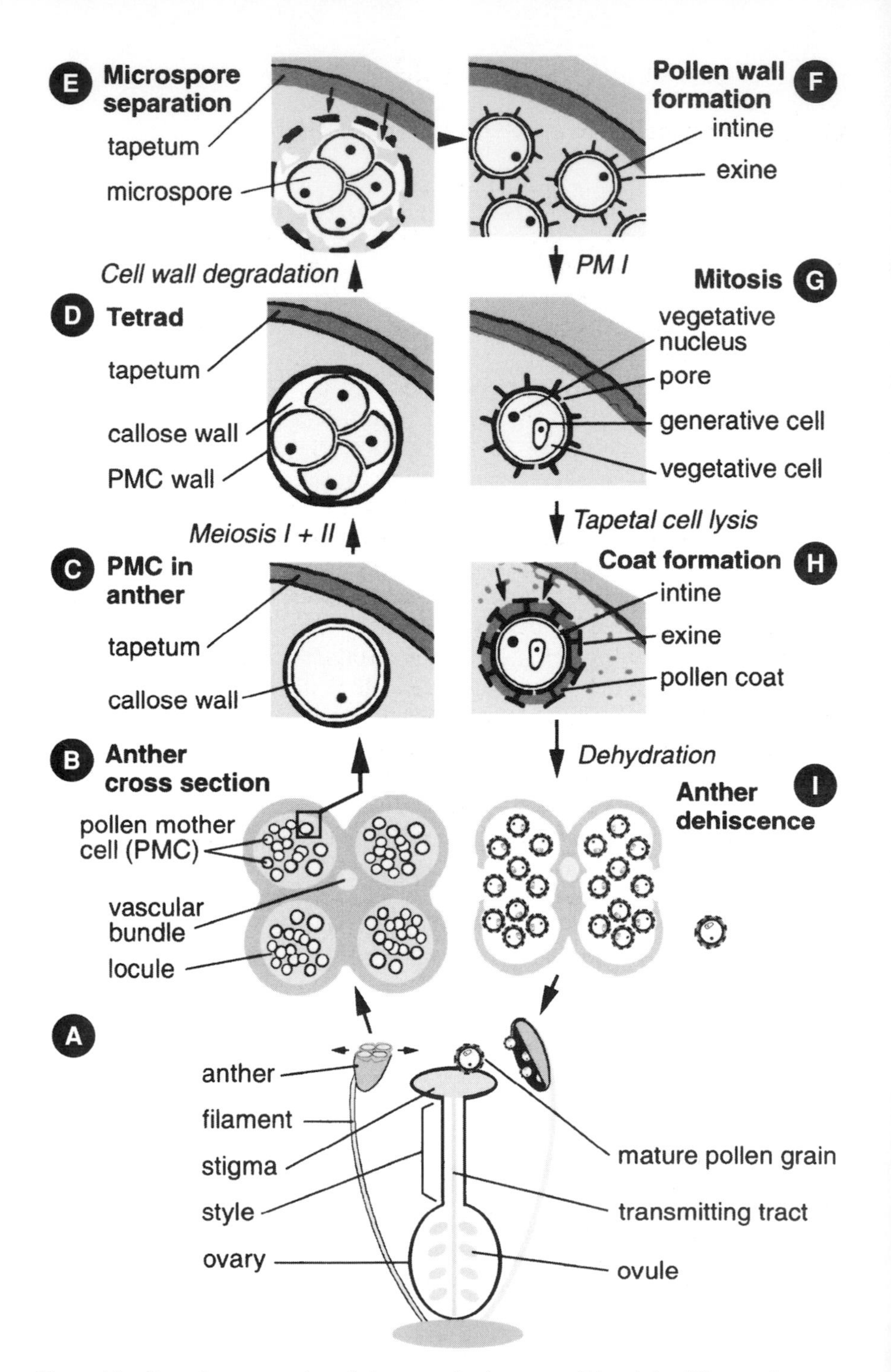

Figure 4.1 General representation of plant reproductive organs (A) and the different microsporogenesis phases (B–I). (B) Diploid pollen mother cells (PMCs) in the anther locules. (C) Locules are

4.2 Pollen formation and microspore separation

4.2.1 Pollen mother cell and tetrad walls

Inside the anthers, diploid pollen mother cells (PMCs) (Figure 4.1A–4.1C) are restricted by an outer wall composed mainly of cellulose, a β-(1,4)-glucan, pectins, a structurally complex polysaccharide family enriched in galacturonosyl acid, rhamnosyl, arabinosyl and galactosyl residues and hemicelluloses (Carpita and McCann, 2000), and an inner wall being made of callose, a β-(1,3)-glucan (Boavida *et al.,* 2005b). After meiosis of the PMCs (Figure 4.1D), the resulting four haploid cells remain associated as tetrads, with individual microspores separated by a callose wall, in which plasmodesmatal connections tend to disappear (Scott *et al.*, 2004).

In most plant cell walls, callose is generally absent, except in specific locations such as cell plates, plasmodesmata or tracheids, but it can be synthesized in response to stresses (Delmer, 1999). In contrast, callose is one of the main wall polysaccharides of the PMCs, microspores and pollen tubes but its role in pollen development is not clear. It was suggested that callose could isolate the microspores and/or serve as a template for the proper formation of exine – the outer wall layer of the mature pollen grains (Worrall *et al.*, 1992; Scott *et al.*, 2004). In pollen tubes, the callose wall was recently implicated in protection against tension and compression stresses (Parre and Geitmann, 2005). Several studies with transgenic tobacco plants (Worrall *et al.*, 1992; Tsuchiya *et al.*, 1995) and Arabidopsis mutants (Dong *et al.*, 2005b) have shown that an accurate callose wall deposition is a prerequisite for the normal development of the future pollen grains. In Arabidopsis, mutations in the *CalS5* gene, encoding a callose synthase in PMCs, tetrads and microspores exhibited a lack of callose in the microspore wall leading to the abnormal formation of the exine wall and degeneration of the microspores (Dong *et al.*, 2005b). This study supports the hypothesis that callose deposition, at the plasma membrane, may serve as a template for the proper synthesis of exine and this exine wall has an important function in pollen development. Nevertheless, the PMC callosic wall is apparently not essential for the proper nuclear meiotic processes (Worrall *et al.*, 1992; Scott *et al.*, 2004).

Figure 4.1 (*Continued*) bordered by the tapetum, and PMCs are composed of an inner callose wall and an outer wall. (D) Meiosis (I + II) of the PMCs leads to the formation of tetrads, composed of four haploid microspores encased in a callose wall. (E) Degradation of the tetrad wall with enzymes produced in the tapetum (arrows). (F) Release of the microspores in the locule and formation of the pollen wall (intine and exine). (G) Asymmetric mitosis (PM I) results in the formation of a large vegetative cell and a small generative cell. The second pollen mitosis (PM II) of the generative cell occurs in the anther during the late pollen development in species such as *Arabidopsis* and *Zea mays* or in most Angiosperms, during pollination. (H) Degeneration of the tapetum and formation of the pollen coat. (I) Dehiscence of anther and release of partially dehydrated pollen grains. Pollen grains land on the stigma surface, starting the pollination process (A).

4.2.2 Microspore separation

In several plant families (Juncacae and Ericaceae), microspore separation is not normally observed and the resulting mature pollen grains, with fused exine walls, are released as tetrads (Scott *et al.*, 2004). In many other plants and in normal pollen development, soon after meiosis, cell separation generally occurs to release the free microspores from the tetrads (Figure 4.1E and F). During this process, a decrease of pH (Izhar and Frankel, 1971; Worrall *et al.,* 1992) and a peak of β-(1,3)-glucanase activity were observed in the anther locular fluid (Stieglitz, 1977) suggesting that this enzyme may be involved in the degradation of the tetrad wall. Degradation of the callose wall requires a complex of endo- and exo-enzymes, called callase. Most of the enzymes are presumably produced in and secreted from the tapetum (Figure 4.1E) – the surrounding nutritive diploid tissue that borders the locules in the anthers (Stieglitz, 1977; Hird *et al.*, 1993). A tight developmental regulation is *de facto* required to synchronize the production of wall-degrading enzymes by the tapetal cells with the maturation of the PMCs. Indeed, the premature dissolution of the callose wall or premature programmed cell death of the tapetum, which normally occurs during the late pollen formation, lead generally to a reduction in male fertility (Worrall *et al.*, 1992; Tsuchiya *et al.*, 1995; Ku *et al.*, 2003). Several possible callase members have been described. In Arabidopsis and *Brassica*, the A6 protein encoded by the tapetum-specific *A6* gene shares some sequence similarities with plant β-(1,3)-glucanases and the *A6* gene is strongly expressed at the time of callose degradation suggesting that this protein may be one member of the callase complexes involved in tetrad separation (Hird *et al.*, 1993). However, functional studies need to confirm this. Indeed, in tobacco plants, reduction of Tag1 level, a specific tapetum β-(1,3)-glucanase, expressed normally from the tetrad to the free microspore stages, did not affect dramatically the microspore separation, suggesting that the Tag1 protein is implicated in other degradation processes or that other enzymes are necessary to complete the degradation process (Bucciaglia *et al.*, 2003).

Interestingly, in genetically modified plants, microspores lacking a callose wall can remain associated at the tetrad stage suggesting that the degradation of other wall components is also required for the cell separation (Worrall *et al.,* 1992). These observations were confirmed with the Arabidopsis *quartet* (*qrt*) mutants that showed abnormal microspore separation from the tetrads whereas the degradation of the callose wall was apparently normal (Rhee and Somerville, 1998). Immuno-localization data on the *qrt3* mutant PMCs pointed out that persistence of pectic polymers in the PMC wall was responsible for this abnormality. Expression of QRT3 protein in yeast showed polygalacturonase (PG) activity and *in planta* QRT3 is secreted from the tapetum into the anther locule at the time of tetrad separation, as demonstrated for other enzymes (Rhee *et al.*, 2003). These data suggest that the degradation of the PMC pectic wall is also necessary for the microspore separation.

The implication of cellulases or other enzymes able to degrade hemicelluloses has not been convincingly demonstrated in this separation mechanism. However, in *Lathyrus*, a peak of β-(1,4)-glucan hydrolase activity was demonstrated in the locular fluid at the time of microspore separation (Sexton *et al.*, 1990; Neelam and Sexton,

1995). An Arabidopsis cDNA macroarray study revealed that out of the 52 anther-specific genes identified, 20% encode putative proteins sharing some sequence identities to known cell wall modifying enzymes (Amagai *et al.*, 2003). For instance, it included the pectin-modifying enzymes, i.e. pectin methylesterases (PME) and PGs, and the cell wall remodeling xyloglucan endotransglucosylase/hydrolase (XTH) able to catalyze the endocleavage and the reconnection of hemicellulosic xyloglucans, implicated in the cellulose–xyloglucan framework modification (Rose *et al.*, 2002). It would not be surprising to find that several of these putative enzymes are involved in the microspore separation but functional studies are necessary to determine the fine-tuning of the events implicated in this process.

4.2.3 Pollen grain wall and pollen coat

At the free microspore stage, pollen maturation proceeds with the formation of the pollen wall (Figure 4.1F–4.1H) and asymmetric mitosis of the microspores (Figure 4.1G). Mature pollen grains are surrounded by multilayered walls (intine and exine) composed mostly of an inner sheath of callose covered with a layer of cellulose, pectins and hemicelluloses and an outer layer (exine) of the acid-resistant sporopollenin; this gives the diverse and structurally attractive aspect of the pollen grains (Edlund *et al.*, 2004). In addition, many molecules present on the pollen grain surface (pollen coat) originate either from the pollen or from the degenerating tapetum (Figure 4.1H), which releases its content, in vector-pollinated plants, via two organelles – tapetosomes and elaioplasts – that adhere non-randomly to the exine, presumably by tapetally derived proteinaceous fibrils (Piffanelli *et al.*, 1998). The amount and the nature of the pollen coating are highly variable and depend on the species, composed mostly of lipids, pigments, aromatic compounds, proteins and enzymes related to the pollinating vectors (Piffanelli *et al.*, 1998; Edlund *et al.*, 2004). The role of pollen coat components is not well defined but they have been implicated in pollen–pollen adhesion, pollen adhesion to vectors, pollen adhesion to the stigma, pollen rejection in self-incompatibility pollination and pollen hydration (Lord and Russell, 2002; Pacini and Hesse, 2005).

After maturation, pollen grains are released by dehiscence of the anthers (Figure 4.1I) and pollination starts when pollen grains land on the surface of the female reproductive organ, i.e. the stigma (Figure 4.1A).

4.3 Pollen–stigma adhesion and pollen tube guidance

4.3.1 Adhesion of pollen grain

The initial step of pollination involves the binding of pollen grains to the female stigma cells that creates *de novo* contacts required in cell adhesion, cell recognition and cell hydration. Such interactions occur in different environments, mainly because of the nature of the stigma surfaces, which have been classified as 'wet' or 'dry' based on the presence (Figure 4.2A) or absence of a copious stigmatic

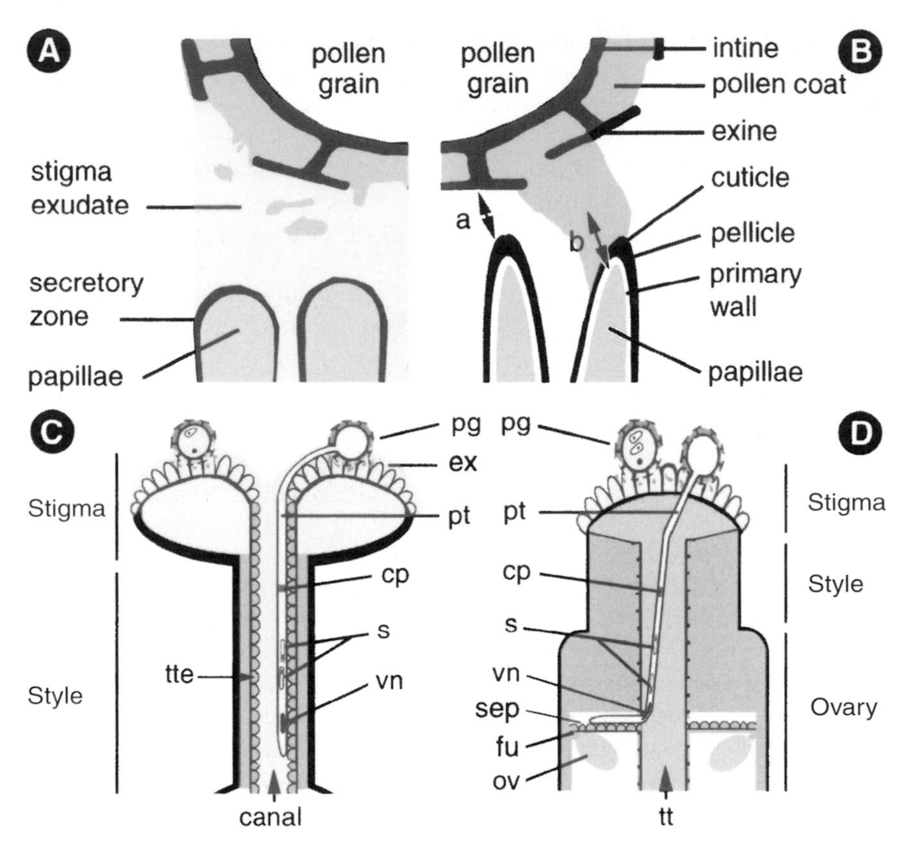

Figure 4.2 Pollen adhesion on stigma (A and B) and path of pollen tube growth in female sporophytic tissues (C and D). (A and B) The presence or absence of exudates on the surface of the papillae characterizes plants with wet and dry stigmas. (A) Wet stigmas present abundant exudates secreted by the papillae cells. This secretion can be enriched in water and carbohydrates (*Lilium*) or lipids (*Nicotiana* and *Petunia*). Pollen adhesion on wet stigmas is permissive and pollen coat mixes with stigma exudates. (B) Dry stigmas are found in Arabidopsis and *Brassica*. They are composed of papillae covered by a proteinaceous pellicle of poorly defined composition and an interrupted waxy cuticle. Initial pollen adhesion is exine-mediated (a) and after migration of the pollen coat on the stigma surface, the "late" adhesion involves pollen coat proteins and papillae cell wall glycoproteins (b). (C and D) Comparison of paths of pollen tube growth in (C) *Lilium*, with wet stigma and hollow style and (D) Arabidopsis, with dry stigma and solid style. (C) Pollens (pg) hydrate and produce pollen tubes (pt), which grow on the surface of the stigma covered with exudates (ex), directed toward the opened style. The pt enter the style and expand by adhering to the ECM of the secretory transmitting tract epidermal cells (tte). The tte borders a canal filled with stylar exudates. (D) Pollen grains (pg) adhere to the dry stigma, hydrate and produce pt that grow through the cell wall of the stigmatic papillae and stylar transmitting tract (tt) cells. In the ovary, pt exit the tt to enter the septum (sep) toward the ovule (ov). The pt grow and adhere on the surface of the septum epidermis and funiculus (fu). During pt expansion, the pollen tube cell, composed of the two sperm cells (s) and the vegetative cell, is maintained in the tip region by deposition of callose plugs (cp). (vn) vegetative nucleus.

secretion (Figure 4.2B). Solanaceae (*Nicotiana*, *Lycopersicum* and *Petunia*), Leguminosae and Liliaceae (*Lilium*) have wet stigmas, with exudates enriched in either lipids or carbohydrates (Figure 4.2A). Adhesion of pollen to mature wet stigmas occurs but it is apparently not specific and more permissive than plants with dry stigmas and relies on liquid surface tension (Heslop-Harrison, 2000; Swanson *et al.*, 2004). Indeed, pollen grains typically can bind to wet stigmas, hydrate and eventually germinate without any species selectivity (Wheeler *et al.*, 2001).

In the more advanced families (i.e. Papaveraceae, Poaceae, Asteraceae and Brassicaceae) with dry stigmas (Figure 4.2B), specific interactions occur between pollen grains and stigmas. Mainly studied in Arabidopsis and *Brassica*, the nature of the pollen–stigma interacting partners was investigated with the use of genetically modified pollens and stigmas and with chemical treatments or competition assays. Adhesion force was assessed by centrifugation-based assays or a cantilever displacement device (Luu *et al.*, 1999; Zinkl *et al.*, 1999). To integrate the results obtained with those two species, Swanson *et al.* (2004) divided the pollen–stigma adhesion in several steps.

The initial adhesion step (i.e. pollen capture) is rapid (30 s) and species specific. Indeed, pollen grains, from 16 different species and the Arabidopsis *cer6-2* (*eceriferum6-2*) mutant with an extreme pollen coat defect, were tested for their abilities to adhere to wild-type Arabidopsis stigmas. After a detergent treatment, pollen grains from only Arabidopsis were still adhering, suggesting that in a natural setting this species-specific affinity would allow plants to capture appropriate pollen (Zinkl *et al.*, 1999). Interestingly, affinities of the *cer6-2* and wild-type pollen grains for the wild-type stigmas were similar, and removal of the pollen coat of wild-type pollen did not affect the adhesion force (Zinkl *et al.*, 1999). Based on this, it was proposed that molecules in the exine wall were responsible for this initial adhesion step (Figure 4.2B-a). This hypothesis was supported by several observations: (1) immediately after pollination, the stigma surface becomes altered at the interface, acquiring a pattern that interlocks with the exine; (2) purified exine fragments remain bound to stigmas under high-stringency conditions, i.e. detergent treatments (Zinkl *et al.*, 1999); and (3) the *less adherent pollen 1* (*lap1*) mutant has reduced exine material (Zinkl and Preuss, 2000). The interacting partner of the pollen exine in the stigma is not known, but it was proposed that a hydrophobic interaction might be involved based on the nature of the chemicals tested to disrupt the pollen-stigma adhesion (Zinkl *et al.*, 1999).

Following this initial exine-mediated contact, the pollen coat migrates from the exine onto the stigma surface called 'coat conversion', where it establishes an 'attachment foot' in the zone of contact between pollen and stigma (Figure 4.2B-b) (Swanson *et al.*, 2004). In *Brassica*, several stigmatic components have been described and the S-locus receptor kinase (SRK) was implicated in the *Brassica* self-incompatibility response, after interaction with the S-locus cysteine-rich protein (SCR/SP11), originating from the pollen coat protein (PCP) (Hiscock and McInnis, 2003). Luu *et al.* (1997, 1999) implicated in pollen adhesion two other stigmatic cell wall molecules, S-locus glycoprotein (SLG) and S-locus related protein 1 (SLR1), also members of this S-multigene family but with an undefined role in pollination.

This adhesion was described as a 'late' event as the maximum binding was obtained 30 min after the initial pollen capture (Heizmann *et al.*, 2000). They also showed that SLG and SLR1 might contribute separately to this adhesion, suggesting that different pollen-interacting partners may exist. *In vitro*, the stigmatic SLG was shown to bind PCP-A1, a member of the Class A PCPs (Hiscock *et al.*, 1995; Doughty *et al.*, 1998) and SLR1 to interact with two other PCP-A proteins, SLR1-binding protein 1 (BP1) and SLR1-BP2 (Takayama *et al.*, 2000). PCP-A1 and SLR1-BP display very low amino acid sequence identity but share common features (Takayama *et al.*, 2000). They are small (6–7 kD), basic (pIs 8.5–10), cysteine-rich (eight cysteine residues expected to form four intramolecular disulfide bonds) secreted proteins related to defensins – peptides with antimicrobial activities (Doughty *et al.*, 2000). Reduction of the disulfide bonds of SLR1-BP abolished the binding to SLR1, suggesting that the three-dimensional structure of SLR1-BP is essential for the adhesion, but the interaction specificity is probably linked to the amino acids located on the surface of the protein (Takayama *et al.*, 2000). Interestingly, whereas SLR1 is present in the stigmas of all *Brassica* species and relatives, the *SLR1-BP* gene was not found in all species suggesting that other PCPs may have a similar function (Takayama *et al.*, 2000).

Results from these two relative species indicate that adhesion between pollen grains and stigmas can become stronger over time with different types of adhesive contacts supplementing each other.

4.3.2 Pollen tube emergence and guidance on the stigma

Following adhesion, hydration of pollen grains occurs and pollen tubes emerge from one of the grain apertures. Pollen tubes grow on the surface of the stigma as in lily (Figure 4.2C), through the intercellular matrix of the secretory cells as in tobacco, or through the 'attachment foot' into the papillae cell wall as in Arabidopsis (Figure 4.2D).

Several molecules from the stigma have been involved in the initial pollen tube guidance. In tobacco with lipid-rich exudates, application of triacylglycerides to transgenic stigmaless plants restored the ability of pollen tubes to grow within the pistil tissues suggesting that the hydrophobic lipids from the stigma are implicated in directional pollen tube growth by controlling water flow (Wolters-Arts *et al.*, 1998; Lush *et al.*, 2000). More recently, in *Lilium* with carbohydrate-rich exudates, a small ($\sim$10 kD) ECM peptide was able *in vitro* to redirect pollen tube growth by creating a gradient (Kim *et al.*, 2003). This protein was named chemocyanin based on its chemotropic function and its deduced amino acid sequence identity with other plantacyanins (Kim *et al.*, 2003, 2004). Interestingly, the effect of chemocyanin on lily pollen tubes was enhanced by the presence of stigma/stylar cysteine-rich adhesin (SCA), another small lily stigma/stylar protein, implicated in lily pollen tube adhesion within the style (Kim *et al.*, 2003), discussed in the following paragraph. The implication of plantacyanin in pollen tube guidance was further investigated in Arabidopsis by overexpression of the unique gene encoding for plantacyanin (Dong *et al.*, 2005a). In addition to a lack of anther dehiscence, wild-type pollen tubes were misguided on the plantacyanin overexpressing stigmas. These plants, compared with wild-type plants, displayed a perturbed plantacyanin concentration gradient in the female tissues along the pollen tube path, which may explain the misguidance of the pollen tubes toward the style (Dong *et al.*, 2005a). Plantacyanins are ECM proteins that belong to the blue copper protein family involved in stress responses and signaling, and they are intriguingly shown to bind the tobacco S-RNase implicated in pollen rejection (Fedorova *et al.*, 2002; Kreps *et al.*, 2002; Cruz-Garcia *et al.*, 2005). The precise mode of action of these proteins in pollen tube guidance is not understood but it was proposed that they might directly or indirectly activate calcium channels at the plasma membrane of the pollen tube apical zone to promote an extracellular calcium influx. This increase of intracellular calcium concentration, at the site of the influx, would control the direction of pollen tube growth (Holdaway-Clarke and Hepler, 2003; Dong *et al.*, 2005a).

4.4 Adhesion and guidance of pollen tubes in the style

In the style, pollen tubes grow in close contact with the components of the specialized ECM of the transmitting tract (TT) in plants with a hollow style as in *Lilium* (Figure 4.2C), or with solid styles as in *Brassica*, *Nicotiana* or Arabidopsis (Figure 4.2D) (Lord, 2000). In addition to a nutritive role, the TT was implicated in pollen tube guidance and several mechanisms, including chemical cues; mechanical constraints and haptotactism (adhesion-mediated guidance) have been proposed (de Graaf *et al.*, 2001; Lord *et al.*, 2001). Based on the biochemical and genetic studies, several adhesion or signaling molecules originating either from the pistil or pollen have been highlighted and are presented.

4.4.1 Proline/hydroxyproline rich glycoproteins

The proline/hydroxyproline-rich glycoprotein/proteoglycan family is abundant in reproductive tissues (Sommer-Knudsen *et al.*, 1997) and is classically divided in four groups: arabinogalactan proteins (AGPs), extensins, proline-rich proteins (PRPs) and Solanaceaous lectins; however, many chimeric molecules with additional domains have been described as well (Wu *et al.*, 2001).

4.4.1.1 Pollen and pistil AGPs

In many species, AGPs are abundantly found in pistil and at the plasma membrane/cell wall interface of pollen tubes (Lord *et al.*, 2000) and their carbohydrate contents (mostly arabinosyl and galactosyl residues) can represent over 95% of their weight. In addition, AGPs can be attached to the plasma membrane via a glycosylphosphatidylinositol (GPI) anchor or released in the cell wall, after cleavage of the GPI anchor with specific phospholipase (Schultz *et al.*, 1998; Majewska-Sawka and Nothnagel, 2000). They have been implicated in many biological processes and are predicted to have adhesive and signaling properties (Majewska-Sawka and Nothnagel, 2000; Schultz *et al.*, 2000).

In *Nicotiana tabacum*, the transmitting tissue-specific (TTS) proteins, with AGP characteristics, are bound to the intercellular matrix of the stylar TT. It was shown *in vitro* and *in vivo* to promote pollen tube growth, to attract pollen tubes, to adhere at the pollen tube tip and to be deglycosylated then incorporated into the pollen tube walls (Cheung *et al.*, 1995; Wu *et al.*, 1995). Moreover, a gradient of glycosylation of TTS was found in the TT, with higher carbohydrate moieties near the ovary (Wu *et al.*, 1995). This glycosylation gradient was suggested to guide the pollen tube growth in the style toward the ovary. Then, after deglycosylation of TTS, the released carbohydrates were used as nutritive compounds (Wu *et al.*, 2001). In *Nicotiana*, other AGP-related stylar glycoproteins were characterized (Sommer-Knudsen *et al.*, 1997; de Graaf *et al.*, 2001) but none showed similar effect as TTS proteins on pollen tube growth, presumably because of their distinct post-translational glycosylation patterns (Wu *et al.*, 2001).

A subclass of AGP, with at least 21 members in the Arabidopsis genome, possesses predicted AGP regions with putative fasciclin-like domains (FLAs) (Borner *et al.*, 2003; Johnson *et al.*, 2003). Proteins with fasciclin domains are wide spread and several of them have been implicated in cell adhesion (Johnson *et al.*, 2003). As an example, the *salt overly sensitive 5* (*sos5*) mutant of Arabidopsis, with alteration of one of the fasciclin domains of FLA4, showed abnormal cell expansion and defects in cell–cell adhesion (Shi *et al.*, 2003). Several *FLAs* are strongly expressed in flowers (Johnson *et al.*, 2003) and promoter–GUS fusion of FLA8 was seen at the top of the style in open flowers (K.L. Johnson, A. Bacic and C.J. Schultz, unpublished data). Specific expression and functional studies focused on reproductive tissues are needed to assess more precisely their presence in pollen tubes or pistil tissues and their possible implication in cell adhesion.

AGPs are also detected at the pollen tube plasma membrane/wall interface in many species suggesting that they might be GPI anchored (Lord *et al.*, 2000). Although GPI-anchored AGPs are apparently abundant in plants, their role in pollen development is not clear (Borner *et al.*, 2003). Recent studies have shown that mutations in genes encoding for enzymes involved in the GPI biosynthesis affect pollen tube growth presumably due to an abnormal cell wall synthesis, assembly and remodeling (Lalanne *et al.*, 2004).

4.4.1.2 Pex, pollen-specific leucine-rich repeat extensin chimeras

Pollen extensin-like (Pex) are pollen-specific members of the leucine-rich repeat extensin chimera (LRX) proteins (Stratford *et al.*, 2001). The mPex1 protein was located in the callosic wall of maize pollen tubes (Rubinstein *et al.*, 1995), and several *Pex* genes were found in tomato and Arabidopsis (Stratford *et al.*, 2001; Baumberger *et al.*, 2003). LRX proteins have C-terminal structural extensin domains that may anchor the protein to the ECM and N-terminal leucine-rich repeat (LRR) domains with protein–protein interaction potentials (Stratford *et al.*, 2001). Despite the presence of Pex in pollen tube walls, their interacting partners and functions are not known yet, but preliminary data, using a yeast two-hybrid screen, identified a Pollen-specific Rapid ALkalinization Factor (PRALF) (Bedinger *et al.*, 2005) closely related to RALF – a peptide able to inhibit root growth (Pearce *et al.*, 2001).

The significance of these findings in terms of pollen tube growth needs further investigations.

4.4.2 *Pollen and pistil cysteine-rich proteins*

4.4.2.1 *SCA–pectin complex*

Using lily as a model plant, an *in vitro* bioassay was developed to mimic the *in vivo* pollen tube adhesion to the ECM of the TT by creating an artificial matrix impregnated with stylar components (Jauh *et al.*, 1997). In this assay, adhesion of pollen tubes occurs near the tip region and pollen tube growth is enhanced compared to the one observed in liquid medium. Two stylar components are required for this adhesion event. The first one, bound to the TT wall, is a 9-kD secreted, heat stable, basic protein known as SCA showing some sequence similarities with plant lipid transfer proteins (LTPs) (Park *et al.,* 2000; Park and Lord, 2003). The second one is a weakly esterified pectic polymer. An ionic interaction between those two stylar ECM molecules is necessary for pollen tube adhesion (Mollet *et al.*, 2000, 2003). Implication of weakly esterified pectins in pollen tube adhesion is not surprising, as homogalacturonans were thought, for a long time, to be involved in cell adhesion at the middle lamella of plant tissues via calcium cross-bridges, but functional evidence was lacking (Lord and Mollet, 2002). The interacting partners of the SCA/pectin matrix in the pollen tube are not known yet. But SCA binds to the pollen tube wall (Ravindran *et al.*, 2005) composed mainly of methylesterified pectins at the tip and weakly esterified pectins back from the tip (Figure 4.3A) (Lord *et al.*, 2001). SCA may bring together the pectic polymers from the stylar ECM and pollen tube, which if linked to a wall-associated kinase, may act directly on a pollen tube receptor or a complex of interacting molecules secreted from pollen tubes or it may act as a signal molecule (Park *et al.*, 2000). LTP-like proteins are widespread in plants and several putative LTP genes are expressed specifically along the pollen tube path of Arabidopsis pistils (Tung *et al.*, 2005). To date, LTPs have been implicated in many biological processes such as cutin biosynthesis, defense against microorganisms, adaptation to environmental changes (Kader, 1997) and pollen tube adhesion (Lord, 2000), and recently used as wall-loosening agents (Nieuwland *et al.*, 2005).

Finally, considering the role of calcium in plant development (Hepler, 2005), the involvement of this cation in this adhesion cannot be ruled out as: (1) it is required for *in vitro* pollen tube growth; (2) it is known to interact with homogalacturonans; and (3) its amount increases in the ECM of the TT after pollination (Zhao *et al.*, 2004) supporting the hypothesis that pollen triggers the changes in the female tissue.

4.4.2.2 *Cys-rich protein's interaction with pollen LRR receptor kinases*

In tomato, the cysteine-rich pollen extracellular protein LATE ANTHER TOMATO52 (LAT52) interacts, before pollen germination, with the extracellular domain of LePRK2, one of the three plasma membrane localized, pollen-specific receptor kinases (Tang *et al.*, 2002). LePRKs have LRR extracellular domains, known to be involved in protein–protein interaction, and an intracellular kinase domain (Wengier *et al.*, 2003). Interestingly, upon *in vivo* pollen germination or *in vitro*

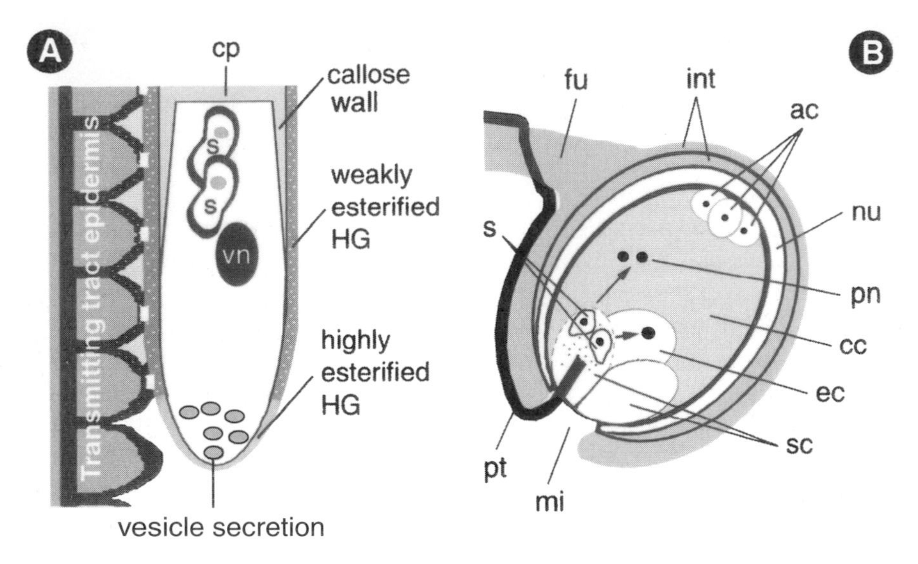

Figure 4.3 Representation of a pollen tube (A) and an ovule (B) found in most flowering plants. (A) Vesicle secretion at the pollen tube tip with membrane and wall materials necessary for tube growth. The tube cell, with the two sperm cells (s) and the vegetative nucleus (vn), is maintained at the tip of the tube by deposition of callose plugs (cp). Based on immuno-localization data: methylesterified homogalacturonans (HG) are found mostly in the tip wall, and back from the tip, weakly esterified HG and callose comprise the wall. Cellulose is apparently absent in the pollen tube tip wall but is found in low amounts, back from the tip. (B) Structure of an ovule and path of a pollen tube (pt) growing on the surface of the funiculus (fu) and integuments (int) of the ovule toward the micropyle (mi) to deliver the two sperm cells (s) and allow double fertilization. The embryo sac is composed of the egg cell (ec), two synergids (sc), a large central cell (cc) containing two nuclei (pn) and three antipodal cells (ac). In order to accomplish the double fertilization, pt penetrate one of the sc, which degenerates, and discharge the two spermatic cells (s). One of them fertilizes the egg cell (ec) (short arrow) giving the zygote and the future embryo, the other one fuses with the two nuclei (pn) (long arrow) of the central cell (cc) that will become the endosperm. (nu) nucellus.

pollen germination in the presence of a tomato stylar extract, interaction between LAT52 and LePRK2 is abolished and LAT52 is replaced by a small pistil-secreted Cys-rich protein, *Lycopersicum esculentum* stigma-specific protein 1 (LeSTIG1), shown to promote pollen tube growth (Tang *et al.*, 2004). These observations indicate that LeSTIG1 may be one of the pistil-interacting partners of the extracellular domain of LePRK and such partners can change throughout pollen development. Indeed, another interactor of the LePRK2 – LeSHY, a tomato ortholog of the *Petunia* germinating pollen SHY – was also described (Guyon *et al.*, 2004). LeSHY is a small pollen-specific protein with an LRR extracellular domain. Transgenic pollen tubes with reduced levels of SHY RNA showed reduced *in vitro* pollen tube growth. *In vivo*, pollen germination was delayed, pollen tube growth was arrested at the apex of the ovary and abnormal deposition of callose in the tubes was observed. These data suggest that LeSHY may facilitate growth through the pistil presumably with interacting partners yet to be determined (Guyon *et al.*, 2004). However, another report in *Petunia* and tobacco suggested that STIG1, a close relative of LeSTIG1 with 72% amino acid identity, was involved in the temporal regulation of the stigma secretion (Verhoeven *et al.*, 2005).

4.4.3 Wall-associated kinases

Another potential candidate that may be involved in adhesion/cell signaling is the wall-associated kinase (WAK)/WAK-like kinases (WAKLs) that can physically link the intracellular compartment and the ECM (Anderson *et al.*, 2001; Baluska *et al.*, 2003). *WAK/WAKL* genes encode transmembrane proteins with cytoplasmic serine/threonine kinases and variable extracellular epidermal growth factor-like (EGF) domains (Verica and He, 2002). Among the five WAKs described in Arabidopsis, WAK1 can interact with pectins with covalent linkages (Anderson *et al.,* 2001) or non-covalently via a calcium-dependent conformation (Decreux and Messiaen, 2005). In Arabidopsis, WAK(1–5) are apparently not expressed in floral organs (Anderson *et al.,* 2001), but several *WAKLs* genes including *WAKL6*, shown to be associated with the cell wall, are strongly expressed in anthers, styles and stigmas (Verica *et al.*, 2003). So far, WAKs and WAKLs have been implicated in plant defense responses and cell elongation (Verica and He, 2002) but specific studies on pollen and pistil WAKLs may unravel interesting mechanisms in the perception of factors originating from either side.

4.5 Cell wall modifying proteins and pollen tube growth in the ECM

For the success of pollination, pollen tubes need to grow, depending on the structure of the female tissue, on or through the stigma and style ECM. To facilitate the growth of the compatible pollen tubes, cell wall modifying proteins, originating from male and/or female tissues, may be implicated in this process.

4.5.1 Cell wall modifying proteins from pollen

Many putative cell wall modifying proteins have been studied at the DNA or mRNA level but only a few proteins have been characterized. Among them, several wall-remodeling proteins have been detected in and/or isolated from the pollen grain surface, the pollen tube wall or the culture medium of *in vitro* grown pollen tubes. In *Zea mays*, several active proteins have been characterized from the pollen grains, i.e. an endoxylanase, a β-glucanase (Suen *et al.*, 2003) and the wall-loosening group 1 pollen allergen β-expansin 1 (Zea m 1) (Cosgrove *et al.*, 1997). All were implicated in the degradation or loosening of the stigmatic cell walls, to facilitate the invasion of the female tissues by the pollen tubes. In addition, a polygalacturonase, isolated from the culture medium of *in vitro* grown maize pollen tubes, was proposed to hydrolyze the pectic components of the TT (Suen *et al.*, 2003).

In *Brassica*, with stigma papillae covered with a cuticle (Figure 4.2B) and a solid style, several serine esterases have been detected in pollen and stigma including a 22-kD, intine-localized pollen protein with cutinase activity (Hiscock *et al.*, 2002).

Application of serine esterase inhibitors on stigmas did not affect pollen germination but reduced significantly the penetration of the pollen tubes into the stigma surface suggesting that the degradation of the cuticle is a prerequisite for pollen tube entry into the stigma (Hiscock *et al.*, 2002). Similarly, immuno-localization of a polygalacturonase at the pollen tube tip growing into the papillar wall suggests a role of this enzyme in pollen tube growth through the female tissues (Dearnaley and Daggard, 2001). The importance of pectin methylesterases (PME), enzymes responsible of the demethyesterification of the pectic homogalacturonans, in pollen tube development was highlighted in the functional studies of the Arabidopsis *VANGUARD1* (*VGD1*) gene encoding for a pollen-specific PME (Jiang *et al.*, 2005). *In vivo*, the *vgd1* pollen mutant was able to germinate, and tube morphology and guidance toward the ovule were normal but growth through the TT was retarded. The results suggest that VGD1 functions in the remodeling of the pollen tube pectic wall but the secretion of VGD1 from the pollen tube into the female tissue for modifying and loosening the ECM was not excluded (Jiang *et al.*, 2005). In addition, β-glucanases (Takeda *et al.*, 2004) have been implicated in pollen tube growth and degradation of female tissues.

4.5.2 Cell-wall-modifying proteins in the pistil

Similarly, the female tissue, whether pollinated or not, could produce components to facilitate the penetration of the pollen tubes by loosening or modifying the TT wall, and/or to act on the pollen tube wall. Several pistil proteins have been identified that may fulfill this role. The β-expansin pistil pollen allergen-like (PPAL) and a lipid transfer protein (TobLTP2) were identified in the stigma of *Nicotiana tabacum* but, unexpectedly, only TobLTP2 displayed the wall-loosening activity (Nieuwland *et al.*, 2005). However, PPAL and LTP-silenced plants did not affect wild-type pollen tube growth, indicating that these proteins may not facilitate pollen tube growth through the female tissues, or that other LTPs, present in the pistil, may have caused functional redundancy (Nieuwland *et al.*, 2005). In Arabidopsis, polygalacturonases have also been detected in stigmas and the TT of styles (Dearnaley and Daggard, 2001; Khosravi *et al.*, 2003) and several specific genes encoding putative secreted proteins with possible wall-modifying properties (expansins, extensin-like, proteases, pectinesterases, LTPs, β-glucanases and cellulases) are expressed along the path of pollen tube growth (Tung *et al.*, 2005).

As expected, many cell wall modifying protein candidates from male and female parts have been described with potential roles in cell separation. However, the activity for most of these cell wall processing proteins needs to be demonstrated and their exact target elucidated to ascertain their functions.

4.6 Pollen tube adhesion, interaction and guidance in the ovary

After traveling through the style, pollen tubes generally emerge from the TT (Figure 4.2D), adhere to the septum and funiculus surface (Lennon *et al.*, 1998) and grow toward the ovules (Figure 4.3B). Near the ovules, pollen tubes turn toward the micropyle, uncovered by integuments, and therefore offer a direct access to the female gametophyte. The emerging hypothesis is that a series of chemical signals (inhibiting or promoting signals) controlled by sporophytic (funicular guidance) and gametophytic (micropylar guidance) cells would limit the access to a single pollen tube near the ovule, preventing polyspermy (Higashiyama *et al.*, 2003).

4.6.1 Pollen tube attraction by sporophytic cells

Most of the progress in this area was obtained by genetic analysis of mutants with pollen tubes that fail to adhere to ovule cells. The *inner no outer* (*ino*) mutant, with ovules lacking the outer integument, failed to properly guide the pollen tubes to the micropyle (Baker *et al.*, 1997). This study clearly implicates the diploid cells of the ovule in pollen tube guidance. Among other mutants, the *pollen-pistil-interaction2* (*pop2*) mutation causes a lack of pollen tube adhesion on the surface of the funiculus, presumably due to an excess of γ-aminobutyric acid (GABA) in the pistil (Wilhelmi and Preuss, 1996). Mutation in the *POP2* gene was associated with a defect in the GABA transaminase that converts GABA to succinic semi-aldehyde. This enzyme, in addition to other activities, is essential for the creation of a GABA gradient at the surface of the ovule, which is proposed to be responsible for guiding the pollen tubes to the micropyle (Palanivelu *et al.*, 2003). However, *in vitro*, a gradient of GABA has no effect on pollen tube guidance; low concentration of GABA can stimulate pollen tube growth but this effect is inhibited with high levels of GABA. These observations suggest that the diploid cells of the ovule are the source of GABA, which is then degraded by the pollen tubes. It remains to determine whether GABA is a metabolite (stimulating the growth of pollen tube) or a signaling molecule acting perhaps with other factors or both (Bouché and Fromm, 2004).

4.6.2 Pollen tube guidance by gametophytic cells

Several studies implicated the female gametophyte in the attraction of the pollen tube near the ovule. Using laser ablation on their model plant *Torenia fournieri*, Higashiyama *et al.* (2001) showed clearly that only one of the two synergids is sufficient for pollen tube attraction before fertilization. Moreover, *in vitro*, the gametophyte is no longer able to attract any pollen tube after fertilization. The nature of the signal (loss of attractive signal or production of repellent) is not known, but recent studies suggest that nitric oxide (NO) may be one of the repellent signal candidates (Prado *et al.*, 2004).

In maize, an attracting signal, *Zea mays* egg apparatus 1 (ZmEA1) produced by the female gametophyte, is required for pollen tube guidance at the vicinity of the ovule (Márton *et al.*, 2005). ZmEA1 is a predicted polypeptide with transmembrane domain that may be cleaved to release ZmEA1 in the micropylar zone of the nucellus. Altered *ZmEA1* expression showed greatly reduced seed set and pollen tubes fail to enter the micropyle due to a random growth. Interestingly, homologs of ZmEA1 are found in other Graminae but not in dicotyledons, suggesting that the ZmEA1 signal could, as part of a multiple checkpoint, be one of the mechanisms supporting species-barriers concept. The perception system of these signals in the pollen tube has to be identified, but it seems to become functional only if pollen tubes have passed through the stigmatic and stylar tissues. This indicates that interactions between male gametophytes and female sporophytic tissues may trigger the expression of guidance signal receptor (Higashiyama *et al.*, 1998).

4.6.3 Interaction during fertilization: female control of male gamete delivery

Once the pollen tube has reached the female gametophyte, two key processes occur for the success of fertilization. One of the two synergids flanking the egg cell degenerates and the pollen tube stops growing. This phenomenon is called pollen tube reception and is followed by the release of the two sperm cells. The controlling factors of this process are not clearly established but the identification of the two Arabidopsis gene products FERONIA (FER) and SIRENE (SIR) will certainly provide important information on the female gametophyte control of pollen tube reception. Indeed, mutation in those genes leads to the same phenotype with pollen tube that failed to rupture and release their sperms (Huck *et al.,* 2003; Rotman *et al.*, 2003).

After the discharge of the sperm cells in the degenerated synergid, the double fertilization can occur; there are new ideas that a preferential fusion of one of the two sperm cells with the egg might be predetermined (Weterings and Russell, 2004). The two separate fusion products will produce the embryo and endosperm of the future seed.

4.7 Conclusions and perspectives

The success of plant reproduction relies on the synchronized maturation of the male and female counterparts, pollination, fertilization and embryogenesis to produce seeds. The diversity of molecules that are presented in this review reflect multiple species-specific checkpoints and mechanisms that restrict the access of foreign or unfit pollens or pollen tubes. The use of genetic and biochemical tools, on model plants, has been successful in elucidating several adhesion and signaling molecules and studies on non-model systems will unravel many other complex signals. Additionally, the design of original *in vitro* or semi-*in vitro* assays is of great interest, to be able to dissect and simplify the complex pollen–pistil interactions. Pollen tubes are a good model to study, after the complete genomic sequence of Arabidopsis, a fully described pollen transcriptome (Boavida *et al.,* 2005a) will undoubtedly give a more integrative view of the different processes. The combination of these approaches should help to clarify the intricately complex mechanisms involved in the production, perception and regulation signals of how plants make seeds.

Acknowledgements

The authors are grateful to Elizabeth Lord (CEPCEB, UC Riverside, USA) and Vanessa Vernoud (ENS-INRA, Lyon, France) for the critical reading of the manuscript. We are also thankful to Carolyn Schultz (University of Adelaide, Australia) for sharing unpublished data.

References

Amagai, M., Ariizumi, T., Endo, M., Hatakeyama, K., Kuwata, C., Shibata, D., Toriyama, K. and Watanabe, M. (2003) Identification of anther-specific genes in a cruciferous model plant, *Arabidopsis thaliana*, by using a combination of Arabidopsis macroarray and mRNA derived from *Brassica oleracea. Sexual Plant Reproduction* **15**, 213–220.

Anderson, C.M., Wagner, T.A., Perret, M., He, Z.H., He, D. and Kohorn, B.D. (2001) WAKs: cell wall-associated kinases linking the cytoplasm to the extracellular matrix. *Plant Molecular Biology* **47**, 197–206.

Baker, S.C., Robinson-Beers, K., Villanueva, J.M., Gaiser, J.C. and Gasser, C.S. (1997) Interactions among genes regulating ovule development in *Arabidopsis thaliana. Genetics* **145**, 1109–1124.

Baluska, F., Samaj, J., Wojtaszek, P., Volkmann, D. and Menzel, D. (2003) Cytoskeleton-plasma membrane-cell wall continuum in plants. Emerging links revisited. *Plant Physiology* **133**, 482–491.

Baumberger, N., Doesseger, B., Guyot, R., Diet, A., Parsons, R.L., Clark, M.A., Simmons, M.P., Bedinger, P.A., Goff, S.A., Ringli, C. and Keller, B. (2003) Whole-genome comparison of LRR extensins (*LRXs*) in *Arabidopsis thaliana* and *Oryza sativa*: a conserved family of cell wall proteins form a vegetative and a reproductive clade. *Plant Physiology* **131**, 1313–1326.

Bedinger, P.A., Covey, P.A., Reeves, A., Arthur-Asmah, R., Parsons, R.L., Clark, M.A., Pearce, G. and Ryan, C.A. (2005) PRALF: a potential pollen growth regulator. *Plant Biology 2005 Seattle*, Washington, Abs #174.

Boavida, L.C., Becker, J.D. and Feijó, J.A. (2005a) The making of gametes in higher plants. *International Journal of Developmental Biology* **49**, 595–614.

Boavida, L.C., Vieira, A.M., Becker, J.D. and Feijó, J.A. (2005b) Gametophyte interaction and sexual reproduction: how plants make a zygote. *International Journal of Developmental Biology* **49**, 615–632.

Borner, G.H., Lilley, K.S., Stevens, T.J. and Dupree, P. (2003) Identification of glycosylphosphatidylinositol-anchored proteins in *Arabidopsis*. A proteomic and genomic analysis. *Plant Physiology* **132**, 568–577.

Bouché, N. and Fromm, H. (2004) GABA in plants: just a metabolite? *Trends in Plant Sciences* **9**, 110–115.

Bucciaglia, P.A., Zimmermann, E. and Smith, A.G. (2003) Functional analysis of a beta-1,3-glucanase gene (*Tag1*) with anther-specific RNA and protein accumulation using antisense RNA inhibition. *Journal of Plant Physiology* **160**, 1367–1373.

Carpita, N.C. and McCann, M. (2000) The cell wall. In: *Biochemistry and Molecular Biology of Plants* (eds Buchanan, B.B., Gruissem, W. and Jones, R.L.), pp. 52–108. American Society of Plant Physiologists, Rockville, MD.

Cheung, A.Y., Wang, H. and Wu, H.M. (1995) A floral transmitting tissue-specific glycoprotein attracts pollen tubes and stimulates their growth. *Cell* **82**, 383–393.

Cosgrove, D.J., Bedinger, P. and Durachko, D.M. (1997) Group I allergens of grass pollen as cell wall-loosening agents. *Proceedings of the National Academy of Sciences of the United States of America* **94**, 6559–6564.

Cruz-Garcia, F., Hancock, C.N., Kim, D. and McClure, B. (2005) Stylar glycoproteins bind to S-RNase *in vitro. Plant Journal* **42**, 295–304.

Dearnaley, J.D.W. and Daggard, G.A. (2001) Expression of a polygalacturonase enzyme in germinating pollen of *Brassica napus. Sexual Plant Reproduction* **13**, 265–271.

Decreux, A. and Messiaen, J. (2005) Wall-associated kinase WAK1 interacts with cell wall pectins in a calcium-induced conformation. *Plant and Cell Physiology* **46**, 268–278.

de Graaf, B.H.J., Derksen, J.W.M. and Mariani, C. (2001) Pollen and pistil in the progamic phase. *Sexual Plant Reproduction* **14**, 41–55.

Delmer, D.P. (1999) Cellulose biosynthesis: exciting times for a difficult field of study. *Annual Review of Plant Physiology and Plant Molecular Biology* **50**, 245–276.

Dong, J., Kim, S.T. and Lord, E.M. (2005a) Plantacyanin plays a role in reproduction in *Arabidopsis*. *Plant Physiology* **138**, 778–789.

Dong, X., Hong, Z., Sivaramakrishnan, M., Mahfouz, M. and Verma, D.P. (2005b) Callose synthase (CalS5) is required for exine formation during microgametogenesis and for pollen viability in *Arabidopsis*. *Plant Journal* **42**, 315–328.

Doughty, J., Dixon, S., Hiscock, S.J., Willis, A.C., Parkin, I.A.P. and Dickinson, H.G. (1998) PCP-Al, a defensin-like *Brassica* pollen coat protein that binds the S locus glycoprotein, is the product of gametophytic gene expression. *The Plant Cell* **10**, 1333–1347.

Doughty, J., Wong, H.Y. and Dickinson, H.G. (2000) Cysteine-rich pollen coat proteins (PCPs) and their interactions with stigmatic *S* (Incompatibility) and *S*-related proteins in *Brassica*: putative roles in SI and pollination. *Annals of Botany* **85**(Suppl. A), 161–169.

Drøbak, B.K., Franklin-Tong, V.E. and Staiger, C. J. (2004) The role of the actin cytoskeleton in plant cell signaling. *New Phytologist* **163**, 13–30.

Edlund, A.F., Swanson, R. and Preuss, D. (2004) Pollen and stigma structure and function: the role of diversity in pollination. *The Plant Cell* **16**(Suppl.), 84–97.

Fedorova, M., van de Mortel, J., Matsumoto, P.A., Cho, J., Town, C.D., Vanden-Bosch, K.A., Gantt, J.S. and Vance, C.P. (2002) Genome-wide identification of nodule-specific transcripts in the model legume *Medicago truncatula*. *Plant Physiology* **130**, 519–537.

Feijó, J.A., Costa, S.S., Prado, A.M., Becker, J.D. and Certal, A.C. (2004) Signalling by tips. *Current Opinion in Plant Biology* **7**, 589–598.

Gu, Y., Wang, Z. and Yang, Z. (2004) ROP/RAC GTPase: an old new master regulator for plant signaling. *Current Opinion in Plant Biology* **7**, 527–536.

Guyon, V., Tang, W.H., Monti, M.M., Raiola, A., Lorenzo, G.D., McCormick, S. and Taylor, L.P. (2004) Antisense phenotypes reveal a role for SHY, a pollen-specific leucine-rich repeat protein, in pollen tube growth. *Plant Journal* **39**, 643–654.

Harwood, A. and Coates, J.C. (2004) A prehistory of cell adhesion. *Current Opinion in Cell Biology* **16**, 470–476.

Heizmann, P., Luu, D.T. and Dumas, C. (2000) Pollen-stigma adhesion in the Brassicaceae. *Annals of Botany* **85**(Suppl. A), 23–27.

Hepler, P.K. (2005) Calcium: a central regulator of plant growth and development. *The Plant Cell* **17**, 2142–2155.

Heslop-Harrison, Y. (2000) Control gates and micro-ecology: the pollen–stigma interaction in perspective. *Annals of Botany* **85**(Suppl. A), 5–13.

Higashiyama, T., Kuroiwa, H., Kawano, S. and Kuroiwa, T. (1998) Guidanc *in vitro* of the pollen tube to the naked embryo sac of *Torenia fournieri*. *The Plant Cell* **10**, 2019–2031.

Higashiyama, T., Kuroiwa, H. and Kuroiwa, T. (2003) Pollen-tube guidance: beacons from the female gametophyte, *Current Opinion in Plant Biology* **6**, 36–41.

Higashiyama, T., Yabe, S., Sasaki, N., Nishimura, S., Kuroiwa, H. and Kuroiwa T. (2001) Pollen tube attraction by the synergid cell. *Science* **293**, 1480–1483.

Hird, D.L., Worrall, D., Hodge, R., Smartt, S., Paul, W. and Scott, R. (1993) The anther-specific protein encoded by the *Brassica napus* and *Arabidopsis thaliana A6* gene displays similarity to beta-1,3-glucanases. *Plant Journal* **4**, 1023–1033.

Hiscock, S.J., Bown, D., Gurr, S.J. and Dickinson, H.G. (2002) Serine esterases are required for pollen tube penetration of the stigma in *Brassica*. *Sexual Plant Reproduction* **15**, 65–74.

Hiscock, S.J., Doughty, J., Willis, A.C. and Dickinson, H.G. (1995) A 7-kD pollen coating-borne peptide from *Brassica napus* interacts with S-locus glucoprotein and S-locus-related glycoprotein. *Planta* **196**, 367–374.

Hiscock, S.J. and McInnis, S.M. (2003) Pollen recognition and rejection during the sporophytic self-incompatibility response: *Brassica* and beyond. *Trends in Plant Sciences* **8**, 606–613.

Holdaway-Clarke, T.L. and Hepler, P.K. (2003) Control of pollen tube growth: role of ion gradients and fluxes. *New Phytologist* **159**, 539–569.

Huck, N., Moore, J.M., Federer, M. and Grossniklaus, U. (2003) The *Arabidopsis* mutant *feronia* disrupts the female gametophytic control of pollen tube reception, *Development* **130**, 2149–2159.

Izhar, S. and Frankel, R. (1971) Mechanism of male sterility in *Petunia*: the relationship between pH, callase activity in the anthers and the breakdown of the microsporogenesis. *Theoretical and Applied Genetics* **41**, 104–108.

Jauh, G.Y., Eckard, K., Nothnagel, E. and Lord, E.M. (1997) Adhesion of lily pollen tubes on an artificial matrix. *Sexual Plant Reproduction* **10**, 173–180.

Jiang, L., Yang, S.L., Xie, L.F., Puah, C.S., Zhang, X.Q., Yang, W.C., Sundaresan, V. and Ye, D. (2005) *VANGUARD1* encodes a pectin methylesterase that enhances pollen tube growth in the *Arabidopsis* style and transmitting tract. *The Plant Cell* **17**, 584–596.

Johnson, K.L., Jones, B.J., Bacic, A. and Schultz, C.J. (2003) The fasciclin-like arabinogalactan proteins of *Arabidopsis*. A multigene family of putative cell adhesion molecules. *Plant Physiology* **133**, 1911–1125.

Kader, J.C. (1997) Lipid-transfer proteins: a puzzling family of plant proteins. *Trends in Plant Sciences* **2**, 66–70.

Khosravi, D., Joulaie, R. and Shore, J.S. (2003) Immunocytochemical distribution of polygalacturonase and pectins of distylous and homostylous Turneraceae. *Sexual Plant Reproduction* **16**, 179–190.

Kim, S., Dong, J. and Lord, E.M. (2004) Pollen tube guidance: the role of adhesion and chemotropic molecules. *Current Topics in Developmental Biology* **61**, 61–79.

Kim, S., Mollet, J.C., Dong, J., Zhang, K., Park, S.Y. and Lord, E.M.(2003) Chemocyanin, a small, basic protein from the lily stigma, induces pollen tube chemotropism. *Proceedings of the National Academy of Sciences of the United States of America* **100**, 16125–16130.

Kreps, J.A., Wu, Y., Chang, H.S., Zhu, T., Wang, X. and Harper, J.F.(2002) Transcriptome changes for *Arabidopsis* in response to salt, osmotic and cold stress. *Plant Physiology* **130**, 2129–2141.

Ku, S., Yoon, H., Suh, H.S. and Chung, Y.Y. (2003) Male-sterility of thermosensitive genic male-sterile rice is associated with premature programmed cell death of the tapetum. *Planta* **217**, 559–565.

Lalanne, E., Honys, D., Johnson, A., Borner, G.H., Lilley, K.S., Dupree, P., Grossniklaus, U. and Twell, D. (2004) SETH1 and SETH2, two components of the glycosylphosphatidylinositol anchor biosynthetic pathway, are required for pollen germination and tube growth in *Arabidopsis*. *The Plant Cell* **16**, 229–240.

Lennon, K.A., Roy, S., Hepler, P.K. and Lord, E.M. (1998) The structure of the transmitting tissue of *Arabidopsis thaliana* (L.) and the path of pollen tube growth. *Sexual Plant Reproduction* **11**, 49–59.

Lord, E.M. (2000) Adhesion and cell movement during pollination: cherchez la femme. *Trends in Plant Sciences* **5**, 368–373.

Lord, E.M., Holdaway-Clarke, T.L., Roy, S.J., Jauh, G.Y. and Hepler, P.K. (2000) Arabinogalactan-proteins in pollen tube growth. In: *Cell and Developmental Biology of Arabinogalactan-proteins* (eds Nothnagel, E.A., Bacic, A. and Clarke, A.), pp. 153–167. Kluwer Academic/Plenum Publishers, New York.

Lord, E.M. and Mollet, J.C. (2002) Plant cell adhesion: a bioassay facilitates discovery of the first pectin biosynthetic gene. *Proceedings of the National Academy of Sciences of the United States of America* **99**, 15843–15845.

Lord, E.M., Mollet, J.C. and Park, S.Y. (2001) *In vivo* pollen tube growth: tube cell adhesion and movement in lily. In: *Cell Biology of Plant and Fungal Tip Growth* (eds Geitmann, A. Cresti, M. and Heath, I.B.), pp. 187–201. NATO Science Series, IOS Press, Amsterdam.

Lord, E.M. and Russell, R.D. (2002) The mechanisms of pollination and fertilization in plants. *Annual Review of Cell and Developmental Biology* **18**, 81–105.

Lush, W.M., Spurck, T. and Joosten, R. (2000) Pollen tube guidance by the pistil of a solanaceous plant. *Annals of Botany* **85**, 39–47.

Luu, D.T., Heizmann, P., Dumas, C., Trick, M. and Cappadocia, M. (1997) Involvement of *SLR1* genes in pollen adhesion to the stigmatic surface in Brassicaceae. *Sexual Plant Reproduction* **10**, 227–235.

Luu, D.T., Marty-Mazars, D., Trick, M., Dumas, C. and Heizmann, P. (1999) Pollen-stigma adhesion in *Brassica spp* involves SLG and SLR1 glycoproteins. *The Plant Cell* **11**, 251–262.

Majewska-Sawka, A. and Nothnagel, E.A. (2000) The multiple roles of arabinogalactan proteins in plant development. *Plant Physiology* **122**, 3–9.

Márton, M.L., Cordts, S., Broadhvest, J. and Dresselhaus, T. (2005) Micropylar pollen tube guidance by Egg Apparatus1 of maize. *Science* **307**, 573–576.

McCormick, S.M. (2004) Control of male gametophyte development. *The Plant Cell* **16**(Suppl.), 142–153.

Mollet, J.C., Park, S.Y. and Lord, E.M. (2003) Interaction of a stylar pectic polysaccharide and a basic protein (SCA) mediates lily pollen tube adhesion. In *Advances in Pectin and Pectinase Research* (eds F. Voragen, H. Schols and R. Visser), pp 1–14. Kluwer Academic Publishers, Dordrecht.

Mollet, J.C., Park, S.Y., Nothnagel, E.A. and Lord, E.M. (2000) A lily stylar pectin is necessary for pollen tube adhesion to an *in vitro* stylar matrix. *The Plant Cell* **12**, 1737–1749.

Neelam, A. and Sexton, R. (1995) Cellulase (endo-β-1,4 glucanase) and cell wall breakdown during anther development in the sweet pea (*Lathyrus odoratus* L.): isolation and characterization of partial cDNA clones. *Journal of Plant Physiology* **146**, 622–628.

Nieuwland, J., Feron, R., Huisman, B.A., Fasolino, A., Hilbers, C.W., Derksen, J. and Mariani C. (2005) Lipid transfer proteins enhance cell wall extension in tobacco. *The Plant Cell* **17**, 2009–2019.

Pacini, E. and Hesse, M. (2005) Pollenkitt – its composition, forms and functions. *Flora* **200**, 399–415.

Palanivelu, R., Brass, L., Edlund, A.F. and Preuss, D. (2003) Pollen tube growth and guidance is regulated by *POP2*, an *Arabidopsis* gene that controls GABA levels. *Cell* **114**, 47–59.

Palanivelu, R. and Preuss, D. (2000) Pollen tube targeting and axon guidance: parallels in tip growth mechanisms. *Trends in Cell Biology* **10**, 517–524.

Park, S.Y., Jauh, G.Y., Mollet, J.C., Eckard, K.J., Nothnagel, E.A., Walling, L.L. and Lord, E.M. (2000) A lipid transfer-like protein is necessary for lily pollen tube adhesion to an *in vitro* stylar matrix. *The Plant Cell* **12**, 151–163.

Park, S.Y. and Lord, E.M. (2003) Expression studies of SCA in lily and confirmation of its role in pollen tube adhesion. *Plant Molecular Biology* **51**, 183–189.

Parre, E. and Geitmann, A. (2005) More than a leak sealant. The mechanical properties of callose in pollen tubes. *Plant Physiology* **137**, 274–286.

Pearce, G., Moura, D.S., Stratmann, J. and Ryan Jr, C.A. (2001) RALF, a 5-kDa ubiquitous polypeptide in plants, arrests root growth and development. *Proceedings of the National Academy of Sciences of the United States of America* **98**, 12843–12847.

Piffanelli, P., Ross, J.H.E. and Murphy, D.J. (1998) Biogenesis and function of the lipidic structures of pollen grains. *Sexual Plant Reproduction* **11**, 65–80.

Prado, A.M., Porterfield, D.M. and Feijó, J.A. (2004) Nitric oxide is involved in growth regulation and re-orientation of pollen tubes. *Development* **131**, 2707–2714.

Ravindran, S., Kim, S., Martin, R., Lord, E.M. and Ozkan, C.S. (2005) Quantum dots as bio-labels for the localization of a small plant adhesion protein. *Nanotechnology* **16**, 1–4.

Rhee, S.Y. and Somerville, C.R. (1998) Tetrad pollen formation in *quartet* mutants of *Arabidopsis thaliana* is associated with persistence of pectic polysaccharides of the pollen mother cell wall. *Plant Journal* **15**, 79–88.

Rhee, S.Y., Osborne, E., Poindexter, P.D. and Somerville, C.R. (2003) Microspore separation in the *quartet3* mutants of *Arabidopsis* is impaired by a defect in a developmentally regulated polygalacturonase required for pollen mother cell wall degradation. *Plant Physiology*, **133** 1170–1180.

Rose, J.K.C., Bramm, J., Fry, S.C. and Nishitani, K. (2002) The XTH family of enzymes involved in xyloglucan endotransglucosylation and endohydrolysis: current perspectives and a new unifiying nomenclature. *Plant and Cell Physiology* **43**, 1421–1435.

Rotman, N., Rozier, F., Boavida, L., Dumas, C., Berger, F. and Faure, J.E.(2003) Female control of male gamete delivery during fertilization in *Arabidopsis thaliana*. *Current Biology* **13**, 432–436.

Rubinstein, A.L., Broadwater, A.H., Lowrey, K.B. and Bedinger, P.A.(1995) *PEX1*, a pollen-specific gene with an extensin-like domain. *Proceedings of the National Academy of Sciences of the United States of America* **92**, 3086–3090.

Sanchez, A.M., Bosch, M., Bots, M., Nieuwland, J., Feron, R. and Mariani, C. (2004) Pistil factors controlling pollination. *The Plant Cell* **16** (Suppl.), 98–106.

Schultz, C.J., Gilson, P., Oxley, D., Youl, J. and Bacic, A. (1998) The GPI-anchors on arabinogalactan-proteins: implication for signalling in plants. *Trends in Plant Sciences* **3**, 426–431.

Schultz, C.J., Johnson, K.L., Currie, G. and Bacic, A. (2000) The classical arabinogalactan protein gene family of *Arabidopsis*. *The Plant Cell*, **12**, 1751–1768.

Scott, R.J., Spielman, M. and Dickinson, H.G. (2004) Stamen structure and function. *The Plant Cell* **16** (Suppl.), 46–60.

Sexton, R., Del Campillo, E., Duncan, D. and Lewis, L.N. (1990) The purification of an anther cellulose (β(1,4)-glucan hydrolase) from *Lathyrus odoratus* L. and its relationship to the similar enzyme found in abscission zones. *Plant Science* **67**, 169–176.

Shi, H., Kim, Y.S., Guo, Y. and Zhu, J.K. (2003) The *Arabidopsis SOS5* locus encodes a putative cell surface adhesion protein and is required for normal cell expansion. *The Plant Cell* **15**, 19–32.

Sommer-Knudsen, J., Clarke, A.E. and Bacic, A. (1997) Proline- and hydroxyproline-rich gene products in the sexual tissues of flowers. *Sexual Plant Reproduction* **10**, 253–260.

Stieglitz, H. (1977) Role of beta-1,3-glucanase in postmeiotic microspore release. *Developmental Biology* **57**, 87–97.

Stratford, S., Barnes, W., Hohorst, D.L., Sagert, J.G., Cotter, R., Golubiewski, A., Showalter, A.M., McCormick, S.M. and Bedinger, P.A. (2001) A leucine rich repeat region is conserved in pollen extensin-like (Pex) proteins in monocots and dicots. *Plant Molecular Biology* **46**, 43–56.

Suen, D.F., Wu, S.S., Chang, H.C., Dhugga, K.S. and Huang, A.H. (2003) Cell wall reactive proteins in the coat and wall of maize pollen: potential role in pollen tube growth on the stigma and through the style. *Journal of Biological Chemistry* **278**, 43672–43681.

Swanson, R., Edlund, A.F. and Preuss, D. (2004) Species specificity in pollen-pistil interactions. *Annual Review of Genetics* **38**, 793–818.

Takayama, S., Shiba, H., Iwano, M., Asano, K., Hara, M., Che, F.S., Watanabe, M., Hinata, K. and Isogai, A. (2000) Isolation and characterization of pollen coat proteins of *Brassica campestris* that interact with S locus-related glycoprotein 1 involved in pollen-stigma adhesion. *Proceedings of the National Academy of Sciences of the United States of America* **97**, 3765–3770.

Takeda, H., Yoshikawa, T., Liu, X.Z., Nakagawa, N., Li Y.Q. and Sakurai, N. (2004) Molecular cloning of two exo-ß-glucanases and their *in vivo* substrates in the cell walls of lily pollen tubes. *Plant and Cell Physiology* **45**, 436–444.

Tang, W., Ezcurra, I., Muschietti, J. and McCormick, S.M. (2002) A cysteine-rich extracellular protein, Lat52, interacts with the extracellular domain of the pollen receptor kinase LePRK2. *The Plant Cell* **14**, 2277–2287.

Tang, W., Kelley, D., Ezcurra, I. and McCormick, S.M. (2004) LeSTIG1, an extracellular binding partner for the pollen receptor kinases LePRK1 and LePRK2, promotes pollen tube growt *in vitro*. *Plant Journal* **39**, 343–353.

Tsuchiya, T., Toriyama, K., Yoshikawa, M., Ejiri, S. and Hinata, K. (1995) Tapetum-specific expression of the gene for an endo-beta-1,3-glucanase causes male sterility in transgenic tobacco. *Plant and Cell Physiology* **36**, 487–494.

Tung, C.W., Dwyer, K.G., Nasrallah, M.E. and Nasrallah, J.B. (2005) Genome-wide identification of genes expressed in *Arabidopsis* pistils specifically along the path of pollen tube growth. *Plant Physiology* **138**, 977–989.

Verhoeven, T., Feron, R., Wolters-Arts, M., Edqvist, J., Gerats, T., Derksen, J. and Mariani, C. (2005) STIG1 controls exudate secretion in the pistil of *Petunia* and tobacco. *Plant Physiology* **138**, 153–160.

Verica, J.A., Chae, L., Tong, H., Ingmire, P. and He, Z.H. (2003) Tissue-specific and developmentally regulated expression of a cluster of tandemly arrayed cell wall-associated kinase-like kinase genes in *Arabidopsis*. *Plant Physiology* **133**, 1732–1746.

Verica, J.A. and He, Z.H (2002) The cell wall-associated kinase (*WAK*) and WAK-like kinase gene family. *Plant Physiology* **129**, 455–459.

Wengier, D., Valsecchi, I., Cabanas, M., Tang, W. and McCormick, S.M. (2003) The receptor kinases LePRK1 and LePRK2 associate in pollen and when expressed in yeast, but dissociate in the presence of style extract. *Proceedings of the National Academy of Sciences of the United States of America* **100**, 6860–6865.

Weterings, K. and Russell, S.D. (2004) Experimental analysis of the fertilization process. *The Plant Cell* **16**(Suppl.), 107–118.

Wheeler, M.J., Franklin-Tong, V.E. and Franklin, F.C.H. (2001) The molecular and genetic basis of pollen-pistil interactions. *New Phytologist* **151**, 565–584.

Wilhelmi, L.K. and Preuss, D. (1996) Self-sterility in *Arabidopsis* due to defective pollen tube guidance. *Science* **274,** 1535–1537.

Wolters-Arts, M., Lush, W.M. and Mariani, C. (1998) Lipids are required for directional pollen-tube growth. *Nature* **392**, 818–821.

Worrall, D., Hird, D.L., Hodge, R., Paul, W., Draper, J. and Scott, R.J. (1992) Premature dissolution of the microsporocyte callose wall causes male sterility in transgenic tobacco. *The Plant Cell* **4**, 759–771.

Wu, H.M., de Graaf, B.H.J., Mariani, C. and Cheung, A.Y. (2001) Hydroxyproline-rich glycoproteins in plant reproductive tissues: structure, functions and regulation. *Cellular and Molecular Life Sciences* **58**, 1418–1429.

Wu, H.M., Wang, H. and Cheung, A.Y. (1995) A pollen tube growth stimulatory glycoprotein is deglycosylated by pollen tubes and displays a glycosylation gradient in the flower. *Cell* **82**, 395–403.

Zhao, J., Yang, H.Y. and Lord, E.M. (2004) Calcium levels increase in the lily stylar transmitting tract after pollination. *Sexual Plant Reproduction*, **16** 259–263.

Zinkl, G.M. and Preuss, D. (2000) Dissecting *Arabidopsis* pollen-stigma interactions reveals novel mechanisms that confer mating specificity. *Annals of Botany* **85** (Suppl. A), 15–21.

Zinkl, G.M., Zwiebel, B.I., Grier, D.G. and Preuss, D. (1999) Pollen-stigma adhesion in *Arabidopsis*: a species-specific interaction mediated by lipophilic molecules in the pollen exine. *Development* **126**, 5431–5440.

5 Cell separation in roots

Fushi Wen, Marta Laskowski and Martha Hawes

5.1 Introduction: cell wall solubilization by microbial enzymes: a framework for understanding cell wall solubilization by plant enzymes

Plant cell walls comprise a major part of the world's biomass. The capacity of organisms to use this vast resource for energy, however, is limited by the availability of enzymes that digest it into its component sugars. Solubilization and turnover is also an integral component of normal cell wall function, and an array of solubilizing enzymes is present within cell walls (Varner and Lin, 1989). Nevertheless, most knowledge about how cell walls are degraded derives from a long history of studying how microbial pathogens solubilize plant cell walls during the process of infection.

DeBary (1886) documented for the first time that sap taken from roots infected by a fungal pathogen causes cellular disintegration when added to plant tissues. By the 1960s, it was clear that the production of cell wall degrading enzymes is a common trait of fungi and bacteria that associate with plants (Brown, 1965; Bateman and Millar, 1966). Most enzymes with the capacity to solubilize cell walls have been classified (Figure 5.1). Because of their presumed key role in causing disease, the primary experimental focus has been on pectolytic enzymes which, when purified, could by themselves cause maceration when added to tissue slices (Collmer and Keen, 1986). In general, only those enzymes that cleave internal linkages within pectic polymers fulfilled these criteria. Enzymes that cleaved monomers or dimers, or those that degraded cellulose and hemicellulose as well as pectin methyl esterase (PME), were inactive in maceration assays.

A systematic analysis of the role of pectolytic enzymes in cell wall solubilization during soft rot pathogenesis unveiled the complexity of the process (Collmer and Keen, 1986; Barras *et al.*, 1987). Using the soft rot pathogen *Erwinia chrysanthemi* as a model, expression of genes encoding individual enzymes was inactivated by marker exchange mutagenesis (Collmer *et al.*, 1985, 1988; Ried and Collmer, 1988; He and Collmer, 1990; Kelemu and Collmer, 1993; Alfano *et al.*, 1995). When grown on a medium containing citrus pectin, *E. chrysanthemi* strain EC16 secretes seven pectolytic enzymes, including a PME, exo-polygalacturonase, exo-pectate lyase and four isozymes of pectate lyase. Each pectate lyase activity, detected by gel overlay assay, is sufficient to confer the ability to macerate potato tubers, when expressed individually in nonpathogenic *E. coli* strains. Yet mutation of individual genes had little effect on pathogenicity as measured by the ability to macerate potato tuber tissue. Eliminating expression of all four pectate lyase genes in strain

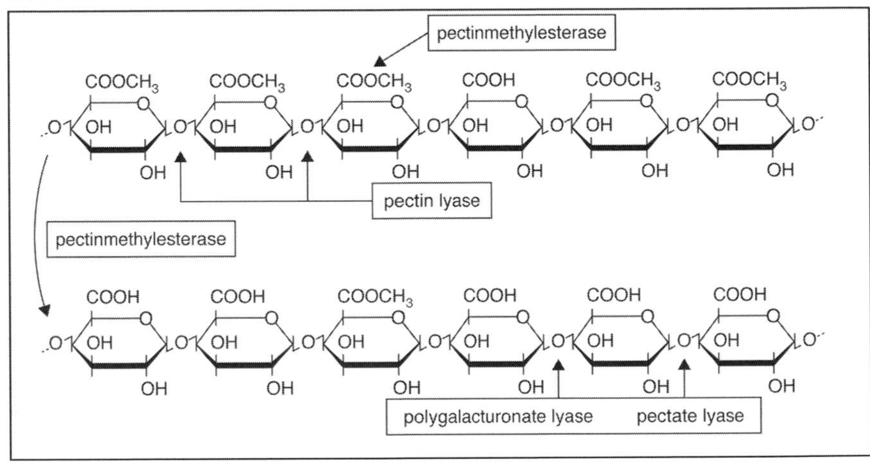

Figure 5.1 Cell wall pectic polymers and points of attack. Pectic polymers within the cell wall include methylated polygalacturonic acid (PGA) polymers (pectin), which are substrates for demethylation (pectinmethylesterase) and cleavage by β-elimination (pectin lyase). After demethylation, PGA is subject to β-elimination (pectate lyase) and hydrolytic cleavage (polygalacturonase) (redrawn from Rombouts and Pilnik, 1980).

EC16, in fact, reduced macerating ability but still did not eliminate it completely. Additional experiments to eliminate six of the seven pectolytic enzymes (all except PME) were carried out. Surprisingly, considering that these seven enzymes were predicted to underlie the strain's macerating activity, some residual activity on potato tubers remained. Moreover, virtually no loss in pathogenicity on the normal host chrysanthemum was observed.

These data suggested that, perhaps, pectolytic enzymes were not as important in plant disease as suggested by previous research. Alternatively, it was also observed that even more enzymes were present in host tissue than in the potato tuber model system. Subsequent experiments revealed that a set of five additional pectate lyase enzymes – a total of nine – was produced in the host tissue (Kelemu and Collmer, 1993). These enzymes were induced when the mutant bacterium was grown in a culture medium with chrysanthemum extracts.

These observations were a landmark in establishing a foundation for appreciating the biochemical complexity of cell wall solubilization (Collmer and Keen, 1986; Barras *et al.*, 1987; Hugouvieux-Cotte-Pattat *et al.*, 1996). Discoveries from several independent laboratories using Erwinia species as a model highlighted several important principles:

(1) Distinct groups of cell wall degrading enzymes are induced under different conditions.
(2) Components of distinct cell walls induce expression of different sets of cell wall degrading enzymes.
(3) Individual enzymes play divergent roles in cell wall solubilization leading to infection; these include modulation of access to substrate, initiation of depolymerization or enzymatic release of sugar or oligosaccharide signals that control regulation of expression of genes encoding other cell wall degrading enzymes.
(4) Individual enzymes may control whether cell wall solubilization is followed by localized or systemic infection.
(5) Individual enzymes may be required for cell wall solubilization in specific hosts or tissues but not others (Beaulieu *et al.*, 1993). Thus, for example, an *E. chrysanthemi* strain with a mutation in a specific pectate lyase gene was developed. When Saintpaulia was inoculated with this strain, infection was reduced by >80%, but infection of pea was reduced by only 27%. Remarkably, the same mutation rendered the strain *more* virulent than the wild type on chicory leaf (122% of wild-type infection) and on pea (153% wild type). Mutation in a single PME gene reduced infection in Saintpaulia by 90%, but only by 10% in chicory leaf. Such effects, presumably, can be attributed to regulatory effects of metabolites released during the cell wall solubilization process and highlight its inherent complexity.

All of these principles, at a minimum, can be assumed to underlie the endogenous metabolism of cell walls during the life of each plant. Available data reveal that the process is at least as complex in plants as in bacteria, with dozens of genes encoding putative cell wall degrading enzymes in the Arabidopsis genome.

5.2 Cell wall solubilization in plants: border cell separation in legumes as a model

In contrast to the elegant predictive framework for the enzymology of microbial pathogenesis, little knowledge beyond descriptive correlations and genome-based predictions is available about plant genes controlling endogenous cell wall solubilization (reviewed in Roberts *et al.*, 2002). It is reasonable to predict that the process is even more complex in plants than in soft rot bacteria. Apart from needing at least as many types of enzymes as pathogenic bacteria, plant cell wall solubilizing activity must be controlled precisely without causing cell death by cytoplasmic leakage or induction of defense responses. Nevertheless, given that the substrate in question is the same, it is reasonable to predict that the fundamental principles will be similar. Results from genomic surveys in recent years are consistent with this prediction: in the *M. truncatula* root cap alone, significant expression of genes encoding putative cell wall degrading enzymes, including 5 PMEs, 4 pectate lyases, 14 polygalacturonases, 5 cellulases (glucanase), 15 galactosidases, 7 xylosidases and 3 arabinofuranosidases, can be detected. The microbial model, therefore, was used as a guide to examine the role of pectolytic enzymes in border cell separation in legumes. The results indicate that altering expression of a single pectolytic enzyme can create comprehensive changes in cell biology.

5.2.1 Border cell separation

Border cell separation involves the complete dissociation of individual cells from each other and from progenitor root tissue. Upon immersion of the root tip into water, border cells begin to disperse within minutes (Figure 5.2A). Vital staining (Figure 5.2A, inset) can be used to confirm viability of the detached cells (Hawes and Wheeler, 1982). On gentle agitation, every cell disperses into the suspension, leaving the root tip free of border cells (Figure 5.2B). Border cell development is therefore similar to maceration that occurs during pathogenesis, except that it occurs without causing functional injury to the cell wall. It is distinct from the process termed 'abscission' in which cell wall solubilization occurs within a specific abscission layer, allowing the separation of two tissues, one of which is commonly composed of dead or dying cells. Instead, during border cell development, the

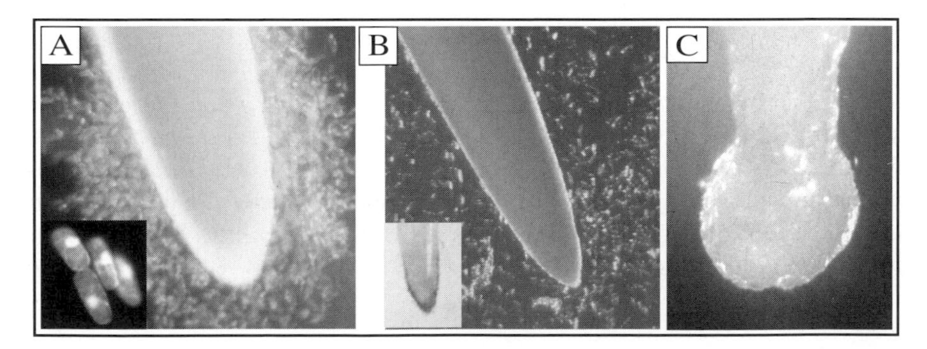

Figure 5.2 Requirement for pectinmethylesterase in border cell separation. (A and B) Border cell separation from legume roots and (C) inhibition of border cell separation by suppression of a root-cap-localized pectinmethylesterase gene (*rcpme1*). (A) Border cells of pea immediately begin to disperse into the suspension upon immersion of the root tip into water; viability of individual cells can be visualized by staining with a fluorescent vital stain (inset). (B) All border cells disperse into suspension with gentle agitation, leaving the root cap periphery free of cells; induction of a *rcpme1* at the cap periphery occurs within minutes after border cells are removed (inset, arrow). (C) In roots expressing *rcpme1* antisense mRNA, border cell separation is prevented, resulting in accumulation of border cells into a ball at the root tip.

product is a population of somatic cells with intact walls that are osmotically stable in distilled water. Unlike pathogenesis, abscission (see Chapter 6) and fruit ripening (see Chapter 8), which are terminally differentiated processes, border cell separation is dynamic and renewable (Hawes *et al.*, 2003). Cell separation can be turned on and off experimentally, and synchronized from plant to plant (Hawes and Lin, 1990). Finally, border cells can be collected nondestructively, without cutting or otherwise damaging tissue, and their numbers can be quantified easily by direct counts (Figure 5.2A). These characteristics make the border cell system ideally suited for exploring how plants control cell separation without lethality.

We used this system as a model to examine the role of pectolytic enzymes in cell separation (Hawes and Lin, 1990; Stephenson and Hawes, 1994; Wen *et al.*, 1999). We predicted that: (1) pectolytic enzymes play a role in the process of border cell separation (Hawes and Lin, 1990); (2) given its key role in cell wall solubilization in soft rot pathogenesis, pectinmethylesterase activity was a likely candidate to play a central role in border cell separation (Stephenson and Hawes, 1994). Our results, as summarized below, revealed that partial inhibition of pectinmethylesterase activity in root caps is correlated with complete inhibition of separation of border cells, which instead accumulate in a ball at the tip of the root (Figure 5.2C) (Wen *et al.*, 1999).

5.2.2 *Pectolytic enzyme activity present in the root cap, and correlated with border cell separation*

The hypothesis that pectin-solubilizing enzymes are present in the root cap while cell separation occurs was tested using the gel overlay assay (Collmer *et al.*, 1988). In this procedure proteins are separated electrophoretically and enzyme activity is detected by overlaying the gel with pectin. At sites where pectate lyase, polygalacturonase or other pectin-solubilizing enzyme activities are present a clearing in the substrate can be detected by staining with ruthenium red. Using this approach, pectolytic enzyme activity could be detected in extracts of root caps but only during the process of cell separation (Hawes and Lin, 1990). Reducing sugar assays confirmed that pectolytic activity was initiated within 1 h after border cell separation was induced, and was no longer detectable after 20 h when cell separation had ceased. No activity was detected in populations of detached border cells. Expression of genes with homology to pectate lyase, polygalacturonase and cellulase (endo-1,4-β-glucanase) is also induced in the root cap and correlates with border cell separation (Figure 5.2B, inset) (Twell *et al.*, 1991; Price, 2002).

5.2.3 *Pectinmethylesterase and border cell separation*

The role of PME in border cell separation has been examined in detail (Stephenson and Hawes, 1994; Wen *et al.*, 1999). This was based on the fact that PME plays a pivotal role in the initiation of cell wall solubilization by microbial pathogens, and therefore might play a similar key role in border cell separation (e.g. Beaulieu *et al.*, 1993). PME activity can be detected by its ability to induce the formation of a gel in the presence of pectin. A crude extract of four pea root caps in 0.1 ml of pectin (1% w/v) was found to induce the formation of a gel within 60 min, and boiling or treatment with protease eliminated gel-inducing activity.

PME can also be measured by a change in pH occurring when methoxy groups are cleaved from esterified pectin. One unit is defined as the amount needed to catalyze the release of 1 μmol carboxyl groups in 1 min. Using this assay, a single pea root cap was found to express nearly one unit of PME activity, sufficient to lower the pH of 3 ml of pectin by an order of magnitude within 5 min.

Levels of PME activity in the root cap are not significantly affected by genotype, circadian rhythms, light or germination temperatures ranging from 10–37°C (Stephenson and Hawes, 1994). However, the levels are correlated strongly with border cell separation. Enzyme activity is high during emergence of the radicle, and remains high until border cell separation is complete. If border cells are removed, PME activity increases rapidly and again remains high until a full set of cells is restored to the cap periphery. Similar levels of PME activity are present in root caps of sunflower, corn and alfalfa, all of which produce large numbers of border cells.

A heterologous probe from a bean PME gene was used to identify a root cap PME (*rcpme1*) gene (Wen *et al.*, 1999). Expression of *rcpme1* was found to be localized to the root cap periphery (Figure 5.2B) and temporally correlated with border cell separation (Figure 5.3A and B). When expression of this gene was inhibited by antisense mRNA expression in transgenic pea hairy roots, border cell separation was prevented (Figure 5.2C). Instead of dispersing into the suspension upon immersion into water, border cells accumulated in a ball at the root cap periphery. Normal root cap and border cell phenotypes were restored when *rcpme1* activity was restored (Wen *et al.*, 1999).

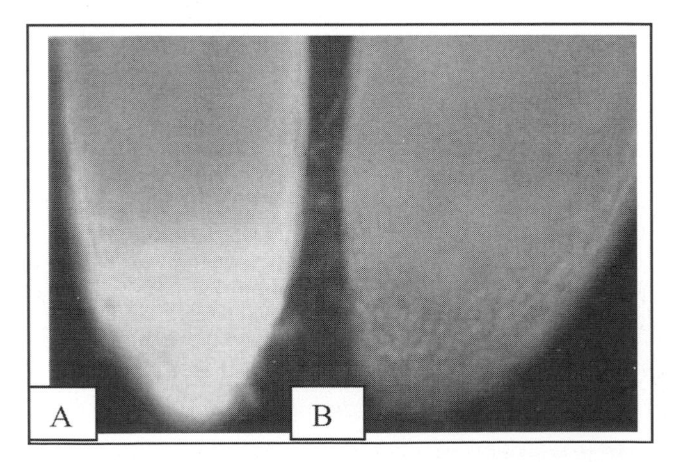

Figure 5.3 Altered extracellular pH throughout the root cap in correlation with *rcpme1* expression at the cap periphery. (A) In wild-type roots, extracellular pH is reduced to <6.0, which allows uptake of fluorescein into the cells throughout the cap. (B) In roots expressing *rcpme1* antisense mRNA, extracellular pH remains >6.5, and so there is no uptake of fluorescein and the root cap remains dark (Dorhout and Kolloffel, 1992; Wen *et al.*, 1999).

Monoclonal antibodies JIM5 and JIM7 react with 0–50% esterified PGA and 35–90% esterified PGA, respectively (Knox, 1997). These were used to demonstrate that changes in the degree of esterification of PGA in the root cap are correlated with PME activity and *rcpme1* expression (Stephenson and Hawes, 1994). In 1-mm roots, before border cells separate from the cap, there is little or no reaction with JIM5 (Figure 5.3C). After border cells begin to separate, JIM5 reactive substrate increases incrementally during the 24-h period of border cell development, in correlation with the expression of *rcpme1*.

High PME activity at the cell wall is thought to be capable of generating a pH gradient as pectin is de-esterified (e.g. Goldberg *et al.*, 1992). This reduction in pH could be an important mode of regulation for other cell wall degrading enzymes such as polygalacturonase, the activity of which is promoted at a low pH. Our data were consistent with this model: extracellular pH was measured using fluorescein, a dye that is taken into the cell when extracellular pH is <6.0 but is not taken up when the pH is >6.5. Uptake of fluorescein into cells within the root cap was correlated with PME activity (Stephenson and Hawes, 1994). In wild-type roots with normal expression of *rcpme1* at the cap periphery, extracellular pH is <6.0 (Figure 5.4A). In roots expressing *rcpme1* antisense mRNA, extracellular pH throughout the cap is increased to >6.5, and no fluorescein uptake occurs (Figure 5.4B).

These findings suggest that, even though dozens of individual enzymes may be required for cell separation, inhibiting activity of a single enzyme – pectinmethylesterase – is a critical evidence to the whole process. Presumably, as in soft rot pathogenesis, this role involves altering the environment by lowering the pH to a

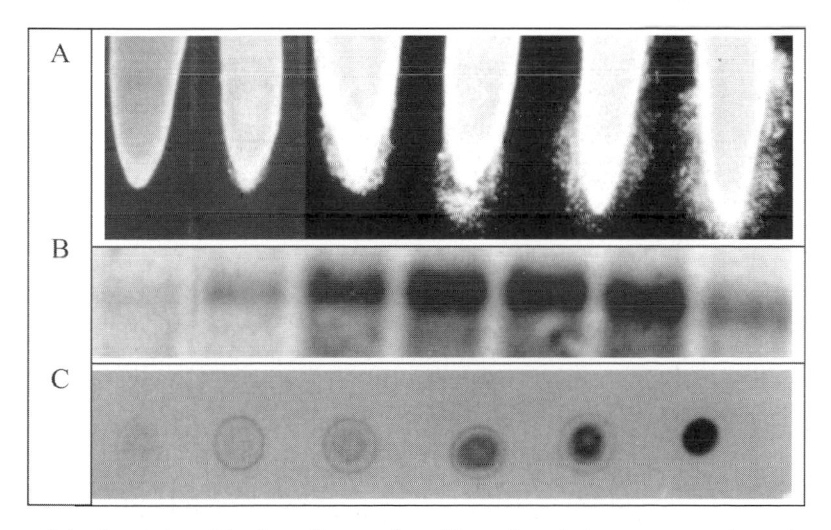

Figure 5.4 Correlation of border cell separation with pectinmethylesterase expression and pectin de-esterification. Over a 24-h period, border cell number increases to a species-specific maximum, in correlation with increased *rcpme1* expression and decreased methylation of pectin (Hawes *et al.*, 2003). Relationship between (A) border cell separation, (B) *rcpme1* expression measured by mRNA blot analysis and (C) increase in demethylated pectin measured by reaction with JIM5, an antibody that recognizes pectin with $<50\%$ esterification (from Stephenson and Hawes, 1994; Wen *et al.*, 1999).

level where the activity of pectolytic enzymes is optimal. Simultaneously, demethylation of pectin yields a substrate accessible to enzymatic solubilization.

5.3 Other mechanisms of cap turnover

With the exception of three families, the border cell separation process that occurs in legumes and cereals is common to all plant species tested (Hawes *et al.*, 2003). Thus, when germinated under conditions in which the cells are not lost due to mechanical abrasion or exposure to water, a species-dependent number of border cells accumulate at the cap periphery. The cell cycle in the root cap meristem is controlled by an extracellular signal secreted by border cells, which thereby regulate their own production (Brigham *et al.*, 1998). The point at which the cell cycle in the cap meristem is suppressed and cap turnover ceases defines the number of border cells per root, which is largely conserved at the family level and varies from a few dozen per root cap to >10,000 (Hawes and Pueppke, 1986). Virtually no variation in cell number occurs within species, unless the seed has been infected or otherwise damaged. Thus, among more than 40 cultivars of maize, each produced 4200 ± 400 cells per root (Hawes, unpublished). Tomato cultivars produced 200 ± 25 cells per root. Similar results were found with pea, bean and other legumes. Significant genotypic variation does occur in the degree of clumping of the cells. Thus, when germinated under identical conditions, roots from some maize cultivars release 4200 cells whose phenotypes range in varying degrees from single cells to clumps of 2–10 cells. In contrast, every root from other cultivars produces 4200 single cells, all of which are completely separate from one another. Such variation can be predicted to stem from cultivar-specific differences in pectolytic enzyme expression in response to endogenous or environmental signals. Responses to environment also are distinct among species. Thus, for example, increased carbon dioxide induces increased border cell production in pea but not in alfalfa (Zhao *et al.*, 2000).

We defined border cells as distinct from the classical term 'sloughed root cap cells' to reflect their unique morphology – populations of detached, metabolically active somatic cells – and their unique patterns of protein and gene expression (Hawes and Lin, 1990; Hawes and Brigham, 1992; Brigham *et al.*, 1995). Synonyms for 'sloughed' are 'putrid' and 'gangrenous', which do not accurately describe cells that exhibit a very high metabolic rate and can survive for months after detachment from the parent tissue (Knudson, 1919; Hawes and Wheeler, 1982). According to this definition, plant species in the Chenopodiacae and Cruciferae families do not produce border cells (Hawes and Pueppke, 1986; Niemira *et al.*, 1996). Thus, when Arabidopsis is germinated on a filter paper, no populations of detached cells can be collected by dipping the root tip into water (Figure 5.5A). When allowed to penetrate solidified medium (e.g. water agar or tissue culture media) over time, the cap does occasionally undergo turnover, yielding a single or multiple files of shed caps (Vicre *et al.*, 2005). In these cases, cells within the cap appear to undergo programmed cell death, followed by sloughing of the dead tissue (Figure 5.5B and C). Cap turnover can be induced by treatment with auxin (Ponce *et al.*, 2005) (Figure 5.5D). However,

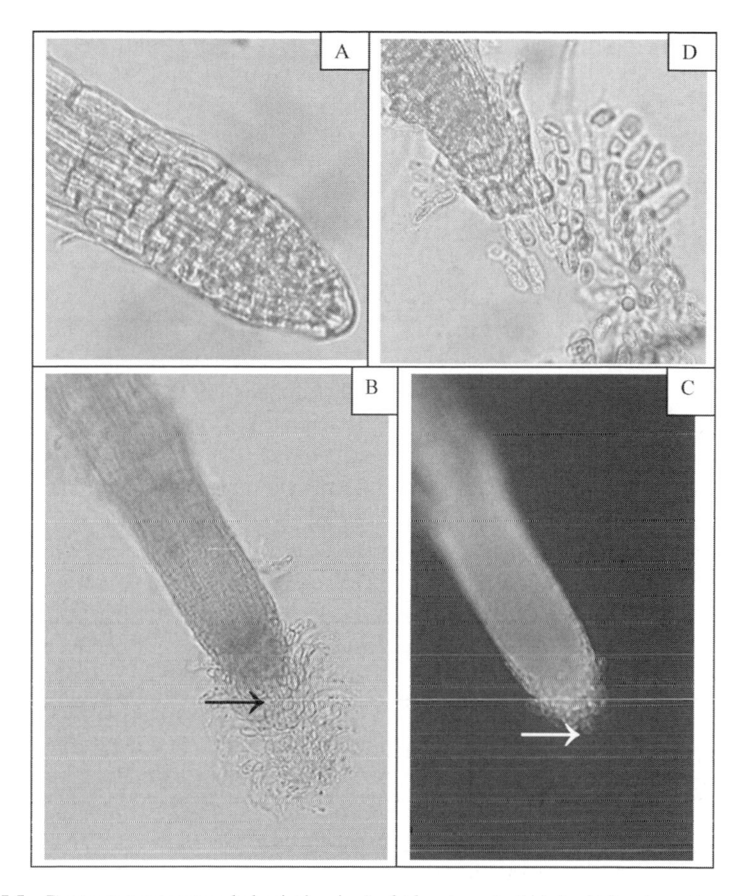

Figure 5.5 Root cap turnover and abscission in *Arabidopsis* roots. (A) *Arabidopsis* roots, such as the typical root shown here, do not have border cells, as defined by populations of living detached cells with unique gene expression patterns: no cells are released when the root tip is immersed into water. (B) Occasional turnover occurs, resulting in detachment or abscission of layers of sloughed cap tissue. (C) When the same root as in (B) is stained with a fluorescent vital stain, only the cells within the cap to the peripheral cap layer (arrow) are viable. None of the multiple layers of sloughed cells are visible in fluorescent light, indicating that they are dead. (D) Induction of root cap turnover in *Arabidopsis*, 24–48 h after treatment with 90 μM IAA (F. Wen, T.J. Kajstura, M. Laskowski and M.C. Hawes, unpublished).

cells within the root cap die before detachment (as in Figure 5.5C). This observation suggests that in these species, the term 'sloughed root cap cells' is accurate. It is possible that border cell production occurs under other, more natural conditions but the families that fail to produce border cells also fail to develop mycorrhizal associations in nature (Niemira *et al.*, 1996). Because living border cells appear to play a role in delivering specific signals required for controlled infection by root-infecting fungi (e.g. Nagahashi and Douds, 2004; Gunawardena *et al.*, 2005a,b), this correlation may reflect an absence of border cells under natural as well as environmental conditions.

How distinct the molecular mechanisms underlying sloughing root caps may be from separating border cells is unknown. However, divergence of root cap phenotypes with regard to cap turnover may be related, solely or in part, to differences in the regulation of expression of cell wall degrading enzymes. A glucanase gene has been partially characterized and proposed to play a role in the sloughing of the Arabidopsis root cap (del Campillo *et al.*, 2004).

5.4 Cortical cell sloughing and lateral root emergence

The mechanism by which lateral roots grow through their parent tissue to reach the soil has been debated for more than 100 years, as yet without resolution. Early experiments focused on the extent to which mechanical pressure (Pond, 1908) and enzymatic degradation (Van Tieghem and Douliot, 1888) may contribute to the process. Microscopic investigations indicate that cortical cells overlying emerging lateral roots separate at the pectin-rich middle lamella. Accordingly, it has been suggested that secretion of pectinases may be part of the process that leads to cell separation, and hence to lateral root emergence (Bell and McCully, 1970). Evidence for involvement of specific enzymes is largely circumstantial, based on a growing body of increasingly strong correlations. As our knowledge of the basic processes driving cell separation progresses, we may further refine our questions, and ask if the mechanisms that allow rapidly expanding pericycle cell derivates to separate from their comparatively quiescent neighboring endodermal cells are similar or distinct from those that allow established primordia to make their way through the cortical and epidermal cell layers.

Cortical cell shedding has been observed in many plants, typically as a result of high auxin concentrations. Radish roots treated with 90 μM IAA activate the entire pericycle to form lateral roots. As this occurs, all of the root cells exterior to the pericycle-cell derivates can slough away (Laskowski *et al.*, 1995). Cortical cell shedding has also been observed in hypocotyls. In Arabidopsis, it occurs at high frequency in superroot (*sur1*) mutants that overproduce auxin (Boerjan *et al.*, 1995). Combining auxin overproduction with auxin insensitivity in *sur1 axr2* double mutants results in plants that do not exhibit cortical cell shedding, reinforcing the connection between auxin response and cell separation. While cortical cell separation typically occurs in regions where adventitious roots form, it can also occur in the absence of root formation, indicating that cell shedding can be initiated independent of lateral root formation (Boerjan *et al.*, 1995). A wide range of enzymes may be involved in the process of cortical cell separation, although none have been proven. Here, we present evidence for the possible involvement of pectolytic enzymes.

5.4.1 Evidence for the involvement of PME

Correlations between PME activity and auxin-induced changes in the cell wall, combined with auxin up-regulation of at least one putative PME family gene, suggest that PME activity may be enhanced downstream of auxin action. Because cortical cell shedding occurs downstream of auxin response, it is possible that pectolytic enzymes involved in the process could be auxin regulated as well. A correlation between PME activity and auxin-induced cell elongation has been observed in tobacco pith (Brian and Newcomb, 1954). A similar correlation was also observed in the third internode of peas (Yoda, 1958). Using microarray analysis, Okushima *et al.* (2005) identified a putative PME family gene (*At3g10720*) that is auxin regulated and differentially expressed in pleiotropic *nph4–1/arf19-1* auxin response mutants that do not make lateral roots until they are more than 2 weeks old. Together, these data raise the possibility that PME activity may be enhanced downstream of auxin action, making it a possible candidate for involvement in cell separation processes.

The fluorescein assay that highlights separating border cells (described above) also highlights emerging lateral roots. Thus, the apoplastic pH of the tissue in regions surrounding emerging lateral roots is lower than that in the surrounding cells (Dorhut and Kolloffel, 1992). When Arabidopsis roots are incubated in fluorescein, the entire epidermis fluoresces, as would be expected due to the presence of de-methylated, and hence acidic, pectin in the outer cell walls. Against this background, increased levels of fluorescence are observed around developing primordia. The youngest primordia, containing 1–2 layers of pericycle cell derivates, were not associated with an increase in fluorescence ($n = 9$), but older primordia were. Eight out of nine primordia that contained at least three cell layers but had not yet emerged showed a substantial increase in fluorescence (Plate 7A). For emerged lateral roots, 79% ($n = 38$) showed substantial increases in fluorescence in the epidermal cells surrounding the root opening (Plate 7B and C). These regions may reflect PME-induced cell wall acidification, or wall acidification due to other factors such as auxin action.

5.4.2 *Evidence for the involvement of polygalacturonase*

Two lines of evidence specifically correlate the presence of polygalacturonase with lateral root emergence. Peretto *et al.* (1992) used immuno-localization with polyclonal antibodies to demonstrate that polygalacturonase accumulates around sites of lateral root emergence in leeks. Tissue prints with anti-polygalacturonase show high levels of signal covering the surface of an emerged lateral root. In comparison, the parent root is nearly devoid of stain. The same antibody used with a fluorescently labeled secondary antibody highlights the perimeter of an emerging lateral root without staining the surrounding tissues. A somewhat different localization was observed in Arabidopsis when polygalacturonase expression was traced with PGAZAT::GFP fusions, and it was found to concentrate in the regions of the parent root immediately surrounding emerging lateral roots as well as in the root cap (Roberts *et al.*, 2002). These results indicate a strong correlation between the location of polygalacturonase and lateral root emergence.

5.4.3 *Evidence for involvement of pectate lyase*

As with PME, auxin induction of pectate lyase family genes may identify candidate genes that could be involved in cell separation. Auxin induction of pectate lyase has been demonstrated in Zinnia (Domingo *et al.*, 1998). In Arabidopsis, where lateral root emergence is accompanied by the breakdown of the pectin-rich middle lamella, auxin-inducible expression of two genes encoding sequence-defined pectate lyases is temporally correlated with the emergence of lateral roots, raising the possibility that they could be related events.

5.5 Aerenchyma

Aerenchyma is a term describing '*plant tissues containing enlarged gas spaces exceeding those commonly found as intracellular spaces*' (Evans, 2003). These spaces may form as a consequence of cell separation, which suggests a potential role of cell wall degrading enzymes. Alternatively, the spaces develop when cells die leaving a 'hole', in which case cell wall degrading enzymes are implicated in the removal of dead cell debris to create the opening. During the last two years, the physiology, cell biology and agronomic importance of aerenchyma formation and function have been detailed in several excellent reviews (Aarts and Fiers, 2003; Colmer, 2003; Evans, 2003; Francis, 2003; Kozela and Regan, 2003; Setter and Waters, 2003; Fukao and Bailey-Serres, 2004; Gunawardena *et al.*, 2005a,b; Igamberdiev *et al.*, 2005; Seago *et al.*, 2005). Specific enzymes that may control cell separation and/or cell wall solubilization, and how their expression may be regulated during aerenchyma formation, remain to be delineated. Rapid changes in pectin esterification during early stages of aerenchyma formation suggest a central role of pectinmethylesterase activity, as in soft rot pathogenesis, border cell separation, and lateral root formation (Gunawardena *et al.*, 2001).

References

Aarts, M.G.M. and Fiers, M.W.E.J. (2003) What drives plant stress genes? *Trends in Plant Science* **8**, 99–102.

Alfano, J.R., Ham, J.H. and Collmer, A. (1995) Use of Tn5tac1 to clone a pel gene encoding a highly alkaline, asparagine-rich pectate lyase isozyme from an *Erwinia chrysanthemi* EC16 mutant with deletions affecting the major pectate lyase isozymes. *Journal of Bacteriology* **177**, 4553–4556.

Barras, F., Thurn, K.K. and Chatterjee, A.K. (1987) Extracellular enzymes and pathogenesis of soft-rot *Erwinia*. *Annual Review of Phytopathology* **32**, 201–234.

Bateman, D.F. and Millar, R.L. (1966) Pectic enzymes in tissue degradation. *Annual Review of Phytopathology* **4**, 119–146.

Beaulieu, C., Boccara, M. and Van Gijsegem, F. (1993) Pathogenic behavior of pectinase-defective *Erwinia chrysanthemi* mutants on different plants. *Molecular Plant Microbe Interactions* **6**, 197–202.

Bell, J.K., McCully, M.E. (1970) A histological study of lateral root initiation and development in *Zea mays*. *Protoplasma* **70**, 179–205.

Boerjan, W., Cervera, M.T., Delarue, M., Beeckman, T., Dewitte, W., Bellini, C., Caboche, M., Van Onckelen, H., Van Montagu, M. and Inze, D. (1995) Superroot, a recessive mutation in Arabidopsis, confers auxin overproduction. *Plant Cell* **7**, 1405–1419.

Brian, W. and Newcomb, E.H. (1954) Stimulation of pectinmethylesterase activity of cultured tobacco pith by indole acetic-acid. *Physiologia Plantarum* **7**, 290–297.

Brigham, L.A., Woo, H.H., Hawes, M.C. (1995) Differential expression of proteins and mRNAs from border cells and root tips of pea. *Plant Physiology* **109**, 457–463.

Brigham, L.A., Woo, H.H., Wen, F. and Hawes, M.C. (1998) Meristem-specific suppression of mitosis and a global switch in gene expression in the root cap of pea by endogenous signals. *Plant Physiology* **118**, 1223–1231.

Brown, W. (1965) Toxins and cell wall dissolving enzymes in relation to plant disease. *Annual Review of Phytopathology* **3**, 1–18.

Colmer, T.D. (2003) Long-distance transport of gases in plants: a perspective on internal aeration and radial oxygen loss from roots. *Plant Cell and Environment* **26**, 17–36.

Collmer, A. and Keen, N.T. (1986) The role of pectic enzymes in plant pathogenesis. *Annual Review of Phytopathology* **24**, 383–409.

Collmer, A., Ried, J.L. and Mount, M.S. (1988) Assay methods for pectic enzymes. *Methods in Enzymology* **161**, 329–335.

Collmer, A., Schoedel, C., Roeder, D.L., Ried, J.L. and Rissler, J.F. (1985) Molecular cloning in *E. coli* of *Erwinia chrysanthemi* genes encoding multiple forms of pectate lyase. *Journal of Bacteriology* **161**, 913–920.

DeBary, A. (1886) Ueber Einige Sclerotinien und Sclerotien-. Krankheiten. *Bot Zeit* **44**, 376–480.

del Campillo, E., Abdel-Aziz, A., Crawford, D. and Patterson, S.E. (2004) Root cap specific expression of an endo-1,4-D-glucanase (cellulase): a new marker to study root development in. *Arabidopsis Plant Molecular Biology* **56**, 309–323.

Domingo, C., Roberts, K., Stacey, N.J., Connerton, I., Riz-Teran, F. and McCann, M.C. (1998) A pectate lyase from *Zinnia elegans* is auxin inducible. *Plant Journal* **13**, 17–28.

Dorhut, R., Kolloffel, C. (1992) Determining apoplastic pH differences in pea roots by use of the fluorescent dye fluorescein. *Journal of Experimental Botany* **43**, 479–486.

Evans, D.E. (2003) Aerenchyma formation. *New Phytologist* **161**, 35–49.

Francis, D. (2003) The interface between the cell cycle and programmed cell death in higher plants: from division unto death. *Advances in Botanical Research Incorporating Advances in Plant Pathology* **40**, 143–181.

Fukao, T. and Bailey-Serres, J. (2004) Plant responses to hypoxia – is survival a balancing act. *Trends in Plant Science* **9**, 449–456.

Goldberg, R., Pierron, M., Durand, L. and Mutaftshiev, S. (1992) In vitro and in situ properties of cell pectinmethylesterases from mung bean hypocotyls. *Journal of Experimental Botany* **43**, 41–46.

Gunawardena, A., Pearce, D.M.E. Jackson, M.B., Hawes, C.R. and Evans, D.E. (2001) Rapid changes in cell wall pectic polysaccharides are closely associated with early stages of aerenchyma formation. *Plant Cell and Environment* **24**, 1369–1375.

Gunawardena, A., Sault, K. and Donnelly, P. (2005a) Programmed cell death and leaf morphogenesis in *Monstera obliqua* (Araceae). *Planta* **221**, 607–618.

Gunawardena, U., Rodriguez, M., Straney, D., VanEtten, H.D. and Hawes, M.C. (2005b) Tissue-specific localization of root infection by *Nectria haematococca*: mechanisms and consequences. *Plant Physiology* **137**, 1363–1374.

Hawes, M.C., Bengough, G.A., Cassab, G. and Ponce, G. (2003) Root caps and rhizosphere. *Journal of Plant Growth Regulation* **21**, 352–367.

Hawes, M.C. and Brigham, L.B. (1992) Impact of root border cells on microbial populations in the rhizosphere. *Advances in Plant Pathology* **8**, 119–148.

Hawes, M.C., Brigham, L.A., Woo, H.H., Zhu, Y. and Wen, F (1997) Root border cells: phenomenology of signal exchange. In: *Radical Biology: Advances and Perspectives on the Function of Plant Roots* (eds Flores, H.E., Lynch, J.P. and Eissenstat, D.), pp. 210–218. American Society of Plant Physiology.

Hawes, M.C. and Lin, H-J. (1990) Correlation of pectolytic enzyme activity with the programmed release of cells from root caps of pea (*Pisum sativum*). *Plant Physiology* **94**, 1855–1859.

Hawes, M.C. and Pueppke, S.G. (1986) Sloughed peripheral root cap cells: yield from different species and callus formation from single cells. *American Journal of Botany* **73**, 1466–1473.

Hawes, M.C. and Wheeler, H.E. (1982) Factors affecting victorin-induced cell death: temperature and plasmolysis. *Physiological Plant Pathology* **20**, 137–144.

He, S.Y. and Collmer, A. (1990) Molecular cloning, nucleotide sequence, and marker-exchange mutagenesis of the exo-poly-a-D-galacturonidase-encoding pehX gene of *Erwinia chrysanthemi* EC16. *Journal of Bacteriology* **172**, 4988–4995.

Hugouvieux-Cotte-Pattat, N., Condemine, G., Nasser, W. and Reverchon, S. (1996) Regulation of pectinolysis in *Erwinia chrysanthemi. Annual Review of Microbiology* **50**, 213–257.

Igamberdiev, A.U., Baron, K. and Manac'h-Little, N. (2005) The haemoglobin//nitric oxide cycle: involvement in flooding stress and effects on hormone signalling. *Annals of Botany* **96**, 557–564.

Kelemu, S. and Collmer, A. (1993) *Erwinia chrysanthemi* EC16 produces a second set of plant inducible pectate lyase isoenzymes. *Applied and Environmental Microbiology* **59**, 1756–1761.

Knox, J.P. (1997) The use of antibodies to study the architecture and developmental regulation of plant cell walls. *International Review of Cytology* **171**, 79–120.

Knudson, L. (1919) Viability of detached root cap cells. *American Journal of Botany* **6**, 309–310.

Kozela, C. and Regan, S. (2003) How plants make tubes. *Trends in Plant Science* **8**, 159–164.

Laskowski, M.J., Williams, M.E., Nusbaum, H.C. and Sussex, I.M. (1995) Formation of lateral root meristems is a two-stage process. *Development* **121**, 3303–3310.

Nagahashi, G. and Douds, D.D. (2004) Isolated root caps, border cells, and mucilage from host roots stimulate hyphal branching of the arbuscular mycorrhizal fungus, *Gigaspora gigantea. Mycological Research* **108**, 1079–1088.

Niemira, B.A., Safir, G.R. and Hawes, M.C. (1996) Arbuscular mycorrhizal colonization and border cell production: a possible correlation. *Phytopathology* **86**, 563–568.

Okushima, Y., Overvoorde, P.J., Arima, K., Alonso, J.M., Chan, A., Chang, C., Ecker, J.R., Hughes, B., Lui, A., Nguyen, D., Onodera, C., Quach, H., Smith, A., Yu, G. and Theologis, A. (2005) Functional genomic analysis of the AUXIN RESPONSE FACTOR gene family members in *Arabidopsis thaliana*: unique and overlapping functions of ARF7 and ARF19. *Plant Cell* **17**, 444–463.

Peretto, R., Favaron, F., Bettini, V., De Lorenzo, G., Marini, S., Alghisi, P., Cervone, F. and Bonfante, P. (1992) Expression and localization of polygalacturonase during the outgrowth of lateral roots in *Allium porrum* L. *Planta* **188**, 164–172.

Pilatzke-Wunderlich, I. and Nessler, C.L. (2001) Expression and activity of cell wall degrading enzymes in the latex of the opium poppy, *Papaver somniferum* L. *Plant Molecular Biology* **45**, 467–576.

Ponce G., Barlow, P.W., Feldman, L.J. and Cassab, G. (2005) Auxin and ethylene interactions control mitotic activity of the quiescent center, root cap size, and pattern of cell differentiation in maize. *Plant, Cell and Environment* **28**, 719–731.

Pond, R.H. (1908) Emergence of Lateral Roots. *Botanical Gazette* **46**, 410–421.

Price, I. (2002) *Characterization of a Rhizosphere Galactosidase Gene Secreted by Root Border Cells.* MS Thesis, University of Arizona, Tucson AZ.

Ried, J.L. and Collmer, A. (1988) Construction and characterization of an *Erwinia chrysanthemi* mutant with directed deletions in all of the pectate lyase structural genes. *Molecular Plant Microbe Interactions* **1**, 32–38.

Roberts, J.A., Elliott, K.A. and Gonzalez-Carranza, Z.H. (2002) Abscission, dehiscence, and other cell separation processes. *Annual Review of Plant Biology* **53**, 131–158.

Rombouts, F.M. and Pilnik, W. (1980) Pectic enzymes. *Economic Microbiology* **5**, 228–282.

Seago, J.L., Marsh, L.C. and Stevens, K.J. (2005) A re-examination of the root cortex in wetland flowering plants with respect to aerenchyma. *Annals of Botany* **96**, 565–579.

Setter T.L., Waters I. (2003) Review of prospects for germplasm improvement for waterlogging tolerance in wheat, barley and oats. *Plant and Soil* **253**, 1–34.

Stephenson, M.B. and Hawes, M.C. (1994) Correlation of pectinmethylesterase activity in root caps of pea with root border cell separation. *Plant Physiology* **106**, 39–45.

Twell, D., Yamaguchi, J., Wing, R.A., Ushiba, J. and McCormick, S. (1991) Promoter analysis of genes that are coordinately expressed during pollen development reveals pollen-specific enhancer sequences and shared regulatory elements. *Genes and Development* **5**, 496–507.

Varner, J.E. and Lin, L. (1989) Plant cell wall architecture. *Cell* **56**, 231–239.

Vicre, M., Santaella, C., Blanchet, S., Gateau, A. and Driouich, A. (2005) Root border-like cells of Arabidopsis. Microscopical characterization and role in the interaction with Rhizobacteria. *Plant Physiology* **138**, 998–1008.

Wen, F., Zhu, Y., Brigham, L.A. and Hawes, M.C. (1999) Expression of an inducible pectin-methylesterase gene is required for border cell separation from roots of pea. *Plant Cell* **11**, 1129–1140.

Van Tieghem, P., Douliot, H. (1888) Recherches comparatives sur l'origine des membres endogènes dans les plantes vasculaires. *Annales des Sciences Naturelles Botanique* **8**, 1–660.

Yoda, S. (1958) Auxin action and pectin enzyme. *Botanical Magazine Tokyo* **71**, 207–213.

Zhao, X., Misaghi, I. and Hawes, M.C. (2000) Stimulation of border cell production in response to increased carbon dioxide levels. *Plant Physiology* **122**, 1–8.

6 Organ abscission

Michelle E. Leslie, Michael W. Lewis
and Sarah J. Liljegren

6.1 Introduction

At certain stages in the life cycle of many multicellular organisms, targeted separation of cells at predetermined sites enables the shedding of entire organs. Plants undergo abscission of many diverse types of organs, such as leaves, flowers, fruit, seeds, bark and branches (Addicott, 1982). Although less common, separation events in animals that involve the loss of entire organs include antler shedding in deer, skin shedding in snakes and tail or arm autotomy in lizards and starfish (Rolf and Enderle, 1999; Alibardi, 2000; Wilkie, 2001; Alibardi, 2005). Each of these events occurs within specific tissue layers or autotomy planes, with some requiring dissolution of cell adhesion similar to organ abscission in plants.

The agronomical benefits of understanding and manipulating organ abscission in plants are well known. Since the advent of agriculture, humans have selected for crop varieties with desirable abscission traits. According to the 'Teosinte Hypothesis', the selection of four key traits within the past 6000–10 000 years was required for the domestication of maize from the primitive teosinte grass (Doebley, 2004). Seed dispersal from teosinte ears is facilitated by the initial separation of eight fruitcases from one another. Each fruitcase then disperses its seed individually. A significant feature of domesticated maize is that the ears remain intact until harvest; thus, it is hypothesized that a mutation(s) disrupting abscission layer formation between teosinte fruitcases was one of the first traits selected by our farming ancestors (Doebley, 2004). More recently, the spontaneously arising *jointless* (*j*) mutation was found to block fruit abscission in tomato (Butler, 1936). Since its discovery, this trait has been introgressed into a variety of tomato cultivars, providing a modern example of human-directed evolution of crop species. Today, we aim to increase our understanding of abscission as a unique developmental process as well as an agronomically important event.

Organ shedding in plants facilitates several key processes, including the following:

(1) *Reproduction*: Cell separation events in plants are orchestrated to drop ripened fruit, disperse seeds from opened fruit, release spores from fern leaf sori, scatter leaves or branches capable of vegetative propagation and abscise leaves that obstruct pollination.
(2) *Recycling*: Many organs in plants are shed after they have served their biological function. For example, senescent leaves fall from deciduous trees and

floral organs are released after fertilization of the flower. These organs are large mineral sinks; thus, shedding of leaves and flowers allows minerals to be redirected to other organs. During the photoperiod-induced senescence of aspen (*Populus tremula*) leaves, many nutrients, such as nitrogen and phosphorous, are redirected out of the leaves prior to their abscission (Keskitalo *et al.*, 2005).

(3) *Remodeling*: Entire branches are often shed to allow for directed growth of a plant.

(4) *Defense*: As a guard against the spreading of disease, pathogen-infected leaves, branches and bark can be shed. Organ shedding can also produce protective scars or spines that deter pathogen attack.

(5) *Competition*: Frequently, a plant will generate more than enough floral buds to be fertilized. Once sufficient fertilization has occurred, a portion of the remaining buds or the developing fruit will be dropped to ensure optimized growth of a subset. In addition, leaves from certain plants contain protective chemicals that can inhibit the growth of other plants in the soil where the 'poisonous' leaves drop and decompose.

Organ abscission events may be diverse in function, yet all events require the proper formation of an abscission zone (AZ) at the base of the organ to be shed. The AZ is morphologically distinct from neighboring cells and is often visible to the naked eye (Addicott, 1982). At the microscopic level, AZ cells are smaller than surrounding cells, and more cytoplasmically dense. The size of the AZ varies from as few as 1–2 rows of cells in the model plant *Arabidopsis thaliana* to as many as 50 rows in *Sambucus nigra* (Taylor *et al.*, 1994; Bleecker and Patterson, 1997). Diverse environmental cues such as drought, nutrient deficiency and pathogen attack can trigger the initiation of abscission at these sites (Addicott, 1982; Taylor and Whitelaw, 2001). Developmentally, the timing of abscission is coordinated with fertilization, fruit maturation and senescence depending on the particular organ to be shed.

Studies of AZ morphology together with physiological, genetic and biochemical experiments have led to a general model of the organ abscission process across plant species (Figure 6.1). First, cells at the future site of detachment must receive and respond to differentiation signals. Differentiation of the AZ is dependent upon the proper developmental patterning of the organ itself. Second, activation of the AZ, allowing the initiation of cell separation, is triggered by developmental and hormonal cues. Third, the organ is abscised when the middle lamella between AZ cells loses integrity due to the activity of cell wall modifying and hydrolytic enzymes. Finally, protective scarring is evident at the site of organ detachment on the plant body. Scarring is associated with expansion of the AZ cells that remain behind – whether cell expansion may itself play a role in the separation process is yet to be determined.

In this chapter, we discuss how genetic analysis in model plants has contributed to our current understanding of the organ abscission process as outlined above (Figure 6.2; Table 6.1). The introduction of *Arabidopsis* as a model organism has been particularly useful for the isolation of genes involved in floral organ abscission

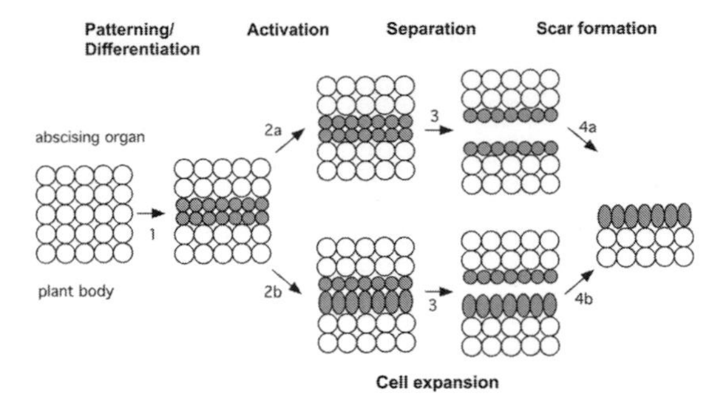

Figure 6.1 Model of the organ abscission process. Morphological changes during organ abscission are conserved across species and between different types of organs that are shed. Cells at the future site of detachment undergo patterning and differentiation within the context of the organ (1) to form distinct layers of small, cytoplasmically dense abscission zone (AZ) cells (shown in dark gray). Next, the AZ is activated for cell separation (2) through the interplay of unknown developmental and hormonal signals. The organ is shed (3) when cell wall modifying enzymes are secreted within the AZ layers, causing dissolution of cell adhesion. Following abscission, a protective scar forms (4) at the site of organ detachment to protect the plant from pathogen attack. It is not yet clear what role cell expansion plays in the abscission process. Cell expansion of a subset of AZ cells during the activation stage (2b) may promote subsequent cell separation; alternatively, cell expansion of the remaining AZ cells on the plant body after abscission (4a) may be solely involved in scar formation.

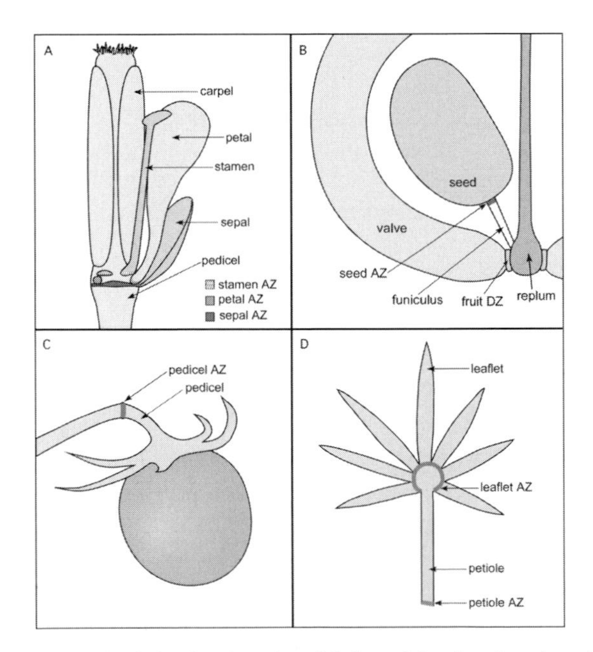

Figure 6.2 Sites of organ abscission in selected model plants. Mutations have been isolated in several model plants that block the abscission of specific organs. Depicted here are the *Arabidopsis thaliana* flower (A) and fruit (B), tomato (*Lycopersicon esculentum*) fruit (C), and *Lupinus angustifolius* compound leaf (D). (A) In *Arabidopsis* flowers, the sepals, petals and stamens are shed at abscission zones (AZs) located at the base of each organ. (B) Within the dry, dehiscent *Arabidopsis* fruit, an AZ forms at the site of seed attachment to the funiculus (seed stalk). Seed abscission occurs after the fruit opens at the dehiscence zones (DZs) along its valves, or walls. (C) In tomato, abscission of either unfertilized flowers or ripened fruit occurs at the pedicel AZ. (D) The entire *L. angustifolius* leaf can be shed by cell separation at the petiole AZ. Alternatively, leaflet AZs allow the shedding of individual leaflets.

Table 6.1 Genes affecting organ abscission in model plants

Gene	Plant	Mutant phenotype	Function	References
Organ patterning and AZ differentiation				
BLADE ON PETIOLE1 (*BOP1*) and *BOP2*	*At*	*bop1 bop2* double LOF[a]: floral organ position disrupted, abscission blocked	Establishment of proximo-distal polarity in lateral organs	Norberg *et al.*, 2005; Hepworth *et al.*, 2005; Ha *et al.*, 2004
SEEDSTICK (*STK*)	*At*	LOF: funicular overgrowth blocks seed release from mature fruits	Funicular patterning, seed AZ differentiation	Pinyopich *et al.*, 2003
development funiculus (*def*)*	*Ps*	Recessive: seed abscission blocked	Unknown	von Stackelberg *et al.*, 2003
JOINTLESS (*J*)	*Le*	LOF: pedicel AZ fails to form, fruit abscission blocked	MADS domain transcription factor	Mao *et al.*, 2000; Syzmkowiak and Irish, 1999, 2005
JOINTLESS-2 (*J-2*)	*Le*	Recessive: pedicel AZ fails to form, fruit abscission blocked	Transcriptional regulator**	Budiman *et al.*, 2004; Yang *et al.*, 2005
Modified stipules (*Mstips*)*	*La*	Recessive: defects in AZ differentiation; delayed senescence	Unknown	Clements and Atkins, 2001
Abscission initiation – ethylene signaling and biosynthesis				
ETHYLENE RECEPTOR1 (*ETR1*)	*At*	Dominant GOF[b]: ethylene insensitivity, delayed abscission	Ethylene perception	Bleecker *et al.*, 1988; Bleecker and Patterson, 1997; Hua *et al.*, 1998
NEVER-RIPE (*LeETR3*)	*Le*	Dominant GOF: ethylene insensitivity, deficient in fruit ripening, and delayed senescence and floral abscission	Ethylene perception	Lanahan *et al.*, 1994; Lashbrook *et al.*, 1998
ETHYLENE INSENSITIVE2 (*EIN2*)	*At*	LOF: ethylene insensitive, delayed abscission	Ethylene signal transduction	Patterson and Bleecker, 2004; Chao *et al.*, 1997; Zimmerman *et al.*, 2004
EIN3-LIKE1 (*EIL1*), *EIL2*, and *EIL3*	*Le*	Nonspecific RNAi: delayed flower abscission	Predicted transcription factors for ethylene signal transduction	Tiemen *et al.*, 2001; Chen *et al.*, 2004
Green-ripe and *Never-ripe 2**	*Le*	Dominant: ethylene-insensitive, deficient in fruit ripening, delayed flower abscission and senescence	Unknown function in ethylene signal transduction	Barry *et al.*, 2005
AUXIN RESPONSE FACTOR2 (*ARF2*), *ARF1*, *ARF7*, *ARF19*	*At*	*arf2* LOF: delayed abscission and senescence; *arf1 arf2* double LOF or *arf2 arf7 arf19* triple LOF: enhanced delay in abscission	Predicted transcriptional repressors and activators; Role in ethylene biosynthesis	Okushima *et al.*, 2005b; Ellis *et al.*, 2005

(*Continued*)

Table 6.1 Genes affecting organ abscission in model plants (*Continued*)

Gene	Plant	Mutant phenotype	Function	References
Abscission initiation – ethylene independent pathway(s)				
delayed abscission1 (*dab1*)*	*At*	Recessive: delayed abscission, ethylene-sensitive	Unknown	Patterson and Bleecker, 2004
*dab2**	*At*	Dominant: delayed abscission, ethylene-sensitive, irregular AZ cell expansion	Unknown	Patterson and Bleecker, 2004
*dab3**	*At*	Recessive: delayed abscission, ethylene-sensitive	Unknown	Patterson and Bleecker, 2004
INFLORESCENCE DEFICIENT IN ABSCISSION (*IDA*)	*At*	LOF: abscission blocked	Predicted signaling ligand	Butenko *et al.*, 2003
HAESA	*At*	AS[c]: abscission blocked	Leucine-rich receptor-like kinase	Jinn *et al.*, 2000
AGAMOUS-LIKE15 (*AGL15*)	*At*	OX[d]: prolonged embryonic development, delayed flowering time, maturation, senescence and abscission, ethylene sensitive; LOF: no phenotype	MADS domain transcription factor	Fernandez *et al.*, 2000; Harding *et al.*, 2003; Lehti-Shiu *et al.*, 2005
ACTIN RELATED PROTEIN7 (*ARP7*)	*At*	LOF: embryonic lethality; RNAi: delayed abscission	Predicted role in chromatin remodeling	Kandasamy *et al.*, 2003, 2005b
Abscission initiation – unknown pathways				
ARP4	*At*	RNAi: early flower maturation, delayed flower senescence and abscission	Predicted role in chromatin remodeling	Kandasamy *et al.*, 2003, 2005a
Delayed abscission (*abs2*)*	*La*	Recessive: cotyledon abscission blocked; reduced frequency of leaflet and flower abscission	Unknown	Clements and Atkins, 2001
Cell wall hydrolysis				
*abs1**	*La*	Recessive: abscission of all organ types blocked, ethylene-sensitive	Predicted role in cell wall dissolution	Clements and Atkins, 2001; Henderson *et al.*, 2001
Cel1, Cel2	*Le*	*Cel1* AS: decreased frequency of fruit abscission. *Cel2* AS: increased force required to remove fruit	Cell wall hydrolysis	Lashbrook *et al.*, 1998; Brummell *et al.*, 1999

[a] Loss-of-function.
[b] Gain-of-function.
[c] Antisense RNA.
[d] Overexpression.
At = *Arabidopsis thaliana*; *La* = *Lupinus angustifolius*; *Le* = *Lycopersicon esculentum* (tomato); *Ps* = *Pisum sativum* (pea). *Gene not yet cloned; **putative gene.

and seed dispersal (Figure 6.2A and B). With the advances in genome analysis in *Lycopersicon esculentum* (tomato), the classic *j* mutants have recently been molecularly characterized, suggesting that tomato will continue to be an ideal plant to study the abscission of fruit (Figure 6.2C). In addition, the recent isolation and characterization of mutations that affect seed abscission in *Pisum sativum* (pea) and flower, leaf and cotyledon abscission in the Australian crop plant *Lupinus angustifolius* point toward the use of these plants in future genetic experiments (Figure 6.2D). Since the abscission process appears to be quite similar across species, genes that are essential for the abscission of particular organs in *Arabidopsis*, tomato, pea and *L. angustifolius* likely control abscission of these organs in other species as well.

6.2 Setting the stage: patterning and differentiation

Differentiation of an organ AZ is thought to be dependent on the proper patterning of plant organs from meristematic tissue. So far, the characterization of mutants in *Arabidopsis*, tomato and *L. angustifolius* that affect both organ patterning and AZ differentiation has provided evidence to support this hypothesis (Mao *et al.*, 2000; Clements and Atkins, 2001; Pinyopich *et al.*, 2003; Hepworth *et al.*, 2005; Norberg *et al.*, 2005). Analysis of a pea mutant in which differentiation of the seed AZ is affected, apparently without altering development of the seed stalk (von Stackelberg *et al.*, 2003), suggests that these two processes can be uncoupled and that proper organ patterning precedes AZ formation.

6.2.1 BLADE ON PETIOLE1 and 2 establish lateral organ polarity

In *Arabidopsis* flowers, the sepals, petals and stamens are shed after pollination at distinct AZs located at the sites of organ attachment to the flower pedicel (see Figure 6.2A). Recent studies suggest that the redundant BLADE-ON-PETIOLE1 (BOP1) and BOP2 proteins may contribute to the patterning of *Arabidopsis* floral organs prior to abscission (Hepworth *et al.*, 2005; Norberg *et al.*, 2005). Single loss-of-function alleles of *BOP1* or *BOP2* do not show any phenotypic differences compared to the wild type; however, floral organ shedding is completely blocked in *bop1 bop2* double mutant flowers. In addition to abscission defects, *bop1 bop2* floral organs and leaves display alterations in organ patterning. In *bop* flowers, the abaxial sepal is replaced by one or two sepal/petal mosaic organs, and the first and second whorl organs cluster around the adaxial side of the flower. In *bop1 bop2* leaves, blade tissue arises aberrantly on the petiole, resulting in multilobed leaves. Although the proximal patterning of these lateral organs is disrupted, the distal regions appear to be unaffected by the loss of *BOP* activity. *bop1 bop2* patterning defects also include ectopic outgrowths at the bases of leaves, cotyledons, sepals and petals.

BOP1 and *BOP2* encode proteins with a BTB/POZ domain and four ankyrin repeats (Ha *et al.*, 2004; Hepworth *et al.*, 2005; Norberg *et al.*, 2005). The BTB/POZ domain, named for the *Drosophila* Broad-Complex, Tramtrak and Bric-a-brac transcriptional regulators and multiple poxvirus zinc-finger proteins in which it was first identified, can facilitate homodimerization as well as interactions with other proteins that may or may not contain a BTB/POZ domain (Stogios *et al.*, 2005). Ankyrin repeats are also predicted to mediate protein–protein interactions. Interestingly, the semi-dominant *bop1-1* allele appears to encode a dominant negative BOP1 isoform with aberrant protein interactions (Ha *et al.*, 2003, 2004). Although abscission defects have not been reported for *bop1-1* plants, similarities in patterning defects between *bop1 bop2* and *bop1-1* plants suggest that the *bop1-1* mutant protein may interfere with the activity of a BOP1–BOP2 heterodimer or perhaps with the activity of shared target proteins (Ha *et al.*, 2004). Of the other 75 BTB/POZ domain-containing proteins in *Arabidopsis*, BOP1 and BOP2 are the most closely related to NONEXPRESSOR OF PR GENES1 (NPR1) (Cao *et al.*, 1997; Ha *et al.*, 2004; Stogios *et al.*, 2005). NPR1, which lacks a canonical DNA-binding domain, mediates transcriptional changes during systemic acquired resistance through interactions with members of the TGA family of bZIP transcription factors (reviewed in Dong, 2004). Thus, it is likely that BOP1 and BOP2 interact with or activate unique transcription factors to promote floral organ patterning and abscission. Indeed, BOP interactions with the TGA transcription factor PERIANTHIA likely control the formation of four sepals rather than five in the first whorl (Hepworth *et al.*, 2005).

Expression analyses indicate that *BOP1* and *BOP2* are expressed in similar patterns during leaf and flower development (Ha *et al.*, 2004; Hepworth *et al.*, 2005; Norberg *et al.*, 2005). During normal vegetative growth, *BOP1* and *BOP2* are expressed in the proximal half of rosette leaves. During reproductive development, *BOP* expression is seen throughout stage 1 and 2 floral primordia. After stage 3, *BOP* expression becomes restricted to the bases of developing sepals, petals, stamens and carpels. In mature flowers, the *BOP* genes are strongly expressed at the bases of sepals, petals and stamens, in domains overlapping the floral organ AZs. This expression profile and the *bop* mutant phenotype described above suggest that BOP1 and BOP2 might function during the organ patterning required for AZ differentiation or during a later stage of the abscission process.

Early BOP activity at the bases of developing lateral organs may specify a proximal fate by restricting the expression domains of two sets of transcription factors (Ha *et al.*, 2004; Norberg *et al.*, 2005). In wild-type plants, BOP activity excludes expression of the *KNOX* transcription factors *BREVIPEDICELLUS* (*BP*), *KNAT2* and *KNAT6* from developing leaves while limiting expression of the C_2H_2 zinc finger transcription factor *JAGGED* (*JGD*) to a distal domain (Ha *et al.*, 2003; Dinneny *et al.*, 2004; Ohno *et al.*, 2004; Norberg *et al.*, 2005). When BOP function is compromised, expression of *BP*, *KNAT2*, *KNAT6* and *JAG* expands into the proximal region of developing leaves (Ha *et al.*, 2003; Norberg *et al.*, 2005). Since the roles of many organ polarity factors are conserved between leaves and floral organs (reviewed in Bowman *et al.*, 2002), it is likely that the BOP function is similarly conserved. Indeed, *JAG* expression was found to expand into the proximal domains of *bop* mutant floral organs as in leaves, although shedding is not restored in a *jag bop1 bop2* triple mutant (Norberg *et al.*, 2005). Another C_2H_2 zinc finger transcription factor, *JAGGED-LIKE* (*JGL*), is thought to act redundantly with *JAG* (Dinneny *et al.*, 2004; Ohno *et al.*, 2004; Norberg *et al.*, 2005), and represents another potential target of BOP regulation. Additional molecular and genetic analyses are now required to determine whether this model of BOP-dependent establishment of proximal–distal polarity also applies to floral organs and can explain the abscission defects of *bop* flowers. It is not yet known whether *BP*, *KNAT2*, *KNAT6* and *JGL* are ectopically expressed in the proximal regions of *bop* floral organs. Characterization of higher order mutant combinations such as the *jag jgl bop1 bop2* quadruple and *bp knat2 knat6 bop1 bop2* quintuple mutants should reveal whether ectopic expression of these two sets of transcription factors prevents the establishment of floral organ boundaries in *bop* mutant flowers, and thus, interferes with AZ development. An alternative, but not exclusive, possibility is that BOP1 and BOP2 may independently act to establish a proximal fate by regulating a third set of candidate target genes with proximal-specific expression patterns in lateral organs, such as *LATERAL ORGAN BOUNDARIES* (*LOB*) and *LATERAL ORGAN JUNCTIONS* (*LOJ*) (Shuai *et al.*, 2002; Prasad *et al.*, 2005). Although *lob* and *loj* single mutants do not display any defects in floral organ shedding, both of the affected genes are predicted to act redundantly. Analysis of *LOB* and *LOJ* expression in *bop* mutant floral organs may reveal yet another link between polarity establishment, pattern formation and organ differentiation, all of which must be properly coordinated for organ abscission to occur.

6.2.2 SEEDSTICK promotes proper development of the seed stalk

In dry dehiscent fruit, like that of *Arabidopsis*, two sequential cell separation events are required for seed dispersal (see Figure 6.2B). First, opening of the fruit along the margins of its walls, or valves, exposes the seed (fruit dehiscence is discussed in greater detail in Chapter 7). Second, cell separation within seed AZs releases the seed from the fruit replum (Figure 6.2B). Seeds are attached to the replum by stalks called funiculi. The seed AZ, a thin layer of distinctly small cells of the funiculus, immediately adjacent to the developing seed, differentiates after fertilization (Pinyopich *et al.*, 2003). A MADS domain transcription factor, *SEEDSTICK* (*STK*), is now known to be required for seed abscission. Characterization of the *stk* mutant revealed that the entire funiculus is enlarged due to cellular overgrowth and over-proliferation. Thus, *STK* may be indirectly required for proper AZ differentiation by regulating development of the funiculus (Pinyopich *et al.*, 2003). Identification of the transcriptional targets of *STK* may lead to more specific regulators of AZ differentiation. In contrast to the *stk* mutant, differentiation of the seed AZ appears to be specifically blocked in the *development funiculus* (*def*) mutation in pea (von Stackelberg *et al.*, 2003). Molecular characterization of the *def* locus should reveal another key factor controlling seed abscission.

6.2.3 *JOINTLESS is essential for differentiation of the pedicel abscission zone*

Tomato is an important model plant to study differentiation of the flower pedicel AZ. Unlike plants such as *Arabidopsis* that only release flower parts, entire tomato flowers and fruit are shed by separation at a single point on the floral pedicel (see Figure 6.2C). Cells in the pedicel AZ are anatomically distinct, and form a visible 'joint'. In *j*, a spontaneous mutation discovered in a tomato crop variety, the pedicel AZ fails to differentiate (Butler, 1936). This mutation has proven to be agronomically desirable and is widely cultivated since the resulting 'stemless' tomato fruit are more easily harvested and transported. In addition to preventing abscission, the *j* mutation affects the transition between vegetative and reproductive growth. In wild-type plants, inflorescence meristems arise from vegetative or sympodial meristems and then transition into floral meristems, whereas *j* inflorescence meristems often revert back to sympodial meristems after initiating the growth of only a few flowers (Szymkowiak and Irish, 1999). Recently, it has been proposed that the lack of AZ differentiation in *j* pedicels may be an indirect effect of disrupted inflorescence development (Szymkowiak and Irish, 2006).

J was found to correspond to *LeMADS,* a MADS domain transcription factor (Mao *et al.*, 2000). Analysis of the *j* locus revealed a 939-basepair (bp) deletion including the 5' UTR and the first 33 bp of the MADS-box region; a tomato anionic peroxidase inverted repeat (TAPIR) transposable element was likely responsible for this deletion (Mao *et al.*, 2000, 2001). Although the *Arabidopsis* MADS box family includes two genes closely related to *J* – *SHORT VEGETATIVE PHASE* (*SVP*/*AGL22*) and *AGAMOUS-LIKE24 (AGL24)* – neither *SVP* nor *AGL24* have known roles in abscission (Hartmann *et al.*, 2000; Yu *et al.*, 2004). *AGL24* is thought to promote an inflorescence meristem fate and *SVP* to act in the autonomous flowering time pathway repressing the transition from vegetative to reproductive growth.

To investigate how J affects both meristem identity and the differentiation of multiple tissue layers into a properly formed pedicel AZ, Szymkowiak and Irish (1999) generated chimeric tomato plants with L1, L2 and L3 meristem layers of wild-type (*J*) and mutant (*j*) origins. Their work demonstrated that a *J* signal from the L3 tissue is sufficient to maintain inflorescence meristem identity, and specify pedicel AZ formation, in plants with overlaying L1 and L2 tissues of *j* origin. This result suggests that the *J* signal can spread radially out of the L3 tissue layer to coordinate formation of the AZ. In addition, the authors found that the *J* signal cannot spread laterally within a tissue layer. When *J* and *j* cells are juxtaposed in the L3 layer, wild-type and mutant sectors are seen in the L1 and L2 layers of the tomato pedicel corresponding to the underlying genotype. Only those tissues overlying wild-type *J* cells differentiate into AZ pedicel tissue. This work clearly indicates that J-mediated cell–cell signaling is required for inflorescence patterning and AZ differentiation, and raises the fundamental question of how the J signal is communicated. Further studies are also necessary to explore the relationship between meristem identity and AZ differentiation, particularly since *J* is expressed at high levels in inflorescence and flower meristems but has not been detected in the pedicel itself (Szymkowiak and Irish, 2006).

6.2.4 *jointless-2 maps near a predicted transcriptional regulator*

A second nonallelic mutation in tomato, *jointless-2* (*j-2*), also prevents differentiation of the pedicel AZ. Extensive linkage mapping and sequencing has established that the *j-2* mutation is located within the centromeric region of chromosome 12 (Budiman *et al.*, 2004; Yang *et al.*, 2005). Yang *et al.* (2005) have recently identified a candidate gene, *ToCPL1*, which is closely related to *Arabidopsis C-terminal phosphatase-like gene 3* (*AtCPL3*). AtCPL3 has been shown to regulate gene expression in response to plant stress, and play a role in growth and development (Koiwa *et al.*, 2002). Confirmation that *J-2* corresponds to *ToCPL1* is currently waiting on complementation of *j-2* mutant plants (Yang *et al.*, 2005).

The predicted ToCPL1 and AtCPL3 proteins contain a conserved catalytic domain characteristic of the CPL family and a BRCA1 carboxy-terminal (BRCT) domain. FCP1, a closely related yeast carboxy-terminal domain (CTD) phosphatase, associates with and dephosphorylates the CTD of RNA polymerase II (Pol II) through the interaction of its BRCT domain with the RAP74 subunit of the TFIIF complex (Archambault *et al.*, 1998). TFIIF is part of the Pol II complex during transcription elongation; dephosphorylation of Pol II by CTD phosphatases serves as a release and recycling signal for the complex (Cho *et al.*, 1999). If the proposed identity of *J-2* is confirmed to be that of *ToCPL1*, it may be required to repress transcription of a specific suite of genes for pedicel AZ development to occur. Alternatively, a region of homology between AtCPL3 and the yeast CAPPING ENZYME SUPPRESSOR protein suggest a link to mRNA capping (Schwer and Shuman, 1996; Koiwa *et al.*, 2002). Identification of the *in vivo* substrate(s) of ToCPL1 should further define its function during pedicel AZ development. In the future, chimeric and expression studies to investigate whether J-2 acts in a cell autonomous or non-autonomous manner, as well as genetic analysis of the interactions between *J* and *J-2*, should be particularly informative.

6.3 Communication involved in cell separation

Around the turn of the twentieth century, scientists were intrigued by the defoliating effect of illuminating gas upon nearby shade trees. Ethylene, a component of this coal-manufactured gas, was soon shown to be the 'poisoning' agent as well as a potent inducer of the 'triple response' in dark-grown (etiolated) seedlings (Neljubow, 1901; Doubt, 1917). When etiolated seedlings are exposed to ethylene, hypocotyls and roots fail to elongate, hypocotyls thicken and the apical hook shows exaggerated curving – a phenotype reminiscent of the stunted and twisted plants growing near street lamps in nineteenth century cities (Neljubow, 1901). Since then, ethylene, a simple plant hormone, has been shown to affect a multitude of plant processes including germination, fruit ripening, leaf and flower senescence, pathogen defense, stress response and abscission (Abeles *et al.*, 1992). In support of an *in vivo* role for ethylene in developmental events, disruption of the endogenous biosynthesis of ethylene has been found to delay abscission, ripening and senescence (Ecker and Theologis, 1994).

The role of small, diffusible hormones such as ethylene in regulating the timing of abscission has been studied in depth. Early experiments with bean and cotton leaf explants, in which controlled amounts of hormones were applied to the stem of the cut leaf, support a model in which ethylene promotes and auxin delays cell separation at the AZ (reviewed in Sexton and Roberts, 1982; Brown, 1997). Following the differentiation of organ AZ cells, there are two distinct phases of hormone sensitivity prior to abscission. During the first phase, a constant flux of auxin into the AZ is thought to inhibit cell separation. Exogenous ethylene application during this time is insufficient to promote abscission, suggesting that AZ cells are largely insensitive to ethylene. Just prior to abscission, during the second phase, hormone sensitivities are reversed – AZ cells become competent to respond to ethylene whereas auxin can no longer inhibit cell separation. Continual exposure to ethylene is required to accelerate cell separation (Sexton and Roberts, 1982; Taylor and Whitelaw, 2001). Studies such as these point out the importance of ethylene in controlling the timing and speed of abscission; however, no experiment has convincingly shown that ethylene perception and signaling are required for abscission (reviewed in Patterson, 2001). The recent identification of several ethylene-independent genes required for the proper initiation of abscission supports a model in which ethylene-dependent and -independent pathways work to promote abscission (Butenko *et al.*, 2003; Patterson and Bleecker, 2004). Thus, the goal of this section is to describe the genes from each of these pathways that are known to be involved in the abscission process.

6.3.1 Involvement of the ethylene-signaling pathway

Screens for *Arabidopsis* seedlings that are defective in the 'triple response' have been extremely successful in identifying genes of the primary ethylene-signaling pathway (reviewed in Alonso and Stepanova, 2004; Guo and Ecker, 2004). A subset of the genes required for ethylene sensitivity, including *ETHYLENE RECEPTOR1* (*ETR1*), *ETHYLENE-INSENSITIVE2* (*EIN2*) and members of the *EIN3* transcription factor family, are implicated in the proper timing of floral organ abscission (Bleecker and Patterson, 1997; Tieman *et al.*, 2001; Patterson and Bleecker, 2004). In *Arabidopsis*, turgid floral organs are normally shed just following pollination. Although signs of senescence are evident in wild-type floral organs prior to abscission (Fang and Fernandez, 2002), dominant gain-of-function and recessive loss-of-function mutations in *ETR1* and *EIN2*, respectively, block floral organs from shedding until sepals and petals have begun to wither (Bleecker and Patterson, 1997; Patterson, 2001; Patterson and Bleecker, 2004). AZ formation is normal in these plants, suggesting that ETR1 and EIN2 play specific roles in the initiation of abscission rather than earlier stages (Patterson and Bleecker, 2004). Tomato homologs of ethylene-signaling genes are also implicated in cell separation; *LeETR3* and the *EIN3-like* genes, *LeEIL1*, *2*, and *3*, are all required for proper coordination of abscission within the timeline of development. It is probable that most, if not all, components of the primary ethylene-signaling pathway will likewise be shown to affect abscission.

6.3.1.1 ETHYLENE RECEPTOR1 links ethylene perception with abscission initiation

ETR1 was the first gene from the ethylene-signaling pathway to be directly implicated in *Arabidopsis* floral organ abscission. Despite the normal appearance of adult *etr1* plants, they are insensitive to ethylene-induced senescence, and flowering and abscission are delayed (Bleecker *et al.*, 1988; Bleecker and Patterson, 1997; Chao *et al.*, 1997). *ETR1* is broadly expressed in *Arabidopsis*, with highest transcript levels in flowers, particularly the anthers and developing carpels (Hua *et al.*, 1998). An *etr1* allele was mapped to a region that encodes a protein with domains similar to both the receiver and sensor proteins of the 'two-component' histidine kinase receptors in bacteria (Chang *et al.*, 1993). Since then, ETR1 has been localized to the endoplasmic reticulum (ER) membrane, where the receptor is predicted to bind ethylene at its N-terminus and transmit a signal to the nucleus through a MAPK-like cascade (Chen *et al.*, 2002; reviewed in Guo and Ecker, 2004). The *Arabidopsis* ethylene receptors have been assigned to two subfamilies based upon their sequence similarity: (1) ETR1 and ERS1 and (2) ETR2, EIN4 and ERS2. In screens for ethylene insensitivity, only dominant gain-of-function mutations have been isolated for these receptor genes, suggesting that genetic redundancy masks any single mutant loss-of-function phenotypes (Hua and Meyerowitz, 1998; Hua *et al.*, 1998). In support of this hypothesis, higher order combinations of loss-of-function ethylene receptor mutants result in stronger ethylene responses than those observed for wild-type or any single loss-of-function allele (Hua and Meyerowitz, 1998; Cancel and Larson, 2002). Both the triple *etr1 etr2 ein4* and quadruple *etr1 etr2 ein4 ers2* loss-of-function mutants display a constitutive response in the absence of ethylene (Hua and Meyerowitz, 1998). The effects of higher order combinations of either gain- or loss-of-function mutations upon the timing of floral organ shedding have not been investigated.

6.3.1.2 The NEVER-RIPE ethylene receptor promotes abscission in tomato

The semi-dominant, ethylene-insensitive *Never-ripe* (*Nr*) mutant was isolated in tomato half a century ago, and it was found to affect the ethylene receptor *LeETR3* (Rick and Butler, 1956; Wilkinson *et al.*, 1995; Lashbrook *et al.*, 1998b). Tomato cultivars with the *Nr* mutation are deficient in fruit ripening, senescence and abscission at the floral pedicel (Rick and Butler, 1956; Lanahan *et al.*, 1994). Whereas 87.5% of unfertilized, wild-type flowers of the Pearson tomato cultivar abscise after 20 days on the plant, less than 4% of unfertilized Pearson *Nr* flowers abscise after the same amount of time. Furthermore, ethylene exposure does little to accelerate this process in flower explants (Lanahan *et al.*, 1994). Antisense inhibition of the *Nr* allele has confirmed that Nr plays an inhibitory role in ethylene signal transduction, akin to its *Arabidopsis* homologs (Hackett *et al.*, 2000). The ethylene receptor family in tomato has four additional genes, *LeETR1*, *2*, *4* and *5*; *LeETR4* appears to be partially redundant with *Nr* in regulating fruit ripening (Tieman *et al.*, 2000).

6.3.1.3 Never-ripe2 and Green-ripe may represent mutations in a novel ethylene signaling gene

Two partially ethylene-insensitive, dominant mutations, *Never-ripe 2* (*Nr-2*) and *Green-ripe* (*Gr*), cause defects in tomato fruit ripening (Barry *et al.*, 2005). Both mutants also show a subtle delay in floral organ shedding. Like *Nr*, *Nr-2* and *Gr* also affect abscission at the floral pedicel; wild-type tomato explants abscise almost all flowers after 48 h of ethylene treatment whereas nearly isogenic *Nr*, *Nr-2* and *Gr* explants abscise only 35%, 60% and 85% of flowers, respectively, after 48 h of ethylene treatment. The roots of *Nr-2* and *Gr* etiolated seedlings exhibit weak ethylene insensitivity, although hypocotyl elongation is not affected, suggesting that the affected gene(s) contributes to a specific subset of ethylene responses involved in fruit ripening, flower senescence, abscission and root elongation. *Nr-2* and *Gr* were both mapped to the same 2-cM region of chromosome 1 and are predicted to disrupt a novel gene in the ethylene-signaling pathway (Barry *et al.*, 2005).

6.3.1.4 ETHYLENE-INSENSITIVE2, a transmembrane protein of unknown function, is required for the proper timing of abscission

To date, *EIN2* is the only gene functioning downstream of the ethylene receptors that has been reported to affect the timing of *Arabidopsis* floral organ abscission (Patterson and Bleecker, 2004). *ein2* mutants exhibit a strong ethylene-insensitive phenotype, as etiolated seedlings are deficient in the 'triple response', and flower senescence and organ abscission are not accelerated by treatment with ethylene (Guzmán and Ecker, 1990; Chao *et al.*, 1997; Patterson and Bleecker, 2004). *EIN2* was found to encode a protein with two functional domains: an N-terminal cluster of 12 predicted transmembrane helices with homology to the Nramp class of metal ion transporters and a hydrophilic C-terminus that may facilitate protein–protein interactions (Alonso *et al.*, 1999). Membrane-bound EIN2 is predicted to function in transmission of the ethylene signal from the ER-bound receptor(s) to transcription factors in the nucleus; however, the specific subcellular localization of EIN2 has not yet been determined (Alonso *et al.*, 1999). Delayed flower senescence has also been reported for antisense lines of an *EIN2* homolog in Petunia (Shibuya *et al.*, 2004). Future studies in *Arabidopsis* should increase our understanding of the EIN2-mediated ethylene response and abscission.

6.3.1.5 EIN3-LIKE1, 2 and 3 may induce the transcription of abscission-related genes

While a delay in floral organ shedding has not yet been reported for the *Arabidopsis ein3* mutant, genes from the tomato *EIN3* family are known to be involved in the ethylene-mediated initiation of abscission (Tieman *et al.*, 2001). Knocking down the expression of all three tomato *EIN3-like* genes – *LeEIL1*, *2* and *3* – by nonspecific antisense RNA (*LeEIL-AS*) causes plants to become ethylene-insensitive. Treatment of wild-type tomato plants with ethylene induces premature abscission of flowers prior to fruit development; however, when *LeEIL-AS* plants with the most severe reduction in *LeEIL* expression are treated with ethylene, flower abscission is delayed until fruit development begins (Tieman *et al.*, 2001). Recently, it was found that overexpression of *LeEIL1-GFP* can partially compensate for the ethylene insensitivity of *Nr* plants (Chen *et al.*, 2004). This result suggests that *LeEIL1* acts downstream of *NR* (*LeETR3*) and that the primary ethylene-signaling pathway is conserved from tomato to *Arabidopsis*. Like the *Arabidopsis* EIN3 family of transcription factors, the predicted LeEIL proteins contain nuclear localization signals that are most likely responsible for the movement of LeEIL1-GFP chimeric protein to the nucleus (Chen *et al.*, 2004). Since *LeEIL1*, *2* and *3* appear to be constitutively expressed in seedlings, leaves, flower buds and fruits (Tieman *et al.*, 2001), interacting proteins may be required to impart specificity to LeEIL activity during abscission.

6.3.2 Ethylene-independent avenues to abscission

6.3.2.1 INFLORESCENCE DEFICIENT IN ABSCISSION encodes a putative signaling ligand essential for abscission

Characterization of an *Arabidopsis* T-DNA mutant led to the discovery of the first gene known to be required for floral organ shedding, *INFLORESCENCE DEFICIENT IN ABSCISSION* (*IDA*; Butenko *et al.*, 2003). Although *ida* mutant flowers develop and senesce normally, the sepals, petals and stamens remain attached indefinitely. In contrast to the ethylene-insensitive mutants, *ida* plants exhibit a wild-type response to ethylene during all stages of development tested except floral organ shedding. Whereas exogenous application of ethylene accelerates the senescence and abscission of wild-type flowers, *ida* flowers treated with ethylene senesce more rapidly but do not shed their organs (Butenko *et al.*, 2003).

IDA encodes the founding member of a new class of putative signaling ligands, which include a predicted signal peptide at the N-terminus and a conserved 12 amino acid sequence of unknown function at the C terminus. Although receptors have yet to be identified for IDA or any of the IDA-like ligands identified in *Arabidopsis* and other plants, it has been shown that a chimeric IDA–GFP fusion protein transiently expressed in onion epidermal cells localizes to the extracellular space and/or cell wall, supporting its proposed function as a secreted ligand (Butenko *et al.*, 2003).

IDA is thought to act after differentiation of the floral AZs, during the stage of active cell separation. An *IDA* promoter::ß-glucuronidase (GUS) reporter gene is strongly expressed in floral AZs, first appearing at stage 15 just prior to organ abscission at stage 16 (Butenko *et al.*, 2003). Measurements of the force required to remove petals from sequentially older flowers of the *ida* mutant show an initial decrease, as in the wild type, followed by a striking increase in the oldest *ida* flowers (Butenko *et al.*, 2003). Scanning electron microscopy (SEM) of *ida* and wild-type flowers after petals are manually removed supports these breakstrength data. Initially, the petal AZ cells of *ida* and wild-type flowers have the same broken appearance; however, at the time of abscission, wild-type cells expand and become fully rounded whereas *ida* cells are less rounded (Butenko *et al.*, 2003). At later timepoints, after abscission has occurred in wild-type flowers, *ida* AZ cells again exhibit a broken appearance. Taken together, these results suggest that IDA signaling enables cell separation within the AZ, either by promoting secretion of hydrolytic enzymes or by inhibiting repair of the middle lamella and cell walls (Butenko *et al.*, 2003).

6.3.2.2 HAESA, a receptor-like kinase, may perceive an abscission signal

HAESA (HAE) belongs to the leucine-rich repeat (LRR) class of receptor-like protein kinases (RLKs). Proteins within this class are highly divergent in sequence; however, all are predicted to contain an extracellular ligand-binding domain, a transmembrane hydrophobic domain and a cytoplasmic serine–threonine protein kinase domain (Shiu and Bleecker, 2001). Appropriate ligand binding is thought to trigger autophosphorylation of the RLK and/or phosphorylation of intracellular proteins, affecting a downstream signaling pathway(s). Approximately 200 LRR–RLKs are annotated within the *Arabidopsis* genome (Tarutani *et al.*, 2004); HAE (previously known as RLK5) was among the first for which biochemical studies were conducted and is localized to the plasma membrane (Horn and Walker, 1994; Stone *et al.*, 1994; Jinn *et al.*, 2000). When *HAE* expression was knocked down in transgenic plants by antisense suppression, floral organ abscission is blocked (Jinn *et al.*, 2000). Since *HAE* antisense RNA (*HAE-AS*) may have also knocked down the expression of closely related RLKs, it will be necessary to analyse single loss-of-function mutants of *HAE* to confirm its function in abscission. As in the *ida* mutant, the floral organs of *HAE-AS* plants senesce normally yet completely fail to abscise (Jinn *et al.*, 2000). Considering the similarities between these phenotypes, it will be intriguing to determine whether these genes function in the same pathway, perhaps with HAE functioning as a receptor for IDA (Butenko *et al.*, 2003).

The expression profile of *HAE* is consistent with its predicted role in floral organ abscission. Jinn *et al.* (2000) used antisense RNA *in situ* hybridization and an *HAE::GUS* reporter construct to show that *HAE* is first expressed at the bases of the floral organs at stage 14, when flowers undergo pollination. Since the expression pattern of *HAE* is not affected by disrupting the ethylene-signaling pathway, it is predicted to function in an ethylene-independent pathway to abscission (Jinn *et al.*, 2000). Expression of *HAE* is also detected in leaves at the base of the petioles, where the leaf attaches to the plant stem. Since *Arabidopsis* rosette leaves are not shed, this may represent an evolutionarily conserved expression pattern (Jinn *et al.*, 2000).

Two additional *Arabidopsis* LRR-RLKs, RLK902 and RLK1, which share 75% amino acid identity overall, may also be involved in floral organ shedding (Tarutani *et al.*, 2004). RNA blot analysis of these genes shows the highest levels of expression in floral tissue, and the promoters of *RLK902* and *RLK1* drive GUS reporter expression in floral AZs. However, single *rlk902* and *rlk1* mutants, as well as the double *rlk902 rlk1* mutant, do not display a detectable phenotype (Tarutani *et al.*, 2004). Since LRR-RLKs have been shown to heterodimerize with other receptor-like proteins (Jeong *et al.*, 1999), it will be interesting to determine whether RLK902 and/or RLK1 interact with HAE. Further characterization of LRR-RLK family members, and the identification of potential HAE-interacting proteins (Jinn *et al.*, 2000), should be particularly informative in dissecting the complex signaling pathways that promote abscission.

6.3.2.3 DELAYED ABSCISSION mutations uncouple the initiation of abscission from ethylene signaling

Recently, a set of three new *Arabidopsis* mutants – *dab1, 2* and *3* – was isolated in a screen for T-DNA lines with delayed floral organ abscission (Patterson and Bleecker, 2004). Each of the *dab* mutants exhibits a normal response to ethylene, suggesting that the initiation of abscission can be uncoupled from ethylene perception. In comparison to the ethylene-insensitive mutants *etr1* and *ein2*, a longer delay in shedding is seen for all three *dab* mutants (Patterson and Bleecker, 2004). Although the sepal, petal and stamen AZs of *dab* flowers appear morphologically normal by light microscopy, SEM has revealed that *dab2* AZ cells elongate irregularly (Patterson and Bleecker, 2004). Since the role of cell expansion during abscission is currently unclear, it should be particularly interesting to investigate this aspect of the *dab2* phenotype. Slight variations among the *dab1*, *2* and *3* phenotypic profiles, as detected by SEM and petal breakstrength calculations, suggest that these mutations disrupt genes with unique functions in cell separation (Patterson and Bleecker, 2004). Thus, mapping and further characterization of *DAB1*, *2* and *3* are expected to enhance our understanding of abscission initiation and the connections between ethylene-dependent and -independent pathways.

6.3.2.4 ACTIN-RELATED PROTEINS may globally regulate transcription during senescence and abscission

In a recent study by Kandasamy *et al.* (2005b), *Arabidopsis* ACTIN-RELATED PROTEIN7 (ARP7) was found to be required for embryogenesis and adult plant development. Complete loss of ARP7 causes embryo lethality, with development arresting just prior to the torpedo stage. When *ARP7* expression was knocked down using RNA interference (RNAi), transgenic plants with the strongest viable phenotype exhibit stunted organs, reduced fertility and defects in flower opening, anther dehiscence and fruit growth. Another striking characteristic of the *ARP7* RNAi plants is a delay in floral organ shedding such that 12 or more open flowers on an inflorescence retain their sepals and petals as compared to 4 or 5 in the wild type

(Kandasamy *et al.*, 2005b). Senescence of the floral organs appears to be correspondingly delayed. Based on sections of *ARP7* RNAi and wild-type flowers just before anthesis, AZ differentiation occurs normally. *ARP7* RNAi plants are partially competent to respond to ethylene; etiolated seedlings from a strong RNAi line show similar morphological changes as wild-type seedlings in the ethylene triple-response assay, yet ethylene exposure does not expedite floral organ shedding in *ARP7* RNAi adult plants as in the wild type. Therefore, the ARP7 abscission pathway may be independent of or downstream of ethylene perception in flowers (Kandasamy *et al.*, 2005b).

Arabidopsis ARP7 belongs to a phylogenetic class of ARPs unique to plants (McKinney *et al.*, 2002). Other eukaryotic ARPs have been implicated in the polymerization of conventional actin filaments, dynein motor function and chromatin remodeling (Schafer and Schroer, 1999; Goodson and Hawse, 2002). Of the eight classes of ARPs that are evolutionarily conserved from humans to budding yeast (*Saccharomyces cerevisiae*), four are localized to the cytoplasm and function in actin polymerization or dynein motor function, and the other four are localized to the nucleus, in association with high molecular weight chromatin remodeling complexes or heterochromatin (Goodson and Hawse, 2002). ARP7 also associates with a high molecular weight complex (Kandasamy *et al.*, 2005b), and it was previously shown to be localized to the nucleus during cytokinesis, interphase and prophase (Kandasamy *et al.*, 2003); therefore, it is predicted to play a role in chromatin remodeling during transcription.

Arabidopsis ARP4, with homologs in both humans and yeast, has also been implicated in floral organ abscission. When *ARP4* expression is knocked down using RNAi, a pleiotropic phenotype results that partially overlaps with that of *ARP7* RNAi plants (Kandasamy *et al.*, 2005a). Paradoxically, *ARP4* RNAi plants with the strongest viable phenotype flower early yet senesce and abscise late (Kandasamy *et al.*, 2005a). Thus, the observed delay in floral organ shedding may be an indirect result of an impaired developmental clock in *ARP4* RNAi plants. A detailed characterization of AZ development in these plants should more clearly define the stage of abscission affected. Like ARP7, ARP4 is localized to the nucleus, and is predicted to function as part of a chromatin-remodeling complex to modulate transcription as has been demonstrated for its yeast and human homologs (Kandasamy *et al.*, 2005a). Both proteins are expressed ubiquitously in all cell types, and ARP4 protein expression is not affected in *ARP7* RNAi plants (Kandasamy *et al.*, 2003, 2005b). Identification of ARP4 and ARP7 interacting proteins as well as their transcriptional targets should reveal more clues about global regulation of genes involved in senescence and abscission.

6.3.2.5 Transcriptional activity of AGAMOUS-LIKE 15 inhibits senescence and abscission

Arabidopsis AGAMOUS-LIKE 15 (AGL15) is a MADS domain transcription factor that appears to promote a juvenile state and/or inhibit senescence-related events, such as abscission, in the adult plant (Fernandez *et al.*, 2000; Harding *et al.*, 2003). Transgenic plants that constitutively express *AGL15* under the control of the viral *35S* promoter show delays in several developmental processes, including embryonic development, flowering time, fruit maturation, floral organ senescence and abscission (Fernandez *et al.*, 2000; Harding *et al.*, 2003). *35S::AGL15* plants are ethylene-sensitive: treatment of adult plants with exogenous ethylene promotes floral organ abscission as in the wild type. Since development of the floral organ AZ occurs normally in these plants, AGL15 appears to inhibit floral organ shedding after AZ differentiation (Fernandez *et al.*, 2000).

Consistent with the proposed role for AGL15 in inhibiting senescence-related events, the *AGL15* promoter drives reporter gene expression in many developing tissues, including the vegetative shoot apex and meristem, leaf and stipule primordia, young leaves and floral buds (stages 4 through 13) – but not in senescing or abscising floral organs. As leaves and floral organs mature, reporter activity declines and becomes concentrated at the base of the organs prior to disappearing completely (Fernandez *et al.*, 2000). To determine the floral stage at which ectopic *AGL15* expression can cause developmental delays, Fang and Fernandez (2002) made use of the glucocorticoid-inducible expression system (Aoyama and Chua, 1997). These experiments demonstrated that AGL15 must be active at the time of flower opening (stage 13) to delay senescence and abscission (Fang and Fernandez, 2002).

Several questions remain concerning the role of AGL15 in flower development. First, genetic redundancy has so far proved an obstacle to the loss-of function analysis of *AGL15*. Single *agl15* mutants or plants carrying double mutant combinations of *AGL15* and *AGL18*, the MADS-box gene most closely related to *AGL15*, do not display any detectable defects in embryonic or adult development (Lehti-Shiu *et al.*, 2005). Genetic analysis of additional related MADS-box genes may be necessary to elucidate *AGL15* function. Second, although several *DOWNSTREAM TARGET OF AGL15* (*DTA1-4*) genes have been identified based on the DNA-binding properties of AGL15 (Wang *et al.*, 2002; Tang and Perry, 2003), to identify the genes regulated by AGL15 that affect the timing of senescence and floral organ shedding, genetic screens for suppressors of the *35S::AGL15* phenotype would be useful. Finally, it is not yet clear whether ectopic expression of *AGL15* delays floral organ abscission through a cell autonomous or non-autonomous pathway. When using the AZ-specific bean chitinase promoter to direct *AGL15* expression instead of the *35S* promoter, Fang and Fernandez (2002) observed no delay in floral organ abscission. These results can be interpreted in one of two ways: either AGL15 inhibits abscission through a cell non-autonomous pathway, or this particular promoter drives *AGL15* expression too late in the developmental process to affect abscission. Earlier AZ-specific promoters should be identified and tested to resolve this question.

6.3.3 *Integrative pathways that promote ethylene biosynthesis and abscission*

6.3.3.1 *AUXIN RESPONSE FACTORS regulate the timing of abscission*

For over half a century, abscission has been known to be affected by auxin levels based on experiments carried out with bean leaf AZ explants (Addicott,

1982). Until recently, however, auxin-related mutants that affect organ abscission had not been identified. Genetic analysis of four members of the *Arabidopsis AUXIN RESPONSE FACTOR* (*ARF*) family – *ARF2*, *ARF1*, *NONPHOTOTROPIC HYPOCOTYL4* (*NPH4*)/*ARF7* and *ARF19* – has now revealed that these transcriptional regulators function with partial redundancy to promote senescence and floral organ shedding (Ellis *et al.*, 2005; Okushima *et al.*, 2005). *arf2* single mutant flowers show a delay in the onset of both senescence and floral organ abscission. Although the single *arf1*, *nph4* and *arf19* single mutants do not demonstrate any shedding defects, *arf1 arf2* double mutants and *arf2 nph4 arf19* triple mutants show an enhanced delay in abscission compared to *arf2* single mutants (Ellis *et al.*, 2005). ARF proteins contain an N-terminal DNA binding domain specific to auxin response elements, and a regulatory middle region that either stimulates or represses transcription (Ulmasov *et al.*, 1999; Tiwari *et al.*, 2003). Since ARF2 and ARF1 have been shown to repress transcription while NPH4 and ARF19 stimulate it (Ulmasov *et al.*, 1997, 1999; Tiwari *et al.*, 2003), the mechanism of how these transcriptional regulators function alone or in complexes to regulate senescence and organ shedding is not yet clear. One possible scenario is that their individual activities could be affected by changes in auxin gradients across floral organ AZs, since such gradients likely affect leaf senescence and abscission (Addicott, 1982; Okushima *et al.*, 2005).

A key role of ARF2 may be to promote ethylene biosynthesis prior to senescence and floral organ shedding. Three members of the 1-aminocyclopropane-1-carboxylate (ACC) synthase family – *ACS2*, *ACS6* and *ACS8* – show decreased transcript levels in *arf2* flowers (Okushima *et al.*, 2005). Expression of each of these *ACS* family members is normally observed in flowers after pollination and can be induced in seedlings by auxin treatment. However, ARF2 must clearly play additional roles in regulating the onset of senescence and abscission, since *ein2 arf2* double mutant flowers show an additive delay (Ellis *et al.*, 2005). An additional point raised by these results is whether *ACS2*, *ACS6* and *ACS8* regulation by ARF2 is mediated by AUX/IAA signaling (Yamagami *et al.*, 2003; Ellis *et al.*, 2005).

6.3.3.2 G-protein signaling promotes ethylene biosynthesis and abscission of citrus leaves

The integration of G-protein signaling with the ethylene biosynthetic pathway of the 'Valencia' orange tree (*Citrus sinensis* L. Osbeck) is unprecedented. In a study based on the application of 2-chloroethylphosphonic acid (ethephon), of which ethylene is a byproduct, to stimulate both fruit and leaf abscission, Yuan *et al.* (2005) found that ethephon specifically increases steady-state RNA levels of the ethylene biosynthesis genes *ACS1* and *ACC Oxidase* (*ACO*) in fruit and leaf AZs. Increased expression of *ACS1* and *ACO* corresponds with increased ethylene evolution. Two G-protein-coupled α_{2A}-adrenoreceptor agonists, guanfacine and clonidine, were found to inhibit ethephon-induced *ACS1* and *ACO* RNA expression in the AZ of the leaf but not in that of the fruit (Yuan *et al.*, 2005). These studies implicate G-protein signaling in the regulation of ethylene biosynthesis related to abscission and demonstrate the feasibility of specifically controlling fruit abscission. In the future, it will be of interest to identify the G-protein-coupled receptor(s) and pathway(s) that are the target of the guanfacine and clonidine agonists.

6.4 Dissolution of the cell wall – cell separation

The late stages of abscission are marked by AZ-specific loosening of the primary cell walls and dissolution of the middle lamella, a pectin-rich connective layer between cells (Osborne, 1989). Cellulose microfibrils, hemicelluloses and pectins are highly cross-linked within the cell wall; therefore, cell separation necessitates the activity of a variety of cell wall modifying and hydrolytic enzymes. Expression studies in multiple plant species underscore the importance of two major types of hydrolytic enzymes for abscission: the endo-β-1,4-glucanase (EGase or cellulase) and the polygalacturonase (PG) families. EGase family enzymes are thought to loosen cell walls by releasing xyloglucan, one of the most abundant hemicellulose species, from the cellulose microfibrils (Cosgrove, 2005). PG family enzymes, which hydrolyze polygalacturons of the pectin class, are most likely required for dissolution of the middle lamella. Other proteins that have been implicated in AZ cell wall modification include the cell wall loosening family of expansins (Belfield *et al.*, 2005; Sampredro and Cosgrove, 2005) and pathogenesis-related (PR) chitinases (reviewed in Roberts *et al.*, 2002).

6.4.1 Hydrolytic enzymes

6.4.1.1 Endo-β-1,4-glucanases

EGase class enzymes are encoded by a multigene family, with evidence of gene duplication and divergence of function. In the *Arabidopsis* genome alone, 25 EGases are annotated with numerous examples in which *Arabidopsis* paralogs are predicted to be more similar to each other than to homologous genes from other species. This phylogenetic arrangement suggests a history of gene duplications (Libertini *et al.*, 2004). The EGase family is divided into three subfamilies, of which the α- and β-EGases are implicated in cell elongation, ripening and abscission. γ-EGases, on the other hand, are predicted to be involved in cell elongation and cellulose biosynthesis at the plasma membrane (Libertini *et al.*, 2004). The first AZ-specific cell wall degrading enzyme to be characterized and cloned was bean abscission-specific cellulase (BAC) from *Phaseolus vulgaris* (Lewis and Varner, 1970; Tucker *et al.*, 1988; Koehler *et al.*, 1996). *BAC* promoter-driven expression of a reporter gene (*pBAC::GUS*) is a commonly used marker associated with abscission across species (Fernandez *et al.*, 2000). In addition, *BAC* sequence has been successfully used to isolate related EGase genes expressed within the AZs of *Glycine max* (soybean;

Kemmerer and Tucker, 1994), *S. nigra* (Taylor *et al.*, 1994) and tomato (Lashbrook *et al.*, 1994).

Whereas a single EGase is known to be expressed in the bean leaf AZ, multiple EGases have been detected in the AZ tissue of *S. nigra* leaves, *Capsicum annuum* (pepper) leaves and flowers, *Prunus persica* (peach) leaves and tomato flowers, suggesting cooperativity and/or redundancy of function (Lashbrook *et al.*, 1994; Taylor *et al.*, 1994; Ferrarese *et al.*, 1995; del Campillo and Bennett, 1996; Trainotti *et al.*, 1997, 1998). EGase expression patterns in tomato and peach suggest overlapping functions in fruit maturation and abscission. In tomato, the EGase genes *Cel1*, *Cel2*, *Cel5* and *Cel6* are expressed to varying extents within pedicel AZs, while *Cel1* and *Cel2* are also expressed in developing fruit (Lashbrook *et al.*, 1994; del Campillo and Bennett, 1996). Knocking down expression of either *Cel1* or *Cel2* with antisense RNA impairs abscission, as measured by decreased frequency of abscission in *Cel1-AS* flower explants and increased force required to remove fruits from *Cel2-AS* plants. However, neither fruit ripening nor softening are affected in *Cel1-AS* and *Cel2-AS* plants (Lashbrook *et al.*, 1998a; Brummell *et al.*, 1999). These results suggest that Cel1 and Cel2 activity is redundant for fruit maturation, and cooperative for abscission. Future analysis of genetic interactions among mutant EGase alleles should reveal more about their cooperative and/or redundant functions.

Interestingly, the spontaneously arising *abs1* mutant of *L. angustifolius*, a grain used for livestock feed in Australia, is deficient in the abscission of leaves, flowers and cotyledons (Clements and Atkins, 2001; see Figure 6.2D). This deficiency corresponds with a lack of EGase expression in ethylene-treated *abs1* AZ cells (Henderson *et al.*, 2001). Otherwise, *abs1* mutants appear like the wild type, with no detectable differences in senescence, root-cap border cell separation or the triple response in ethylene-treated seedlings. Although AZ differentiation appears to be unaffected, imaging of *abs1* cotyledon AZs by transmission electron microscopy reveals a lack of middle lamella dissolution in comparison to the wild type (Clements and Atkins, 2001). In contrast to tomato, in which multiple EGases are expected to play a role in cell wall dissolution within the pedicel AZ, *abs1* may represent a single genetic lesion that disrupts the expression of an EGase necessary for all types of *L. angustifolius* abscission. Another possibility is that *abs1* mutants lack a crucial regulator of multiple cell wall modifying and hydrolytic enzymes. Mapping of the *abs1* mutation will definitely enhance our understanding of the mechanism and requirements of cell wall dissolution during abscission.

6.4.1.2 Polygalacturonases

Plants are well equipped with enzymes for the modification of pectin in the primary cell wall and middle lamella. In *Arabidopsis*, approximately 150 proteins are predicted to have pectin-modifying activity, of which 52 belong to the PG family of enzymes (The *Arabidopsis* Genome Initiative, 2000). PGs are implicated in the hydrolysis of the polygalacturon class of pectins during many developmental processes, including fruit ripening, abscission, dehiscence, pollen tube growth and pollen grain maturation (reviewed in Hadfield and Bennett, 1998). Based upon the phylogenetic relationship of 43 PGs from multiple plant species, the enzymes were divided into five clades (A–E) with unique intron–exon structures (Torki *et al.*, 2000). All PGs contain a signal peptide for secretion of the enzyme and four functional domains (I–IV). Domains I–III are thought to be required for catalytic activity, and domain IV may facilitate interaction with the polygalacturon substrate (reviewed in Torki *et al.*, 2000). PGs of clade B are unique in that they contain an additional N-terminal prosequence that may act as a novel protein-sorting signal (Dal Degan *et al.*, 2001). Although initial expression studies and sequence analysis of the first PGs isolated suggested that all PGs acting during organ abscission belong to clade A (Torki *et al.*, 2000), recent studies of a few *Brassica napus* and *Arabidopsis* PGs belonging to clade B (see below) suggest that this simple classification is not absolute.

PG family genes were originally implicated in abscission by expression studies in a variety of plants, including tomato, peach and *S. nigra* (Tucker *et al.*, 1984; Bonghi *et al.*, 1992; Taylor *et al.*, 1993). The first AZ-specific PG mRNA to be isolated was *TAPG1* in tomato (Kalaitzis *et al.*, 1995), suggesting that PG expression is more specific to the abscission process than the more broadly expressed EGase tomato genes. Using *TAPG1* as a probe of a tomato leaf AZ cDNA library, three additional PG sequences were identified: *TAPG2*, *TAPG4* and *TAPG5* (Kalaitzis *et al.*, 1997; Hong and Tucker, 2000). In general, tomato abscission PG genes are more related to each other (76–95% identity) than to tomato fruit PG genes (38–41% identity) (Kalaitzis *et al.*, 1997). All four tomato PGs are expressed in the AZs of the leaf and the flower pedicel, but not in stems, petioles or fruit (Kalaitzis *et al.*, 1997; Hong and Tucker, 2000), and RNase protection assays indicate that expression is sevenfold higher in flower AZs than leaf AZs (Kalaitzis *et al.*, 1997). PG expression is also detected within the AZ of the corolla floral organs at the base of the flower (Hong *et al.*, 2000). Furthermore, promoter–GUS fusion constructs for *TAPG1* and *TAPG4* indicate that *TAPG4* is expressed earlier than *TAPG1*, suggesting that the enzymes function at different stages of cell wall dissolution (Hong *et al.*, 2000).

Whereas numerous PG transcripts have been isolated from the AZ tissue of tomato, the small number of cells within *Arabidopsis* separation zones makes RNA and protein isolation from these sites difficult. To circumvent this problem, González-Carranza *et al.* (2002) selected a closely related species, *B. napus* (oilseed rape), which has a larger AZ. A PG cDNA isolated from *B. napus* leaf AZs was used as a probe to identify corresponding genomic clones in *B. napus* (*PGAZBRAN*) and *Arabidopsis* (*PGAZAT*). *PGAZAT* expression was analysed using *PGAZAT::GUS* and *PGAZAT::GFP* promoter fusion constructs. PGAZAT::GUS expression is first detected in a ring pattern at the anther bases (González-Carranza *et al.*, 2002), similar to that seen for *TAPG1* and *TAPG4* promoter-driven GUS expression in tomato pedicel AZs (Hong *et al.*, 2000). This expression spreads throughout the floral AZ prior to separation, with ring patterns at the sites of petal abscission also becoming apparent. Ethylene exposure accelerates this process, so that the expression of the reporter correlates with the accelerated timing of abscission. Reporter expression was not detected in the seed AZ or the dehiscence zones (DZs) of anthers and fruits, suggesting that PGAZAT functions primarily in floral organ abscission (González-Carranza *et al.*, 2002).

As in tomato, functional redundancy is expected among PGs during *Arabidopsis* and *B. napus* abscission. Another *Arabidopsis* PG, ADPG1, shares 69% amino acid identity with PGAZAT, suggesting that these two PGs of clade B might have overlapping activities. While *ADPG1* expression has only been reported for fruit DZs, expression of a *B. napus* ortholog, *RDPG1*, is also detected in floral AZs by RT-PCR (Sander *et al.*, 2001). In addition, a partial *RDPG1* promoter drives reporter expression in *Arabidopsis* at several sites of cell separation, including fruit and anther DZs and floral AZs. Thus, a functional analysis of both *PGAZAT* and *ADPG1* in *Arabidopsis* may be required to dissect their respective roles in abscission.

6.4.2 Expansins

6.4.2.1 Expansins positioned to affect cell wall loosening during abscission

In addition to the hydrolytic enzymes necessary for cell wall dissolution, a family of enzymes that facilitate cell wall extension, known as expansins, have recently been implicated in abscission. Using an *in vitro* biochemical assay, Belfield *et al.* (2005) observed that similar levels of expansin activity are present in *Sambucus nigra* (elder) leaflet AZ and non-AZ tissue prior to ethylene treatment; however, following ethylene treatment, expansin activity increases sevenfold exclusively in AZ cells (Belfield *et al.*, 2005). Two expansins, *SniEXP2* and *SniEXP4*, were found to be enriched in a cDNA library constructed from ethylene-treated leaflet AZ cells, and RNA blot analysis showed that transcription of both genes is induced in AZ cells 12–24 h after ethylene exposure. RT-PCR amplification of two additional expansins, *SniEXP1* and *SniEXP3,* suggested that they are enriched in leaflet AZ cells as well (Belfield *et al.*, 2005).

Expansins are a multigene family, with 38 predicted *Arabidopsis* members (Li *et al.*, 2003). Proteins of this family have two characteristic domains: (1) a predicted active site with homology to the family-45 EGases and (2) a putative polysaccharide-binding region with homology to a grass pollen allergen. Expansins are thought to weaken the hydrogen bonds between cell wall polysaccharides, particularly cellulose microfibrils and xyloglucans, easing the cell wall tension that impedes growth. Although cellulose and xyloglucans are also targeted by EGases, *in vitro* biochemical experiments have not shown that expansins are capable of hydrolyzing cell wall material (Cosgrove *et al.*, 2002; Li *et al.*, 2003). The localization of expansins is consistent with their role in cell wall expansion. Immunogold labeling of cucumber and maize epidermal cells with an antibody for cucumber EXP1 shows that expansins are dispersed throughout all layers of the cell wall and are occasionally found in Golgi-derived vesicles, suggesting that these enzymes are targeted to the cell wall through the secretory system (Cosgrove *et al.*, 2002).

Historically, expansins have been shown to function in events that require extension of the cell wall, such as polarized growth and general cell expansion – not in events that require softening and breakdown of the cell wall, such as fruit ripening and abscission. Although functional analysis of the *Arabidopsis* expansin family has not yet revealed any single mutants with defects in floral organ shedding, the likelihood of genetic redundancy will require construction of higher order mutants to fully address their potential role in cell separation (reviewed in Cosgrove *et al.*, 2002; Li *et al.*, 2003). In addition to the induction of expansin activity in ethylene-treated *S. nigra* leaflet AZs, a few other observations in *Arabidopsis* flowers suggest that certain expansins might be involved in the abscission process. First, it has been shown that changes in the expression of *AtEXP10*, which is normally found at the bases of flower pedicels, affect the degree of pedicel breakage when flowers are forcibly removed from the inflorescence stem. Overexpression of *AtEXP10* in transgenic plants appeared to increase the frequency of complete breakage at this apparently vestigial AZ site compared to the wild type, whereas antisense *AtEXP10* expression reduced the frequency of complete breakage (Cho and Cosgrove, 2000). Second, SEM of floral organ AZs over time has revealed that proximal AZ cells begin to expand and become rounded at the time of abscission (Patterson and Bleecker, 1997; Butenko *et al.*, 2003). An intriguing possibility is that expansion of these AZ cells, facilitated by expansin activity, might be a prerequisite for abscission, perhaps by affecting the ability of cells within the AZ to adhere to one another (Figure 6.1).

6.5 Discussion and future directions

The organ abscission studies described in this chapter represent only a few enticing episodes of a much more intricate story yet to be told. Since abscission mutants are currently categorized based on our superficial understanding of the cell separation process, as we gain a deeper knowledge of the complex genetic interactions that underlie the morphological changes that take place, the stages of organ abscission are certain to be more evident (Figure 6.1). In the future, sensitized genetic screens and protein interaction experiments should dramatically expand the cast of abscission players, and begin to reveal the structure and intersections of the pathways in which they act. Additional insights into the functional mechanisms of those proteins already known to be involved in organ abscission are likely to be provided by subcellular localization approaches such as fluorescent tagging and immunofluorescence analysis. For instance, the nuclear localization of ARP4 and ARP7 lends strong support to their predicted functions in chromatin remodeling and global regulation of gene networks involved in senescence and abscission. Research in model plants with sequenced genomes also facilitates several powerful approaches including: (1) map-based cloning of abscission mutants, (2) systematic analysis of gene families in which genetic redundancy is expected, (3) cross-species comparisons to identify candidate genes in agronomically important species, (4) microarray studies of AZ-specific gene expression using laser capture microdissection and (5) ChIP-chip analysis to identify downstream targets of transcription factors that regulate abscission. Together with more traditional genetic screens and reverse genetic strategies, these techniques are certain to uncover new tales of genetic interactions required for organ abscission over the next decade.

In addition to investigating the shared characteristics of cell separation across plant species, a parallel goal of abscission research is the ability to manipulate the shedding of specific organs in important crop plants. Characterization of new mutants such as *abs2*, which is nonallelic with *abs1* and completely blocks cotyledon abscission yet only delays flower and leaf shedding in *L. angustifolius* (Clements and Atkins, 2001), should increase our understanding of the regulatory differences between various types of organ abscission. Since ethylene promotes the abscission of all lateral organs in 'Valencia' orange trees, there is ongoing research to identify compounds that affect the timing of mature fruit abscission without inducing citrus trees to drop their flowers, immature fruit or leaves. One such compound, 5-chloro-3-methyl-4-nitro-1*H*-pyrazole (CMNP), marketed as 'Release' 30 years ago, specifically induces the abscission of mature fruit. When applied to citrus peel, CMNP appears to promote membrane breakdown and the activation of lipid-signaling pathways (Alferez *et al.*, 2005). This study suggests that lipid signaling is yet another component of particular pathways controlling plant separation, and nicely illustrates the feedback potential between the knowledge gained from applied and basic research of organ abscission.

Acknowledgments

We thank Lalitree Darnielle for her fabulous illustrations, and José Alonso, Lalitree Darnielle, Zinnia González-Carranza, Adrienne Roeder and Chandra Tucker for their critical reading of the manuscript. M.E.L. is supported by a William R. Kenan Graduate Research Fellowship and the UNC Curriculum in Genetics and Molecular Biology; M.W.L. is supported by the National Institutes of Health-funded UNC Developmental Biology training program in the Carolina Center for Genome Sciences. Research in the laboratory is funded by a National Science Foundation grant to S.J.L.

References

Abeles, F.B., Morgan, P.W. and Saltveit, M.E. (1992) *Ethylene in Plant Biology*, 2nd edn. Academic Press, San Diego, CA.

Addicott, F.T. (1982) *Abscission*. University of California Press, Berkeley, CA.

Alferez, F., Singh, S., Umbach, A.L., Hockema, B. and Burns, J.K. (2005) Citrus abscission and *Arabidopsis* plant decline in response to 5-chloro-3-methyl-4-nitro-1*H*-pyrazole are mediated by lipid signaling. *Plant, Cell and Environment* **28**, 1436–1449.

Alibardi, L. (2000) Ultrastructural localization of alpha-keratins in the regenerating epidermis of the lizard *Podarcis muralis* during formation of the shedding layer. *Tissue and Cell* **32** (2), 153–162.

Alibardi, L. (2005) Differentiation of snake epidermis, with emphasis on the shedding layer. *Journal of Morphology* **264** (2), 178–190.

Alonso, J.M., Hirayama, T., Roman, G., Nourizadeh, S. and Ecker, J.R. (1999) EIN2, a bifunctional transducer of ethylene and stress responses in *Arabidopsis*. *Science* **284** (5423), 2148–2152.

Alonso, J.M. and Stepanova, A.N. (2004) The ethylene signaling pathway. *Science* **306** (5701), 1513–1515.

Aoyama, T. and Chua, N.H. (1997) A glucocorticoid-mediated transcriptional induction system in transgenic plants. *Plant Journal* **11** (3), 605–612.

Archambault J., Pan, G., Dahmus, G.K., Cartier, M., Marshall, N., Zhang, S., Dahmus, M.E. and Greenblatt, J. (1998) FCP1, the RAP74-interacting subunit of a human protein phosphatase that dephosphorylates the carboxyl-terminal domain of RNA polymerase IIO. *Journal of Biological Chemistry* **273** (42), 27593–27601.

Barry, C.S., McQuinn, R.P., Thompson, A.J., Seymour, G.B., Grierson, D. and Giovannoni, J.J. (2005) Ethylene insensitivity conferred by the *Green-ripe* and *Never-ripe 2* ripening mutants of tomato. *Plant Physiology* **138** (1), 267–275.

Belfield, E.J., Ruperti, B., Roberts, J.A. and McQueen-Mason, S. (2005) Changes in expansin activity and gene expression during ethylene-promoted leaflet abscission in *Sambucus nigra. Journal of Experimental Botany* **56** (413), 817–823.

Bleecker, A.B., Estelle, M.A., Somerville, C. and Kende, H. (1988) Insensitivity to ethylene conferred by a dominant mutation in *Arabidopsis thaliana. Science* **241** (4869), 1086–1089.

Bleecker, A.B. and Patterson, S.E. (1997) Last exit: senescence, abscission, and meristem arrest in *Arabidopsis. Plant Cell* **9** (7), 1169–1179.

Bonghi, C., Rascio, N., Ramina, A. and Casadoro, G. (1992) Cellulase and polygalacturonase involvement in the abscission of leaf and fruit explants of peach. *Plant Molecular Biology* **20** (5), 839–848.

Bowman, J.L., Eshed, Y. and Baum, S.F. (2002) Establishment of polarity in angiosperm lateral organs. *Trends in Genetics* **18** (3), 134–141.

Brown, K.M. (1997) Ethylene and abscission. *Physiologia Plantarum* **100**, 567–576.

Brummell, D.A., Hall, B.D. and Bennett, A.B. (1999) Antisense suppression of tomato endo-1,4-beta-glucanase Cel2 mRNA accumulation increases the force required to break fruit abscission zones but does not affect fruit softening. *Plant Molecular Biology* **40** (4), 615–622.

Budiman, M.A., Chang, S.-B., Lee, S., Yang, T.J., Zhang, H.-B., de Jong, H. and Wing, R.A. (2004) Localization of *jointless-2* gene in the centromeric region of tomato chromosome 12 based on high resolution genetic and physical mapping. *Theoretical and Applied Genetics* **108** (2), 190–196.

Butenko, M.A., Patterson, S.E., Grini, P.E., Stenvik, G.E., Amundsen, S.S., Mandal, A. and Aalen, R.B. (2003) *INFLORESCENCE DEFICIENT IN ABSCISSION* controls floral organ abscission in *Arabidopsis* and identifies a novel family of putative ligands in plants. *Plant Cell* **15** (10), 2296–2307.

Butler, L. (1936) Inherited characters in the tomato. II. Jointless pedicel. *The Journal of Heredity* **37**, 25–26.

Cancel, J.D. and Larsen, P.B. (2002) Loss-of-function mutations in the ethylene receptor ETR1 cause enhanced sensitivity and exaggerated response to ethylene in Arabidopsis. *Plant Physiology* **129** (4), 1557–1567.

Cao, H., Glazebrook, J., Clarke, J.D., Volko, S. and Dong, X. (1997) The Arabidopsis *NPR1* gene that controls systemic acquired resistance encodes a novel protein containing ankyrin repeats. *Cell* **88** (1), 57–63.

Chang, C., Kwok, S.F., Bleecker, A.B. and Meyerowitz, E.M. (1993) *Arabidopsis* ethylene-response gene *ETR1*: similarity of product to two-component regulators. *Science* **262** (5133), 539–544.

Chao, Q., Rothenberg, M., Solano, R., Roman, G., Terzaghi, W. and Ecker, J.R. (1997) Activation of the ethylene gas response pathway in *Arabidopsis* by the nuclear protein ETHYLENE-INSENSITIVE3 and related proteins. *Cell* **89** (7), 1133–1144.

Chen, G., Alexander, L. and Grierson, D. (2004) Constitutive expression of EIL-like transcription factor partially restores ripening in the ethylene-insensitive *Nr* tomato mutant. *Journal of Experimental Botany* **55** (402), 1491–1497.

Chen, Y.-F., Randlett, M.D., Findell, J.L. and Schaller, G.E. (2002) Localization of the ethylene receptor ETR1 to the endoplasmic reticulum of *Arabidopsis. Journal of Biological Chemistry* **277** (22), 19861–19866.

Cho, H.T. and Cosgrove, D.J. (2000) Altered expression of expansin modulates leaf growth and pedicel abscission in *Arabidopsis thaliana*. *Proceedings of the National Academy of Sciences of the United States of America* **97** (17), 9783–9788.

Cho, S., Jang, S., Chae, S., Chung, K.M., Moon, Y.-H., An, G. and Jang, S.K. (1999) Analysis of the C-terminal region of *Arabidopsis thaliana* APETALA1 as a transcriptional activation domain. *Plant Molecular Biology* **40** (3), 419–429.

Clements, J. and Atkins, C. (2001) Characterization of a non-abscission mutant in *Lupinus angustifolius*. I. Genetic and structural aspects. *American Journal of Botany* **88** (1), 31–42.

Cosgrove, D.J. (2005) Growth of the plant cell wall. *Nature Reviews. Molecular Cell Biology* **6** (11), 850–861.

Cosgrove, D.J., Li, L.C., Cho, H.-T., Hoffmann-Benning, S., Moore, R.C. and Blecker, D. (2002) The growing world of expansins. *Plant and Cell Physiology* **43** (12), 1436–1444.

Dal Degan, F., Child, R., Svendsen, I. and Ulvskov, P. (2001) The cleavable N-terminal domain of plant endopolygalacturonases from clade B may be involved in a regulated secretion mechanism. *The Journal of Biological Chemistry* **276** (38), 35297–35304.

del Campillo, E. and Bennett, A.B. (1996) Pedicel breakstrength and cellulase gene expression during tomato flower abscission. *Plant Physiology* **111** (3), 813–820.

Dinneny, J.R., Yadegari, R., Fischer, R., Yanofsky, M.F. and Weigel, D. (2004) The role of *JAGGED* in shaping lateral organs. *Development* **131** (5), 1101–1110.

Doebley, J. (2004) The genetics of maize evolution. *Annual Review of Genetics* **38**, 37–59.

Dong, X. (2004) NPR1, all things considered. *Current Opinion in Plant Biology* **7** (5), 547–552.

Doubt, S.L. (1917) The response of plants to illuminating gas. *Botanical Gazette* **63** (3), 209–224.

Ecker, J.R. and Theologis, A. (1994) Ethylene: a unique plant signaling molecule, in *Arabidopsis* (eds Meyerowitz, E.M. and Somerville, C.R.), pp. 485–521. Cold Spring Harbor Press, Plainview, NY.

Ellis, C.M., Nagpal, P., Young, J.C., Hagen, G., Guilfoyle, T.J. and Reed, J.W. (2005) *AUXIN RESPONSE FACTOR1* and *AUXIN RESPONSE FACTOR2* regulate senescence and floral organ abscission in *Arabidopsis thaliana*. *Development* **132** (20), 4563–4574.

Fang, S.-C. and Fernandez, D.E. (2002) Effect of regulated overexpression of the MADS domain factor AGL15 on flower senescence and fruit maturation. *Plant Physiology* **130** (1), 78–89.

Fernandez, D.E., Heck, G.R., Perry, S.E., Patterson, S.E., Bleecker, A.B. and Fang, S.-C. (2000) The embryo MADS domain factor AGL15 acts postembryonically. Inhibition of perianth senescence and abscission via constitutive expression. *Plant Cell* **12** (2) 183–198.

Ferrarese, L., Trainotti, L., Moretto, P., Polverino De Laureto, P., Rascio, N. and Casadoro, G. (1995) Differential ethylene-inducible expression of cellulase in pepper plants. *Plant Molecular Biology* **29** (4), 735–747.

González-Carranza, Z.H., Whitelaw, C.A., Swarup, R. and Roberts, J.A. (2002) Temporal and spatial expression of a polygalacturonase during leaf and flower abscission in oilseed rape and Arabidopsis. *Plant Physiology* **128** (2), 534–543.

Goodson, H.V. and Hawse, W.F. (2002) Molecular evolution of the actin family. *Journal of Cell Science* **115** (Pt 13), 2619–2622.

Guo, H. and Ecker, J.R. (2004) The ethylene signaling pathway: new insights. *Current Opinion in Plant Biology* **7** (1), 40–49.

Guzmán, P. and Ecker, J.R. (1990) Exploiting the triple response of *Arabidopsis* to identify ethylene-related mutants. *Plant Cell* **2** (6), 513–523.

Ha, C.M., Jun, J.H., Nam, H.G. and Fletcher, J.C. (2004) *BLADE-ON-PETIOLE1* encodes a BTB/POZ domain protein required for leaf morphogenesis in *Arabidopsis thaliana*. *Plant and Cell Physiology* **45** (10), 1361–1370.

Ha, C.M., Kim, G.T., Kim, B.C., Jun, J.H., Soh, M.S., Ueno, Y., Machida, Y., Tsukaya, H. and Nam, H.G. (2003) The *BLADE-ON-PETIOLE1* gene controls leaf pattern formation through modulation of meristematic activity in *Arabidopsis*. *Development* **130** (1), 161–172.

Hackett, R.M., Ho, C.-W., Lin, Z., Foote, H.C.C., Fray, R.G. and Grierson, D. (2000) Antisense inhibition of the *Nr* gene restores normal ripening to the tomato *Never-ripe* mutant, consistent with the ethylene receptor-inhibition model. *Plant Physiology* **124** (3), 1079–1086.

Hadfield, K.A. and Bennett, A.B. (1998) Polygalacturonases: many genes in search of a function. *Plant Physiology* **117** (2), 337–343.

Harding, E.W., Tang, W., Nichols, K.W., Fernandez, D.E. and Perry, S.E. (2003) Expression and maintenance of embryogenic potential is enhanced through constitutive expression of *AGAMOUS-Like 15*. *Plant Physiology* **133** (2), 653–663.

Hartmann, U., Hohmann, S., Nettesheim, K., Wisman, E., Saedler, H. and Huijser, P. (2000) Molecular cloning of *SVP*: a negative regulator of the floral transition in *Arabidopsis*. *Plant Journal* **21** (4), 351–360.

Henderson, J., Lyne, L. and Osborne, D.J. (2001) Failed expression of an *endo*-beta-1,4-glucanhydrolase (cellulase) in a non-abscinding mutant of *Lupinus angustifolius* cv Danja, *Phytochemistry* **58** (7), 1025–1034.

Hepworth, S.R., Zhang, Y., McKim, S., Li, X. and Haughn, G.W. (2005) *BLADE-ON-PETIOLE1* dependent signaling controls leaf and floral patterning in *Arabidopsis*. *Plant Cell* **17** (5), 1434–48.

Hong, S.-B., Sexton, R. and Tucker, M.L. (2000) Analysis of gene promoters for two tomato polygalacturonases expressed in abscission zones and the stigma. *Plant Physiology* **123** (3), 869–881.

Hong, S.-B. and Tucker, M.L. (2000) Molecular characterization of a tomato polygalacturonase gene abundantly expressed in the upper third of pistils from opened and unopened flowers. *Plant Cell Reports* **19** (7), 680–683.

Horn, M.A. and Walker, J.C. (1994) Biochemical properties of the autophosphorylation of RLK5, a receptor-like protein kinase from Arabidopsis thaliana. *Biochimica et Biophysica Acta* **1208** (1), 65–74.

Hua, J. and Meyerowitz, E.M. (1998) Ethylene responses are negatively regulated by a receptor gene family in *Arabidopsis thaliana*. *Cell* **94** (2), 261–271.

Hua, J., Sakai, H., Nourizadeh, S., Chen, Q.G., Bleecker, A.B., Ecker, J.R. and Meyerowitz, E.M. (1998) *EIN4* and *ERS2* are members of the putative ethylene receptor gene family in Arabidopsis. *Plant Cell* **10** (8), 1321–1332.

Jeong, S., Trotochaud, A.E. and Clark, S.E. (1999) The Arabidopsis *CLAVATA2* gene encodes a receptor-like protein required for the stability of the CLAVATA1 receptor-like kinase. *Plant Cell* **11** (10), 1925–1934.

Jinn, T.-L., Stone, J.M. and Walker, J.C. (2000) *HAESA*, an *Arabidopsis* leucine-rich repeat receptor kinase, controls floral organ abscission. *Genes and Development* **14** (1), 108–17.

Kalaitzis, P., Koehler, S.M. and Tucker, M.L. (1995) Cloning of a tomato polygalacturonase expressed in abscission. *Plant Molecular Biology* **28** (4), 647–656.

Kalaitzis, P., Solomos, T. and Tucker, M.L. (1997) Three different polygalacturonases are expressed in tomato leaf and flower abscission, each with a different temporal expression pattern. *Plant Physiology* **113** (4), 1303–1308.

Kandasamy, M.K., Deal, R.B., McKinney, E.C. and Meagher, R.B. (2005a) Silencing the nuclear actin-related protein AtARP4 in *Arabidopsis* has multiple effects on plant development, including early flowering and delayed floral senescence. *Plant Journal* **41** (6), 845–858.

Kandasamy, M.K., McKinney, E.C., Deal, R.B. and Meagher, R.B. (2005b) Arabidopsis ARP7 is an essential actin–related protein required for normal embryogenesis, plant architecture and floral organ abscission. *Plant Physiology* **138** (4), 2019–2032.

Kandasamy, M.K., McKinney, E.C. and Meagher, R.B. (2003) Cell cycle-dependent association of *Arabidopsis* actin-related proteins AtARP4 and AtARP7 with the nucleus. *Plant Journal* **33** (5), 939–948.

Kemmerer, E.C. and Tucker, M.L. (1994) Comparative study of cellulases associated with adventitious root initiation, apical buds, and leaf, flower, and pod abscission zones in soybean. *Plant Physiology* **104** (2), 557–562.

Keskitalo, J., Bergquist, G., Gardestrom, P. and Jansson, S. (2005) A cellular timetable of autumn senescence. *Plant Physiology* **139** (4), 1635–48.

Koehler, S.M., Matters, G.L., Nath, P., Kemmerer, E.C. and Tucker, M.L. (1996) The gene promoter for a bean abscission cellulase is ethylene-induced in transgenic tomato and shows high sequence conservation with a soybean abscission cellulase. *Plant Molecular Biology* **31** (3), 595–606.

Koiwa, H., Barb, A.W., Xiong, L., Li, F., McCully, M.G., Lee, B.-H., Sokolchik, I., Zhu, J., Gong, Z., Reddy, M., Sharkhuu, A., Manabe, Y., Yokoi, S., Zhu, J.K., Bressan, R.A., Hasegawa, P.M. (2002) C-terminal domain phosphatase-like family members (AtCPLs) differentially regulate *Arabidopsis thaliana* abiotic stress signaling, growth, and development. *Proceedings of the National Academy of Sciences of the United States of America* **99** (16), 10893–10898.

Lanahan, M.B., Yen, H.-C., Giovannoni, J.J. and Klee, H.J. (1994) The *Never ripe* mutation blocks ethylene perception in tomato. *Plant Cell* **6** (4), 521–530.

Lashbrook, C.C., Giovannoni, J.J., Hall, B.D., Fischer, R.L. and Bennett, A.B. (1998a) Transgenic analysis of tomato endo-beta-1,4-glucanase gene function. Role of *cel1* in floral abscission. *The Plant Journal* **13** (3), 303–310.

Lashbrook, C.C., Gonzalez-Bosch, C. and Bennett, A.B. (1994) Two divergent endo-beta-1,4-glucanase genes exhibit overlapping expression in ripening fruit and abscising flowers. *Plant Cell* **6** (10), 1485–1493.

Lashbrook, C.C., Tieman, D.M. and Klee, H.J. (1998b) Differential regulation of the tomato *ETR* gene family throughout plant development. *Plant Journal* **15** (2), 243–252.

Lehti-Shiu, M.D., Adamczyk, B.J. and Fernandez, D.E. (2005) Expression of MADS-box genes during the embryonic phase in *Arabidopsis*. *Plant Molecular Biology* **58** (1), 89–107.

Lewis, L.N. and Varner, J.E. (1970) Synthesis of cellulase during abscission of *Phaseolus vulgaris* leaf explants. *Plant Physiology* **46** (2), 194–199.

Li, Y., Jones, L. and McQueen-Mason, S. (2003) Expansins and cell growth. *Current Opinion in Plant Biology* **6** (6), 603–610.

Libertini, E., Li, Y. and McQueen-Mason, S.J. (2004) Phylogenetic analysis of the plant endo-beta-1,4-glucanase gene family. *Journal of Molecular Evolution* **58** (5), 506–515.

Mao, L., Begum, D., Chuang, H.W., Budiman, M.A., Szymkowiak, E.J., Irish, E.E. and Wing, R.A. (2000) *JOINTLESS* is a MADS-box gene controlling tomato flower abscission zone development. *Nature* **406** (6798), 910–913.

Mao, L., Begum, D., Goff, S.A. and Wing, R.A. (2001) Sequence and analysis of the tomato *JOINTLESS* locus. *Plant Physiology* **126** (3), 1331–1340.

McKinney, E.C., Kandasamy, M.K. and Meagher, R.B. (2002) Arabidopsis contains ancient classes of differentially expressed actin-related protein genes. *Plant Physiology* **128** (3), 997–1007.

Neljubow, D.N. (1901) Über die horizontale nutation der stengel von *Pisum sativum* und einiger anderen pflanzen. *Beitrage und Botanik Zentralblatt* **10**, 128–139.

Norberg, M., Holmlund, M. and Nilsson, O. (2005) The *BLADE ON PETIOLE* genes act redundantly to control growth and development of lateral organs. *Development* **132** (9), 2203–2213.

Ohno, C.K., Reddy, G.V., Heisler, M.G.B. and Meyerowitz, E.M. (2004) The *Arabidopsis JAGGED* gene encodes a zinc finger protein that promotes leaf tissue development. *Development* **131** (5), 1111–1122.

Okushima, Y., Mitina, I., Quach, H.L. and Theologis, A. (2005) AUXIN RESPONSE FACTOR 2 (ARF2): a pleiotropic developmental regulator. *Plant Journal* **43** (1), 29–46.

Osborne, D.J. (1989) Abscission. *Critical Reviews in Plant Sciences* **8** (2), 103–129.

Patterson, S.E. (2001) Cutting loose. Abscission and dehiscence in Arabidopsis. *Plant Physiology* **126** (2), 494–500.

Patterson, S.E. and Bleecker, A.B. (2004) Ethylene-dependent and independent processes associated with floral organ abscission in *Arabidopsis*. *Plant Physiology* **134** (1) 194–203.

Pinyopich, A., Ditta, G.S., Savidge, B., Liljegren, S.J., Baumann, E., Wisman, E. and Yanofsky, M.F. (2003) Assessing the redundancy of MADS-box genes during carpel and ovule development. *Nature* **424** (6944), 85–88.

Prasad, A.M., Sivanandan, C., Resminath, R., Thakare, D.R., Bhat, S.R. and Srinivasan (2005) Cloning and characterization of a pentatricopeptide protein encoding gene (*LOJ*) that is specifically expressed in lateral organ junctions in *Arabidopsis thaliana*. *Gene* **353** (1), 67–79.

Rick, C.M., and Butler, L. (1956) Phytogenetics of the tomato. *Advances in Genetics* **8**, 267–382.

Roberts, J.A., Elliott, K.A. and González-Carranza, Z.H. (2002) Abscission, dehiscence, and other cell separation processes. *Annual Review of Plant Biology* **53**, 131–158.

Rolf, H.J. and Enderle, A. (1999) Hard fallow deer antler: a living bone till antler casting? *Anatomical Record* **255** (1), 69–77.

Sampredro, J. and Cosgrove, D.J. (2005) The expansin superfamily. *Genome Biology* **6** (12), 242.

Sander, L., Child, R., Ulvskov, P., Albrechtsen M. and Borkhardt, B. (2001) Analysis of a dehiscence zone endo-polygalacturonase in oilseed rape (*Brassica napus*) and *Arabidopsis thaliana*: evidence for roles in cell separation in dehiscence and abscission zones, and in stylar tissues during pollen tube growth. *Plant Molecular Biology* **46** (4), 469–479.

Schafer, D.A. and Schroer, T.A. (1999) Actin-related proteins. *Annual Review of Cell and Developmental Biology* **15**, 341–363.

Schwer, B. and Shuman, S. (1996) Multicopy suppressors of temperature-sensitive mutations of yeast mRNA capping enzyme. *Gene Expression* **5** (6), 331–344.

Sexton, R. and Roberts, J.A. (1982) Cell Biology of Abscission. *Annual Review of Plant Physiology* **33,** 133–162.

Shibuya, K., Barry, K.G., Ciardi, J.A., Loucas, H.M., Underwood, B.A., Nourizadeh, S., Ecker, J.R., Klee, H.J. and Clark, D.G. (2004) The central role of PhEIN2 in ethylene responses throughout plant development in petunia. *Plant Physiology* **136** (2), 2900–2912.

Shiu, S.H. and Bleecker, A.B. (2001) Receptor-like kinases from *Arabidopsis* form a monophyletic gene family related to animal receptor kinases. *Proceedings of the National Academy of Sciences of the United States of America* **98** (19), 10763–10768.

Shuai, B., Reynaga-Peña, C.G. and Springer, P.S. (2002) The *LATERAL ORGAN BOUNDARIES* gene defines a novel, plant-specific gene family. *Plant Physiology* **129** (2), 747–761.

Stogios, P.J., Downs, G.S., Jauhal, J.J., Nandra, S.K. and Prive, G.G. (2005) Sequence and structural analysis of BTB domain proteins. *Genome Biology* **6** (10), R82.

Stone, J.M., Collinge, M.A., Smith, R.D., Horn, M.A. and Walker, J.C. (1994) Interaction of a protein phosphatase with an *Arabidopsis* serine-threonine receptor kinase. *Science* **266** (5186), 793–795.

Szymkowiak, E.J. and Irish, E.E. (1999) Interactions between *jointless* and wild-type tomato tissues during development of the pedicel abscission zone and the inflorescence meristem. *Plant Cell* **11** (2), 159–175.

Szymkowiak, E.J. and Irish, E.E. (2006) *JOINTLESS* suppresses sympodial identity in inflorescence meristems of tomato, *Planta* **223** (4), 646-658.

Tang, W. and Perry, S.E. (2003) Binding site selection for the plant MADS domain protein AGL15: an *in vitro* and *in vivo* study. *Journal of Biological Chemistry* **278** (30), 28154–28159.

Tarutani, Y., Morimoto, T., Sasaki, A., Yasuda, M., Nakashita, H., Yoshida, S., Yamaguchi, I. and Suzuki, Y. (2004) Molecular characterization of two highly homologous receptor-like kinase genes, *RLK902* and *RKL1*, in *Arabidopsis thaliana. Bioscience, Biotechnology, and Biochemistry* **68** (9), 1935–1941.

Taylor, J.E., Coupe, S.A., Picton, S. and Roberts, J.A. (1994) Characterization and accumulation pattern of a messenger-RNA encoding an abscission-related beta-1,4-glucanase from leaflets of *Sambucus nigra. Plant Molecular Biology* **24** (6), 961–964.

Taylor, J.E., Webb, S.T.J., Coupe, S.A., Tucker, G.A. and Roberts, J.A. (1993) Changes in polygalacturonase activity and solubility of polyuronides during ethylene-stimulated leaf abscission in *Sambucus nigra. Journal of Experimental Botany* **258,** 93–98.

Taylor, J.E. and Whitelaw, C.A. (2001) Signals in abscission. *New Phytologist* **151**, 323–339.

The Arabidopsis Genome Initiative (2000) Analysis of the genome sequence of the flowering plant *Arabidopsis thaliana, Nature* **408** (6814), 796–815.

Tieman, D.M., Ciardi, J.A., Taylor, M.G. and Klee, H.J. (2001) Members of the tomato LeEIL (EIN3-like) gene family are functionally redundant and regulate ethylene responses throughout plant development. *Plant Journal* **26** (1), 47–58.

Tieman, D.M., Taylor, M.G., Ciardi, J.A. and Klee, H.J. (2000) The tomato ethylene receptors NR and LeETR4 are negative regulators of ethylene response and exhibit functional compensation within a multigene family. *Proceedings of the National Academy of Sciences of the United States of America* **97** (10), 5663–5668.

Tiwari, S.B., Hagen, G. and Guilfoyle, T. (2003) The roles of auxin response factor domains in auxin-responsive transcription. *Plant Cell* **15** (2), 533–543.

Torki, M., Mandaron, P., Mache, R. and Falconet, D. (2000) Characterization of a ubiquitous expressed gene family encoding polygalacturonase in Arabidopsis thaliana. *Gene* **242** (1–2), 427–436.

Trainotti, L., Ferrarese, L. and Casadoro, G. (1998) Characterization of cCel3, a member of the pepper endo-beta-1,4-glucanase multigene family. *Hereditas* **128** (2), 121–126.

Trainotti, L., Spolaore, S., Ferrarese, L. and Casadoro, G. (1997) Characterization of ppEG1, a member of a multigene family which encodes endo-beta-1,4-glucanase in peach. *Plant Molecular Biology* **34** (5), 791–802.

Tucker, G.A., Schindler, C.B. and Roberts, J.A. (1984) Flower abscission in mutant tomato plants. *Planta* **160**, 164–167.

Tucker, M.L., Sexton, R., del Campillo, E. and Lewis, L.N. (1988) Bean abscission cellulase. *Plant Physiology* **88** (4), 1257–1262.

Ulmasov, T., Hagen, G. and Guilfoyle, T.J. (1997) ARF1, a transcription factor that binds to auxin response elements. *Science* **276** (5320), 1865–1868.

Ulmasov, T., Hagen, G. and Guilfoyle, T.J. (1999) Activation and repression of transcription by auxin-response factors. *Proceedings of the National Academy of Sciences of the United States of America* **96** (10), 5844–5849.

von Stackelberg, M., Lindemann, S., Menke, M., Riesselmann, S. and Jacobsen, H.J. (2003) Identification of AFLP and STS markers closely linked to the *def* locus in pea. *Theoretical and Applied Genetics* **106** (7), 1293–1299.

Wang, H., Tang, W., Zhu, C. and Perry, S.E. (2002) A chromatin immunoprecipitation (ChIP) approach to isolate genes regulated by AGL15, a MADS domain protein that preferentially accumulates in embryos. *Plant Journal* **32** (5), 831–843.

Wilkie, I.C. (2001) Autotomy as a prelude to regeneration in echinoderms. *Microscopy Research and Technique* **55** (6), 369–396.

Wilkinson, J.Q., Lanahan, M.B., Yen, H.C., Giovannoni, J.J., and Klee, H.J. (1995) An ethylene-inducible component of signal transduction encoded by *Never-ripe*. *Science* **270** (5243), 1807–1809.

Yamagami T., Tsuchisaka A., Yamada K., Haddon W.R., Harden L.A. and Theologis A. (2003) Biochemical diversity among the 1-amino-cyclopropane-1-carboxylate synthase isozymes encoded by the *Arabidopsis* gene family. *Journal of Biological Chemistry* **278** (49), 49102–49112.

Yang, T.-J., Lee, S., Chang, S.-B., Yu, Y., de Jong, H. and Wing, R.A. (2005) In-depth sequence analysis of the tomato chromosome 12 centromeric region: identification of a large CAA block and characterization of pericentromere retrotransposons. *Chromosoma* **114** (2), 103–117.

Yu, H., Ito, T., Wellmer, F. and Meyerowitz, E.M. (2004) Repression of AGAMOUS-LIKE 24 is a crucial step in promoting flower development. *Nature Genetics* **36** (2) 157–161.

Yuan, R., Wu, Z., Kostenyuk, I.A., Burns, J.K. (2005) G-protein-coupled alpha2A-adrenoreceptor agonists differentially alter citrus leaf and fruit abscission by affecting expression of ACC synthase and ACC oxidase. *Journal of Experimental Botany* **56** (417), 1867–1875.

7 Dehiscence

Lars Østergaard, Bernhard Borkhardt and Peter Ulvskov

7.1 Introduction

Flowering plants characteristically produce their seeds within, protected by, and nourished through, a carpel-derived structure called a fruit. Although the word fruit traditionally refers to fleshy, edible and usually sweet, carpellate organs, botanically identifiable fruits include a wide range of forms and sizes ranging from the fruit of *Wolffia angusta*, which is smaller than the size of a grain of table salt, to the giant pumpkin that can weigh more than 500 kg. Some fruits are neither sweet nor fleshy, and become dry by the time they mature. In common with fleshy fruits, many of the dry fruits represent obvious adaptations for seed dispersal. The forcible or explosive manner in which many dry fruits open up, or dehisce, to scatter their seeds is one such type of adaptation. Dehiscence thus refers to the opening of dry fruits to release their seeds. Fruit dehiscence occurs along one or more dehiscence zones consisting of a narrow file of highly specialized cells that differentiate, in the case of Brassicaceae, at the valve/replum border (see examples in Plate 8). These cells are believed to produce and secrete cell wall-degrading enzymes that allow the valves to separate. Opening takes place when the maternal tissue is dead, and thus requires an external force. The biochemical processes that are prerequisite for fruit opening are separated in time from the opening itself by days in Arabidopsis and by weeks in oilseed rape (*Brassica napus*). This separation in time raises interesting regulatory problems. Dehiscence may be viewed as a specialized fruit softening in which the softening has been confined to the carpel margins through the interaction of transcription factors to be detailed below. We shall review some hormonal regulatory aspects that also highlight similarities to softening of fleshy fruits (see Chapter 8). Certain aspects, such as that the same polysaccharide hydrolase is involved in both fruit dehiscence and seed abscission, point to dehiscence being closely related to abscission (see Chapter 6). Similarly, anther dehiscence appears to be related to both fruit softening and abscission. Dehiscence has almost invariably been reviewed in the context of abscission, most recently by Jarvis *et al.* (2003) encompassing intercellular adhesion with a focus on biomechanical aspects and the use of antibody probes in the study of load-bearing structures. Roberts *et al.* (2002) cover both basic and applied aspects of dehiscence in depth. Overlap with these reviews is unavoidable, although emphasis will be on the most recent advances in our understanding of the dehiscence process.

7.2 Shatter resistance

Although pod shattering provides the plants with a dispersal advantage in nature, unsynchronized pod shattering is a problem for oilseed rape farmers and research into dehiscence mechanisms has to a large extent been stimulated by the agronomical significance of the trait *shatter resistance*. The major cereals are all shatter resistant, i.e. seeds are not dispersed when they become mature and dehydrated. A similar trait is not found in modern cultivars of oilseed rape, nor to a sufficient degree in crops like soybean, sesame and lupin. Thus, in tropical and sub-tropical regions, pod shattering in soybean may cause serious yield losses (Tiwari and Bhatnagar, 1991). Oilseed rape is a young crop plant that still retains many traits of a wild plant, including unsynchronized flowering, which aggravates the problem of seed shattering. Early harvest reduces seed losses but leads to the occurrence of seeds containing chlorophyll. Even as little as 2% of green seed can cause the price received by farmers to be discounted. Hence, optimal harvest time is a compromise, and leads typically to a seed loss of 10%. In addition to this yield loss, prematurely released seeds fall to the ground and germinate to become weeds (volunteers) that contaminate the following year's harvest. This adversely affects the crop rotation practice used by many farmers, and is consequently damaging to the environment. Morgan *et al.* (2003) discuss the different measures taken to minimize seed loss, and their implications.

Shatter resistance occurs in various species of Brassicaceae including *B. oleracea* and *B. rapa*, the parents of *B. napus* (Morgan *et al.*, 1998). The limited variability in shatter resistance among modern cultivars reflects the importance of the so-called double-low phenotype (low in the antinutritional constituents erucic acid and aliphatic glucosinolates). The low glucosinolate trait of all modern cultivars originates from the Polish cultivar Bronowski (Kondra and Steffansson, 1970) and the low erucic acid phenotype from the cultivar Liho and its derivatives (Seyis *et al.*, 2004). Modern oilseed rape has, therefore, a relatively narrow genetic base. Thus, transgenic approaches to engineering shatter resistance are attractive. Despite the problems encountered in introducing new variability in shatter resistance, using conventional breeding techniques, without losing the double-low phenotype or other advanced traits, Morgan *et al.* (1998) resynthesized oilseed rape in order to do this. A line, DK142, was selected for its better shatter resistance and characterized in detail using cv. Apex as the reference cultivar. Child *et al.* (2003) found that the increased resistance to shatter was not due to reduced or delayed cell separation in the dehiscence zone of the silique, but due to two other factors: the width of the dehiscence zone and the strength of a thicker main vascular bundle that passes through the dehiscence zone at a shallower angle than in cv. Apex (Figure 7.1B). Of these two factors, the latter was found to be quantitatively more important. DK142 has reduced fertility as well as other undesirable characters, and it will probably require a comprehensive breeding program to turn it into a competitive, modern cultivar. We shall return to the prospects of engineering shatter resistance into oilseed rape in the end of this chapter.

The natural variation in shatter resistance is much larger in soybean than in oilseed rape. Pod anatomy has been studied in soybean varieties with different levels of pod shattering in an attempt to identify the structural basis of resistance

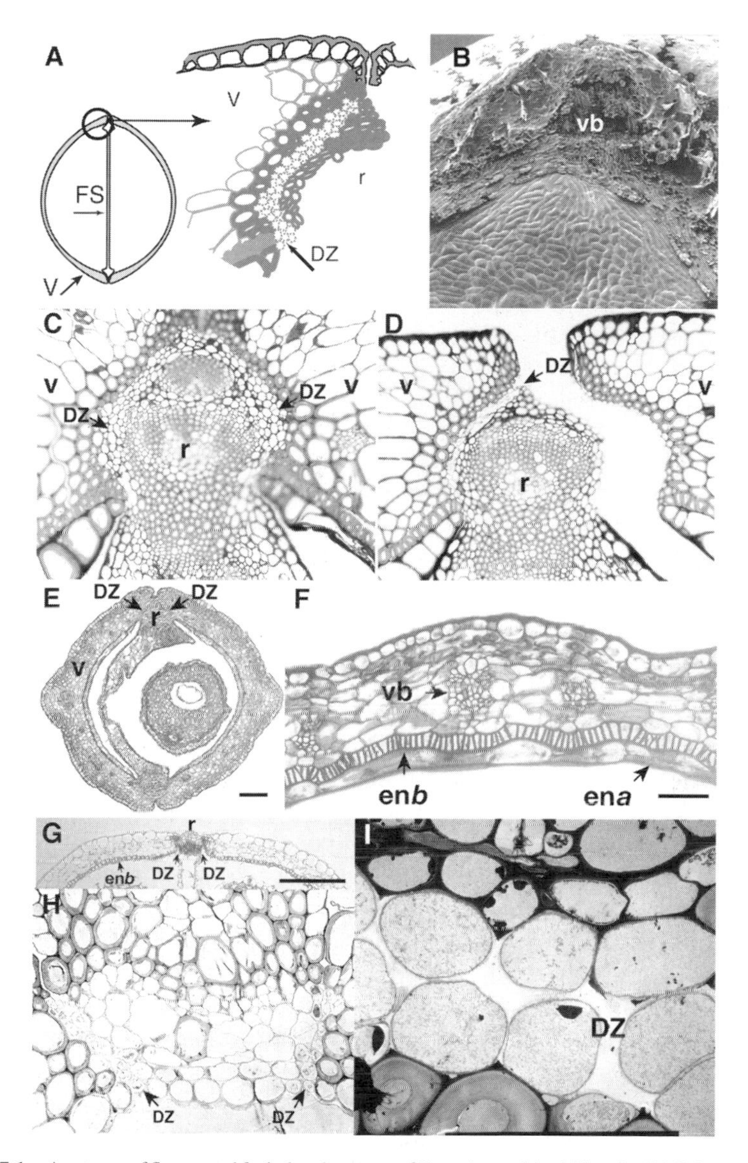

Figure 7.1 Anatomy of flower and fruit development of *Brassica* and Arabidopsis. (A) Schematic cross section of a silique. The encircled area is enlarged to visualize the area containing one of the dehiscence zones. (B) SEM image showing the fracture plane of the main vascular bundle that transverse the valve/replum region in cv. Apex. Dehiscence zone structure in *B. napus* (C) five WAA and (D) seven WAA, respectively. (E) Cross section of young *B. juncea* fruit two WAA stained with alcian blue/safranin O. (F) Cross section of valve from fruit of *B. juncea* six WAA. A vascular bundle (vb) and the lignified en*b* layer as well as en*a* are indicated. (G) Cross section of stage 17 Arabidopsis fruit stained for lignin with phloroglucinol. The lignified en*b* layer is indicated. (H) Cross section of an Arabidopsis silique showing the dehiscence zone located between lignified valve cells and the replum. (I) *B. napus* silique seven WAA with separated, rounded dehiscence zone cells. Scale bars: (E and F) 500 μm and (G) 100 μm. C, carpel; DZ, dehiscence zone; FS, false septum; R, replum. (Panel (A) adapted from Christiansen *et al.*, 2002; panels (B, C, D, H, I) by Robin Child.)

to this phenomenon (Tiwari and Bhatia, 1995). High negative correlation of seed shattering with bundle cap length, bundle cap thickness and pod wall thickness was demonstrated.

7.3 Morphogenesis and anatomy of the dehiscence zone

The species mentioned above for which shatter resistance is agronomically relevant are not phylogenetically closely related: these include, for example, oilseed rape, which belongs to the family Brassicaceae, lupin and soybean (members of the family Fabaceae) and sesame (Pedaliaceae). A recent observation by Chandler *et al.* (2005) that a *Brassica* ortholog of the Arabidopsis *FRUITFULL* gene can functionally inhibit dehiscence of tobacco capsules when introduced into *Nicotiana tabacum* (Solanaceae) suggests that dehiscence of capsules, siliques and pods has not developed by convergent evolution but is based on principles that predate the diversification of the plant families mentioned above. Some members of the family Resedaceae, which like Brassicaceae is included in the order Capparales, have dry dehiscent fruits in which the carpels are only weakly fused and often open at the capsule tip. Carpels are initially free, and fusion at sutures occurs as a secondary event. Thus, the simplest assumption is that no particular developmental program is required to form a dehiscence zone. All that is required is that fusion of carpels is arrested in time to form a weakened zone along which the fruit can split open as it dehydrates. However, species of Fabaceae have only one carpel but both a dorsal and a ventral dehiscence zone. The dorsal dehiscence zone has not developed on the basis of a preceding fusion event and we must conclude that dehiscence zones can develop *de novo.* It is therefore of interest to compare the Brassicaceae dehiscence zone to that of the Fabaceae and further look for differences between the ventral and dorsal dehiscence zones in the legume pod.

7.3.1 Brassicaceae fruit morphology

Arabidopsis thaliana and *Brassica* species are members of the family *Brassicaceae* and produce inflorescences with cruciform flowers (Plate 8A). Despite a significant size difference, the morphologies of Arabidopsis and *Brassica* fruits are remarkably similar (compare Plate 8C with 8D and Plate 8G with 8H). Upon fertilization, the gynoecium develops into a silique in these species (Plate 8B).

In flowering plants, fertilization depends on contact of pollen with the stigmatic tissue of the gynoecium (Plate 8B–8D). The pollen grains germinate on the stigma and produce pollen tubes, which grow down the style, through the transmitting tract and enter the ovules through a small opening in the integument layers, called the micropyle (reviewed in Lord, 2003). After fertilization, Brassicaceous fruits elongate and form a number of specialized tissues including valves or seed pod walls, a central replum, a false septum and valve margins with the dehiscence zones that form at the valve/replum borders (Plate 8D and E and 7.1E and G). The cell types that comprise each of these tissues differ strikingly, forming sharp boundaries along the valve margin (Ferrándiz *et al.*, 1999).

The valves exhibit an adaxial/abaxial symmetry and are composed of an outermost (abaxial) epidermal layer with long and broad cells interspersed with guard cells (stomata). Underneath the epidermal layer, 3–6 layers of mesophyll cells follow (Figure 7.1F). Two specialized adaxial cell layers are formed in the valves: the innermost endocarp *a* (en*a*) layer is composed of large cells that undergo cell death before the fruit has completely matured, whereas cells in the en*b* layer remain small and develop lignified cell walls (Figure 7.1F). It is believed that the rigidity of the lignified en*b* layer assists in the process of fruit opening by providing significant tension in the fruit (Spence *et al.*, 1996). The replum is connected to the septum and contains the main vascular bundles. The epidermis is composed of long and slender cells. Whereas the Arabidopsis outer replum is relatively broad, siliques from *Brassica* species have a comparatively smaller outer replum (Figures 7.1C, D, and G and 7.7). Implications of this difference are discussed below. At the valve margin, a narrow file of specialized dehiscence zone cells differentiates along the entire length of the fruit late in development, allowing the valves to detach from the replum, which causes seed dispersal to occur by a shattering mechanism.

The valve margin consists of two distinct cell layers: a layer where cell separation will take place, and a layer of lignified cells, whose rigidity appears to allow the valves to actively dislocate from the replum in a spring-like action (Spence *et al.*, 1996). Dehiscence zone formation in oilseed rape (*B. napus*) was characterized in detail by Meakin and Roberts (1990a). During the first three weeks after anthesis (WAA), the siliques elongate rapidly and attain full size before any significant seed filling takes place. During the period from three to seven WAA seeds are filled and the cells on either side of the dehiscence zone are reinforced by deposition of secondary cell walls followed by lignification. A cross section of a mature oilseed rape silique of five WAA is shown in Figure 7.1C. The dehiscence zone cells begin to show signs of organelle disintegration from approximately six WAA, but remain alive for another couple of weeks. During this period, the middle lamella and probably much of the primary wall are dissolved in the dehiscence zone leading to cell separation (Figure 7.1D). The dehiscence zone cells appear turgid and round up following middle lamella dissolution and hence cannot yet have undergone cell death at this stage (Figure 7.1I). Eventually, the carpel cells senesce, dehydrate and die. As the lignified cells dry out, tensions develop in the silique, but not sufficiently for the silique to open. An external mechanical force is required, for example, from wind or harvesting equipment.

Arabidopsis is closely related to oilseed rape and its silique is built similarly. While the dehiscence zone is 2–3 cells wide in oilseed rape, it is only one cell layer in Arabidopsis (cf Figure 7.1C and H). The replum, on the other hand, is relatively more developed in Arabidopsis; we shall return to this discussion later in the section dealing with the Arabidopsis *replumless* mutant. The most prominent difference is in timing as progression from anthesis to seed dispersal in Arabidopsis occurs within only two weeks (Smyth *et al.*, 1990).

7.3.2 *Fabaceae fruit morphology*

In contrast to the bilateral two-carpel fruit of *Brassica* species, the basis of the soybean fruit is a single carpel fused at its edges (Esau, 1977; Figure 7.2). The soybean pod wall consists of an exocarp composed of the outer epidermis and hypodermis, both with thickened cell walls, a mesocarp composed of parenchyma and an endocarp, which includes layers of schlerenchyma and the inner epidermis. The thickened cells of the hypodermis and the schlerenchymata of the endocarp are elongated, but in mutually perpendicular planes. As a result, the outer and inner layers of the pod wall shrink in different directions and the developing stresses promote the opening of the dehydrated mature fruit (Esau, 1977). Microscopy of the soybean pod opening zones has demonstrated the presence of a dehiscence zone beneath both the dorsal and ventral sutures (Tiwari and Bhatia, 1995; Christiansen *et al.*, 2002). The two dehiscence zones are functionally equivalent to those found in *Brassicaceae* species, but are not exact copies, as the dorsal dehiscence zone in soybean does not span the entire pod wall. In fully grown green pods, the ventral dehiscence zone extends from the endocarp through the mesocarp, where it passes

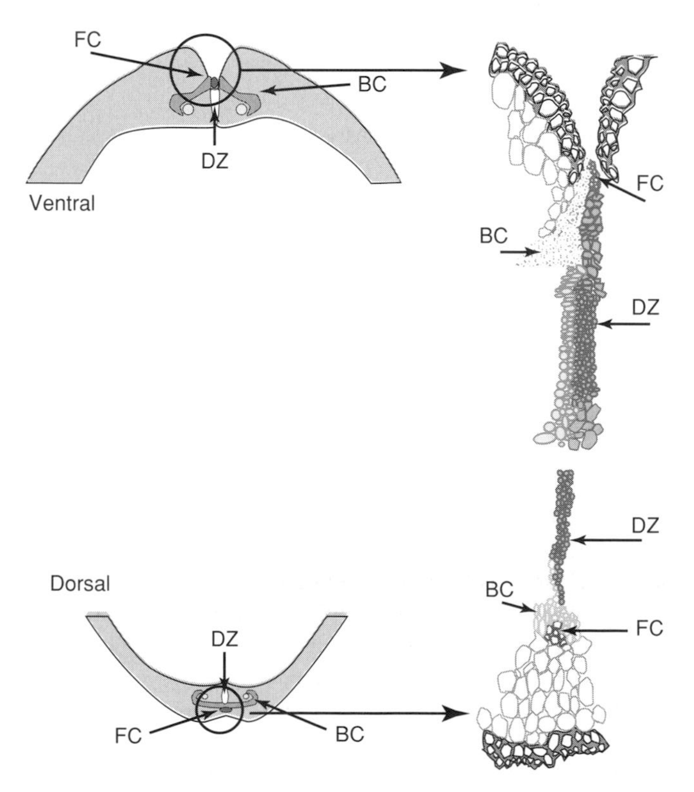

Figure 7.2 Schematic representation of a cross section of a soybean pod. The areas around the dorsal and ventral sutures are depicted and the dehiscence zones are enlarged. BC, bundle cap; DZ, dehiscence zone; FC, fiber cap cells.

through the bundle cap and connects to a bundle of fiber cap cells located at the surface of the suture. In contrast, the dorsal dehiscence zone does not extend through the mesocarp but terminates at the fiber cap cells located at the border between the bundle cap and the mesocarp. This difference in dehiscence zone structure agrees with the notion that pods have evolved from a single carpel where the margins have merged at the ventral suture, while the dorsal suture is a reminiscent of the leaf midrib. In mature green soybean pods, the dehiscence zone cell layer is two cells wide and characterized by having large vacuoles and thick primary walls separated by a thin middle lamella. At this stage, the dehiscence zone cells are easily recognized since the valve edge cells of the bundle cap have thick secondary walls and little vacuolation.

Dehiscence at both the dorsal and ventral dehiscence zones initiates from the interior of the pod and expands through the bundle cap to the lignified fiber cap cells, which probably require additional physical force to break. The mesocarp region located between the fiber cap cells and the outer epidermis at the dorsal suture is likely to need a physical force to break as it lacks a dehiscence zone. In contrast to the situation in oilseed rape, where the separated cells of the dehiscence zone retain their form surrounded by a thinned cell wall, the cells of the soybean dehiscence zone collapse, probably due to a weakening of their cells walls, and adhere to the thick-walled lignified cells of the bundle cap. Christiansen *et al.* (2002) suggested that a programmed cell death might be part of the dehiscence process in the soybean pod. Programmed cell death has also been found to operate in dehiscing anthers (Goldberg *et al.*, 1993; Wang *et al.*, 1999).

7.4 Cell wall dissolution and its hormonal regulation

Cellulase, systematically known as β-1,4-glucanase (EC 3.1.2.4), was probably the first polysaccharide hydrolase to be implicated in cell separation in the dehiscence zone (Meakin and Roberts, 1990b). The authors examined enzyme activity in the dehiscence zone region and in carpels during silique development and observed that a low-p*I* glucanase was expressed during silique elongation, while the expression of a high-p*I* glucanase correlated well with the onset of cell separation. The probable substrate of the glucanase activity is hemicellulosic polysaccharides. The primary cell wall microfibrils are normally not easy to distinguish under the electron microscope, but become clearly visible as cell separation progresses (Figure 7.3A and B). This suggests that not only the middle lamella and primary cell wall pectins but also the hemicellulosic polymers are dissolved. It may be speculated that the removal of the hemicellulosic polysaccharides significantly weakens the vascular bundles that pass through the dehiscence zone and confer much of the strength that remains in fully mature siliques. The involvement of β-1,4-glucanase activity has also been implicated in soybean pod dehiscence (Agrawal *et al.*, 2002; Christiansen *et al.*, 2002).

Meakin and Roberts (1990b) as well as Johnson-Flanagan and Spencer (1994) observed that β-1,4-glucanase activity increased immediately following a peak of

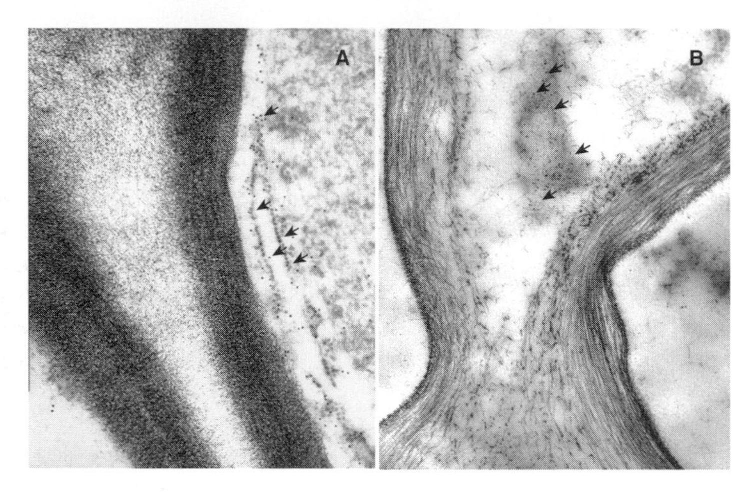

Figure 7.3 Electron micrograph of cross sections of the dehiscence zone of oilseed rape silique and immunogold labeling of RDPG1 (arrows). (A) At five WAA, RDPG1 is located intracellularly only and the cell walls are intact. (B) At seven WAA, RDPG1 is present in the cell walls between separated cells. *CPL*, cytoplasm; *CW*, cell wall; *ML*, middle lamella (adapted from Dal Degan *et al.*, 2001).

climacteric ethylene evolution in the silique. However, treating siliques with ethylene had only limited ability to promote dehiscence, and treatment with ethylene biosynthesis inhibitors delayed dehiscence only very marginally (Meakin and Roberts, 1990b; Child *et al.*, 1998). Furthermore, dehiscence and the rise in β-1,4-glucanase activity in parthenocarpic siliques were slightly delayed but generally similar to the situation in seeded siliques, although the former lacks a clear climacteric peak of ethylene production (Child *et al.*, 1998). Seeds are the main source of ethylene, and the synthesis of the ethylene precursor amino-cyclopropane carboxylic acid (ACC) is maintained throughout silique development. However, the ability to convert ACC into ethylene is lost as the seeds mature and start to desiccate (Rodriguez-Gacio and Matilla, 2001). These data seem to indicate that ethylene does not directly regulate the expression of the β-1,4-glucanase, nor dehiscence in general. Ethylene promotes senescence and dehydration and hence influences silique dehiscence only weakly and indirectly.

The role of ethylene is only slightly different in anther dehiscence. Rieu *et al.* (2003) used tobacco that was rendered ethylene insensitive either by introgression of the Arabidopsis *etr1-1* ethylene receptor mutant gene or by treatment with the ethylene-perception inhibitor methylcyclopropene. Anther development progressed in these plants as in the controls, but dehiscence – specifically degeneration of stomium cells and dehydration – was delayed compared to flower development in general. This led the authors to conclude that ethylene specifically regulates the opening of the stomium. They also note however, that anthers of ethylene-insensitive Arabidopsis mutants are not indehiscent unlike those of some jasmonic acid mutants. They suggest that the involvement of ethylene and jasmonic acid in anther dehiscence is overlapping and partly redundant (Rieu *et al.*, 2003).

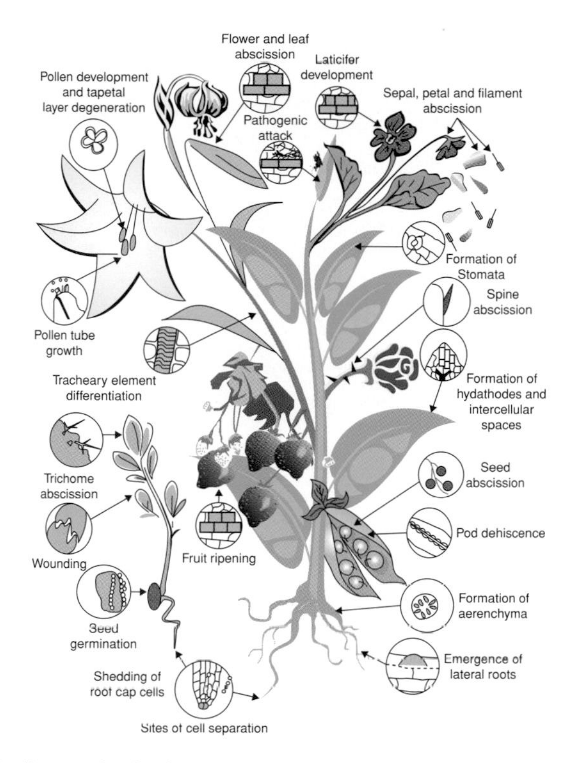

Plate 1 Sites of cell separation in plants.

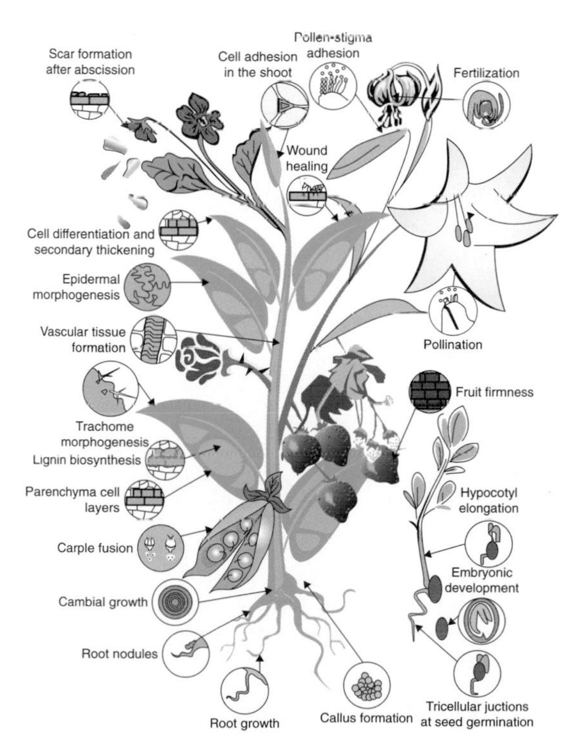

Plate 2 Sites of cell adhesion in plants.

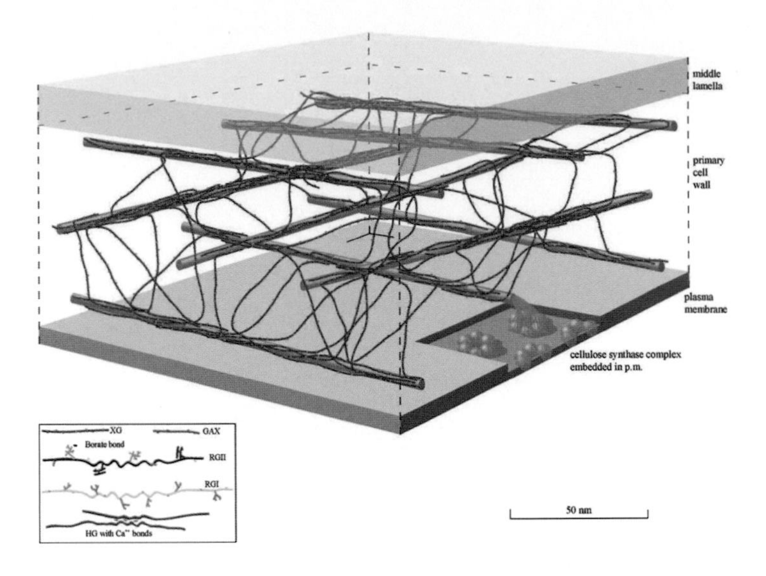

Plate 3A Scale model of the organization of polysaccharides in an Arabidopsis leaf cell. (Panel A was taken with permission from Somerville *et al.*, 2004.)

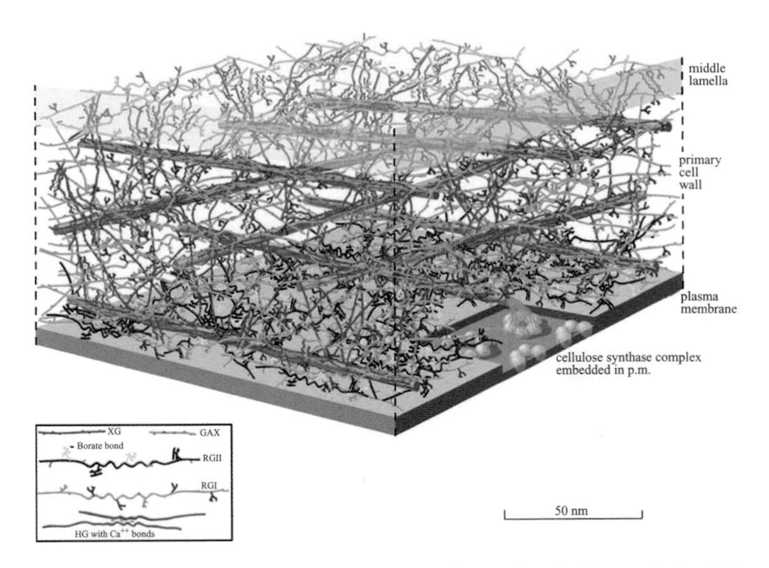

Plate 3B (Panel B was provided by Chris Somerville and reproduced with permission.) The amount of cellulose was reduced for clarity, and therefore, the length of the cross-links generated by xyloglucan and arabinoxylan are exaggerated.

Plate 4 Transdifferentiation processes in *Z. elegans* L. cell culture. The time course of transdifferentiation from mesophyll to TE cells is divided into three distinct stages. Following mechanical isolation, *Zinnia* leaf mesophyll cells undergo dedifferentiation at stage I. Cellular events at this stage involve the transient expression of wounding- and infection-inducible genes. The cells then enter stage II, where they approach the differentiation of vascular cells by expressing procambium-related genes. Finally, about 40% of cells produce secondary wall materials and hydrolytic enzymes to complete TE morphogenesis and reach stage III. Cultured *Zinnia* cells complete this course of events within about 96 h. The 4-day incubation of isolated *Zinnia* single mesophyll cells (A) with appropriate concentrations

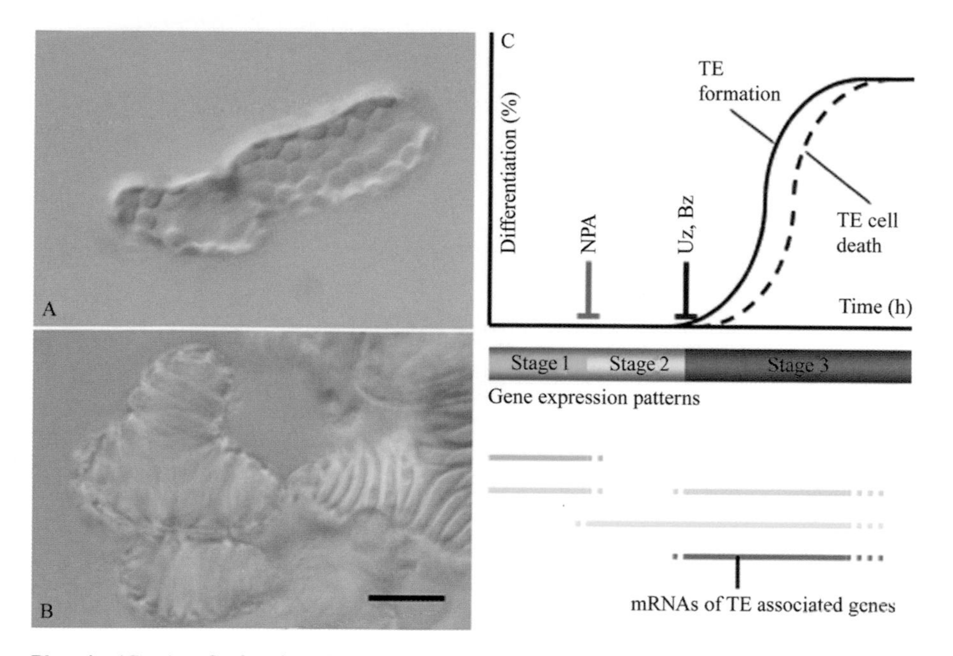

Plate 4 (*Continued*) of auxin and cytokinin results in their transdifferentiation into TEs (B). A maximum of about 40% of cells can become TEs. In this culture, the course of TE differentiation can be subdivided into three distinct stages based on characteristic changes in cellular morphology, biochemical properties, and gene expression patterns (C). When cell culture is initiated with single cell preparation, the mechanical breakage of the first leaves imposes a severe stress on mesophyll cells, leading to the activation of wounding responses. At stage I, the expression of several wound-inducible genes occurs. *ZePI1*, *ZePI2* and *ZePR1* genes, which are homologues of potato proteinase inhibitor (PI)1, PI2 and pathogenesis-related (PR) protein 1 gene, respectively, are expressed at the beginning and continue to be expressed for 24 h. After this time, there is a rapid decrease in the expression levels of these genes. In parallel, the rearrangement of cytoskeletons proceeds. Three-dimensional orientations of microtubules and actin filaments change from transverse to longitudinal relative to the cellular axis (Fukuda, 1992, 1996). These events result in the reorganization of organelle distribution patterns, which is most conspicuously represented by drastic changes in the cellular distribution of chloroplasts (Fukuda, 1992). Consequently, these initial mesophyll cells become dedifferentiated cells until the completion of stage I. The rearrangement of actin filaments, tubulin synthesis and repair-type DNA synthesis proceeds in cells at this stage. At stage II (from about 24 to 48 h after the start of TE-differentiation-inductive culture), the expression of vascular-cell-related genes becomes evident. Expression of these genes cannot be induced by application of single hormones; that is, neither auxin nor cytokinin alone can induce expression of these genes. These findings strongly suggest that some complicated internal cues that regulate gene expression arise as a result of the co-ordinated actions of two such plant hormones. Fukuda (2004) considers that this stage constrains plant cell pluripotency to a certain direction in vascular cells. Cells become xylem cell precursors at the end of stage II. Increases in housekeeping activities (such as RNA and protein synthesis), organelles, cytoskeletons, calmodulin-related molecules, cell volume and the frequency of cell division are also observed at this stage. At stage III (from about 48 h after the start of culture), some cultured cells start TE morphogenesis. TE-differentiation-induced cells (up to about 40% of all cells) construct characteristic secondary walls and undergo PCD. The synthesis of various hydrolytic enzymes, which may work for autolysis, occurs concomitantly with the production of enzymes putatively involved in secondary wall thickening. The disruption of the central vacuole then ensues. Vacuolar disruption immediately leads to TE cell death. Vacuole-localized hydrolytic enzymes in the cytoplasm degrade cellular contents, resulting in the formation of hollow tubules that form components of vessel and tracheal conduits. Inhibitors of polar auxin transport, *N*-(1-naphthyl)phthalamic acid (NPA); Yoshida *et al.*, 2005) and brassinosteroid synthesis, uniconazole (Uz; Yamamoto *et al.*, 1997) and brassinazole (Bz; Asami *et al.*, 2000; R. Yamamoto, unpublished) can block the progression of stages II and III, respectively. The bar indicates 20 μm.

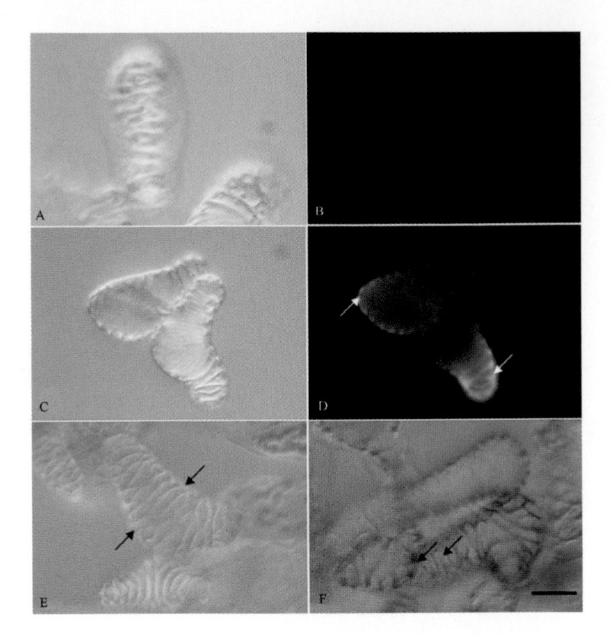

Plate 5 Staining of various cell wall materials in cultured *Zinnia* TEs. (A and B) Calcofluor white strongly stains TE secondary walls. (C and D) WGA-FITC staining occurs in TE secondary walls at specific regions (arrows). (E) Alcian blue staining was observed in the primary walls of *Zinnia* cultured cells including premature TEs (arrows). However, the extent of staining becomes weaker as TEs mature following the degradation of pectic materials. (F) Detection of secondary wall lignin by staining with phloroglucinol. The secondary walls of mature TEs exhibit heavy staining. The bar indicates 20 μm.

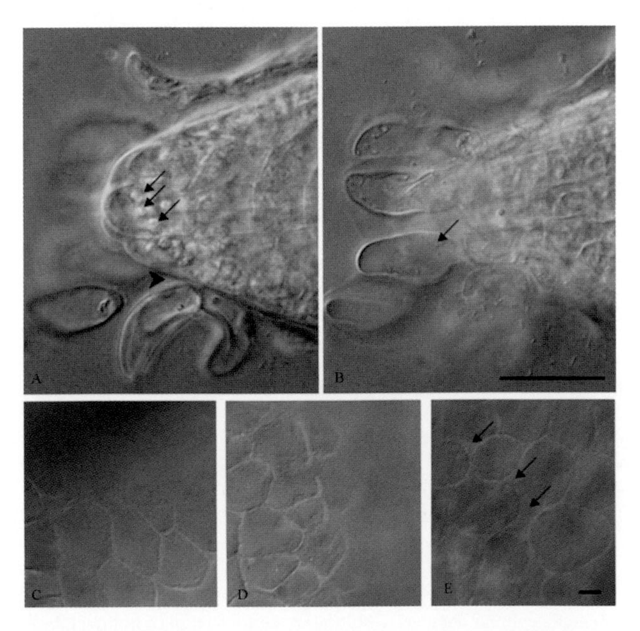

Plate 6 Cell separation in *Arabidopsis* root cap and aleurone cells. (A and B) Root cap cells separate from root at the tip of the root (arrowhead). These cells lose their characteristic amyloplasts and become vacuolated (arrows). The bar indicates 40 μm. (C–E) Aleurone cells separate from each other and become spherical. Intercellular space becomes apparent (arrows). The bar indicates 20 μm.

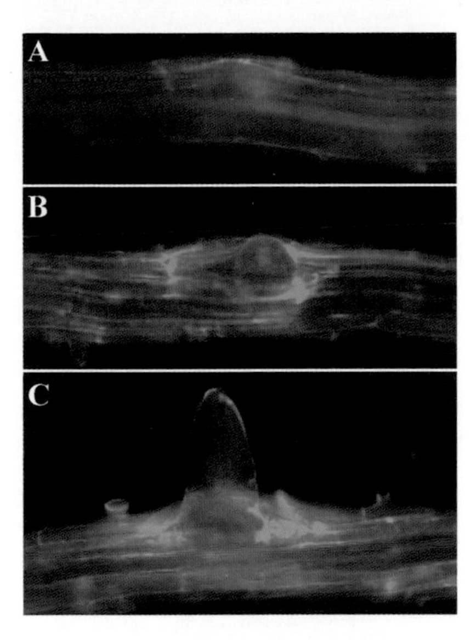

Plate 7 Localized changes in extracellular pH in correlation with lateral root emergence in *Arabidopsis*. Epidermal cells of the parent root show increasing levels of cell wall acidification in regions surrounding emerging lateral roots, as measured by fluorescein fluorescence: (A) prior to lateral root emergence (arrow); (B) shortly after emergence; and (C) as the lateral root develops. Fresh roots were incubated in 0.025 mg/ml fluorescein in 0.1 mM HEPES pH 7–7.2 for 15 min, washed three times in water for 20 min each, with inversion, followed by an overnight wash prior to viewing on a fluorescence microscope fitted with an FITC filter set. Photo credit: T.J. Kajstura.

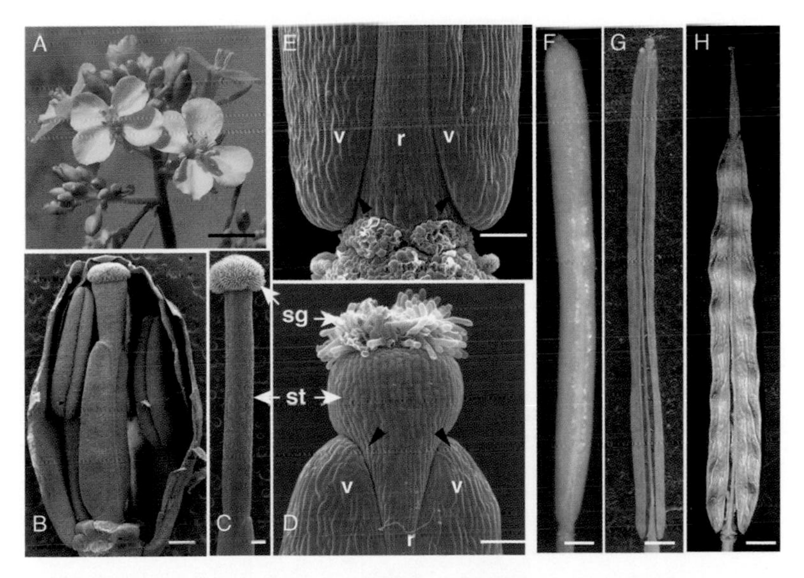

Plate 8 Morphology of flower and fruit development of *Brassica* and Arabidopsis. (A) *B. juncea* inflorescence. (B) SEM image of *B. juncea* flower shortly before fertilization. Two sepals and two petals have been manually removed. (C) SEM image of apical parts of *B. juncea* fruit two WAA. SEM image of basal (D) and apical (E) parts of stage 16 Arabidopsis fruit. (F) Fully elongated stage 17 Arabidopsis silique. (G) Dehiscing Arabidopsis fruit. (H) Dehiscing *B. juncea* fruit. v, valves; r, replum; st, style; and sg, stigma. Black arrows in (D) and (E) point to valve margin region. Scale bars: (A and H) 5 mm, (B, F, and G) 1 mm, (C) 250 μm and (D and E) 100 μm.

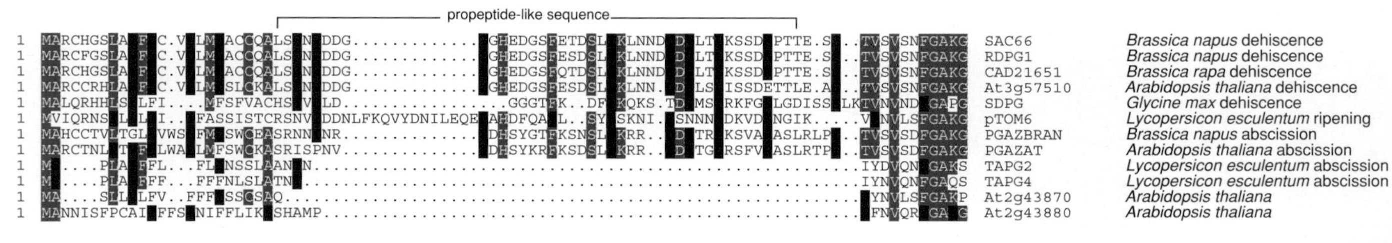

Plate 9 Alignment of endo-PGs from tomato, soybean, *Brassica* and Arabidopsis. Shown are sequences with and without a propeptide-like sequence between the signal peptide and the first conserved block of the mature enzyme (alignment using Muscle (Edgar, 2004) for OS X with default settings, and amino acid residue shading for chemical properties using TeXShade (Beitz, 2000)).

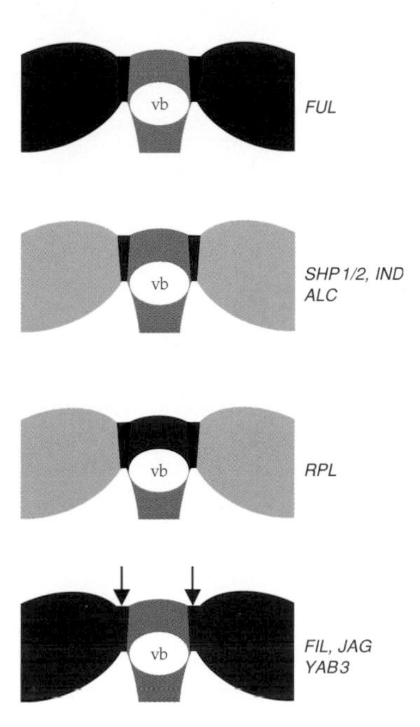

Plate 10 Expression patterns of genes involved in Arabidopsis fruit development. Schematic cross sections of the valve margin regions are shown. Areas in red correspond to the expression pattern of the gene noted to the right of each panel. Light grey indicates valves, dark grey symbolizes replum and septum whereas valve margins are in black. vb denotes vascular bundle.

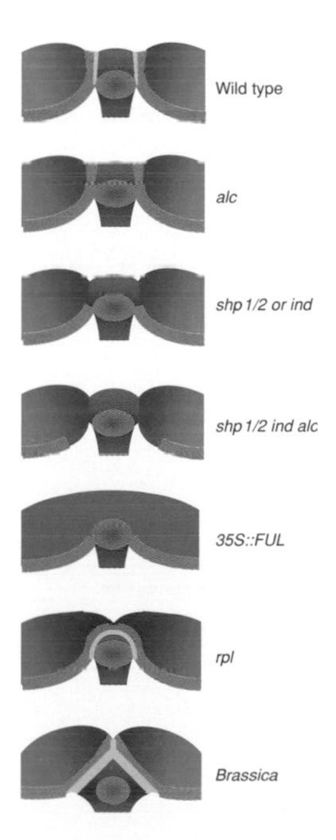

Plate 11 Phenotypic consequences of altered activity of Arabidopsis fruit development genes. Schematic cross sections of the valve margin region are shown for each. Areas in green correspond to valve tissue, orange indicates lignified tissue (lignified valve margin layer, inner valve layer and the vascular bundle of the replum), light blue represents separation layer whereas blue symbolizes replum and septum.

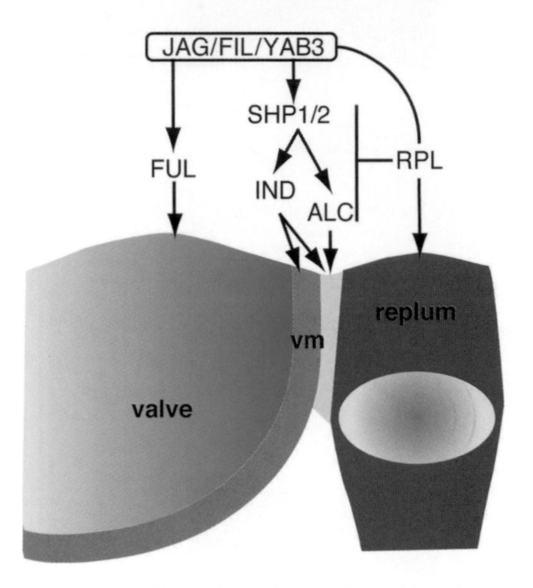

Plate 12 Genetic interactions controlling fruit development in Arabidopsis. Orange indicates lignified tissue (lignified valve margin layer, inner valve layer and the vascular bundle of the replum), light blue represents separation layer whereas blue symbolizes replum and septum.

Although the role of β-1,4-glucanase in silique dehiscence is well established and it has frequently been used as a biochemical marker of dehiscence, attempts to identify and clone the particular glucanase are yet to be made successful. Several of the 24 Arabidopsis glucanases located in the CAZy[1] family GH9 possess a high p*I* value. A survey of the expression pattern of these high-p*I* glucanases using the Genevestigator database (Zimmermann *et al.*, 2004) does not at present point unambiguously to a glucanase being expressed exclusively or predominantly in the Arabidopsis silique. This raises the possibility that either it is recruited into other cellular events or it takes only a small quantity of the transcript to make the enzyme that degrades the dehiscence zone cell walls. However, it cannot be excluded that the activity is encoded by a gene belonging to a family other than GH9. Of other hemicellulytic enzymes with a possible role in dehiscence, only the xyloglucan endotransglycosylases have been investigated.

Dissolution of the middle lamella in the dehiscence zone is the main event leading to cell separation. The middle lamella is believed to be made mostly of homogalacturonan, but it is not homogenous. Tension between cells is not uniform: it is higher near the corners formed where three cells meet, and it has been shown by Willats *et al.* (2001) that pectin composition is distinct in the 'corners of the corners' where tension is the highest (Iwai *et al.*, 2001). This has not been shown to apply to dehiscence zone cells, but it is a reasonable assumption, and hence it is likely that more than one pectin-degrading enzyme contributes to middle lamella dissolution. Only one of these has been identified so far – an endo-polygalacturonase (endo-PG, EC 3.2.1.15). The endo-PG was first cloned and characterized from oilseed rape by Jenkins *et al.* (1996) and Petersen *et al.* (1996), and named SAC66 and RDPG1, respectively. The two groups later identified the single Arabidopsis ortholog in the ecotype Landsberg *erecta* and Colombia, respectively (Jenkins *et al.*, 1999; Sander *et al.*, 2001).

Jenkins *et al.* (1999) set out to characterize the Arabidopsis promoter by transforming oilseed rape with a promoter-ß-glucuronidase (promoter-GUS) fusion, while Sander *et al.* (2001) transformed Arabidopsis with an oilseed rape promoter-GUS fusion. Both combinations direct expression in the dehiscence zone, not only of the siliques but also of the anthers. Expression was also observed in seed abscission zones and in the stigmatic tissue as pollen germinates. Furthermore, inducibility by mechanical stimulation throughout the plant was observed in both systems (unpublished observations). It is biologically reasonable that the same enzymes are involved in dehiscence of anthers and siliques and in abscission of seeds. The findings are also in accordance with the observation that seed shattering is stimulated by stress, but stress inducibility in tissues other than the dehiscence zone and seed abscission zones is puzzling. It is important to consider, of course, that the upstream sequences used in these experiments may not include all relevant regulatory elements.

The lack of silique dehiscence zone specificity of the endo-PG promoter has thus prevented the engineering of shatter resistance using some of the more straightforward approaches, e.g. silencing of the endo-PG or expression of a toxic gene product under the control of the endo-PG promoter in order to arrest development of cells in the dehiscence zone. Jenkins *et al.* (1999) used a ribonuclease as a toxic gene product and indeed observed that male sterility was part of the phenotype, but also that silique breakstrength was increased in hand-pollinated transformants.

Christiansen *et al.* (2002) cloned the soybean endo-PG ortholog and transformed Arabidopsis with a soybean promoter-GUS construct. Based on visual assessment, dehiscence zone specificity in Arabidopsis appeared to be less stringent than in the oilseed rape construct, but expression in the dehiscence zone was clear. Despite the differences at the anatomical level described above, the general conclusion was that silique and pod dehiscence both rely on the endo-PG for middle lamella dissolution and cell separation in the dehiscence zone. However, Agrawal *et al.* (2002) have reached a different conclusion that the endo-PG in soybean plays a role in fruit softening during pod maturation to relieve tension in the pod and thus prevent premature pod opening.

The classical model of auxin involvement in cell separation is well documented for leaf abscission. Morris (1993), for example, provided clear evidence, using cotton as a model, that an auxin source distal to an abscission zone inhibits ethylene-promoted abscission while an auxin source on the proximal side of the abscission zone stimulates abscission. Chauvaux *et al.* (1997) set out to test whether a similar mechanism would apply to dehiscence in oilseed rape. They found that endogenous auxin in the dehiscence zone declined steeply just prior to the rise in β-glucanase activity, and that spraying with an auxin analog suppressed β-glucanase accumulation and delayed dehiscence. The source of auxin in the dehiscence zone was shown to be the seeds, the only organs that can be regarded as distal to the dehiscence zone. Therefore, it is possible that a decline in auxin production by the seeds at maturity can induce cell separation at the dehiscence zone in a manner similar to that observed during leaf abscission. The textbook model may thus hold for dehiscence also, and potentially provides an elegant mechanism for the essential synchronization between seed development and fruit opening. A direct test of this model has been carried out by removing auxin from the dehiscence zone by conjugation. This was accomplished in transgenic Arabidopsis by expressing the bacterial *iaaL* gene under the control of a dehiscence zone-specific promoter. These plants displayed premature and explosive silique opening (Østergaard, unpublished).

Expression of the endo-PG does not appear to depend on the auxin profile as endo-PG synthesis occurs much earlier than the decline in endogenous auxin concentration. This raises the question as to how premature middle lamella dissolution is prevented. Dal Degan *et al.* (2001) showed that the oilseed rape endo-PG was stored initially in the cytoplasm within a membrane compartment until the onset of dehiscence when it was released into the cell wall (Figure 7.3A and B). They investigated the functional significance of the longer N-terminal sequence and showed that it is cleaved off during maturation of the endo-PG. Possible roles in correct folding of the enzyme, in protein stabilization or in modulation of activity were eliminated. They proposed that the N-terminal sequence was a propeptide-like structure possibly involved in storage and secretion of the enzyme. Treatment of the siliques with an auxin analog further delayed secretion (Dal Degan *et al.*, 2001). Plate 9 shows an alignment of endo-PGs from tomato, soybean, *Brassica* and Arabidopsis, comparing sequences with and without a propeptide-like sequence between the signal sequence and the first conserved block of the mature enzyme.

These observations raise the question whether it is likely that the endo-PG alone is subject to regulated secretion. The family GH9 glucanases comprises five enzymes with N-terminal extensions: three membrane-anchored KOR-β-glucanases and two enzymes, At1g19940 and At1g75680, predicted to be secreted. The latter two may feature propeptide-like sequences, but both are more strongly expressed in young than in old siliques (Roberts, personal communication).

The microheterogeneity of the middle lamella near the main load-bearing structures suggests the possible involvement of as yet unidentified pectin-degrading enzymes. The possible role, in silique or anther dehiscence, of pectate lyases, pectin-methyl or -acetyl esterases, or enzymes that degrade rhamnogalacturonan-I and its side chains, remains to be investigated.

7.5 Factors regulating Arabidopsis fruit development

The study of mutant plants with defects in normal development has been and continues to be essential for our understanding of plant biology. Subsequent cloning of the mutated genes responsible for the phenotypes has provided insights into the molecular processes guiding almost all aspects of plant growth and development (Østergaard and Yanofsky, 2004). Such mutant characterizations have, for example, been successful in identifying key regulators of fruit development in Arabidopsis, some of which are described here.

7.5.1 The FRUITFULL gene

The *FRUITFULL* (*FUL*) gene encodes a MADS box transcription factor and is expressed early during flower development in carpel primordial cells. At approximately floral stage 8 (floral stages defined by Smyth *et al.*, 1990), *FUL* expression becomes restricted to the carpel and excluded from the other tissues as they develop (Plate 10; Mandel and Yanofsky, 1995).

Loss-of-function mutants in the *FUL* gene result in failure of the fruit to elongate following fertilization. The replum of *ful* mutant fruits is enlarged and organized in a zigzag pattern. These fruits produce completely viable seeds that become crowded in the small siliques. Since dehiscence does not occur normally, the tight seed packing occasionally leads to fruit rupture (Gu *et al.*, 1998). Scanning electron microscopy (SEM) and histological staining of cells in *ful* fruits revealed that the short-fruit phenotype is in large part due to failure of the cells to elongate into the characteristic long valve cells. Moreover, whereas Arabidopsis valves normally consist of six cell layers as described above, *ful* mutant valves only have 2–3 layers. These valve cells are furthermore ectopically lignified, indicating the possibility of at least a partial adaptation of valve margin cell identity. Interestingly, transgenic plants expressing *FUL* under control of the cauliflower mosaic virus (CaMV) 35S promoter produced completely indehiscent fruit with valve cells surrounding the entire fruit, and therefore there was no formation of valve margin cells or replum (Plate 11; Ferrándiz *et al.*, 2000).

7.5.2 The SHATTERPROOF genes

During flower development, two closely related MADS box transcription factor-encoding genes become expressed in the developing carpel in cells that will later give rise to replum and valve margin cells. These genes have been dubbed *SHATTERPROOF1* (*SHP1*) and *SHP2* (Liljegren *et al.*, 2000). After fertilization, their expression pattern becomes restricted to valve margin cells (Plate 10; Savidge *et al.*, 1995). Mutations in either of these genes have no detectable phenotype on their own. However, the combined *shp1 shp2* double mutant produces indehiscent fruit in which valve margin cells fail to differentiate revealing a redundant function of these genes in valve margin development (Plate 11; Liljegren *et al.*, 2000). The *SHP* genes are also expressed in developing ovules (Savidge *et al.*, 1995), where a redundant function with another MADS box gene, *SEEDSTICK* (*STK*), in proper ovule development was revealed (Pinyopich *et al.*, 2003).

7.5.3 The INDEHISCENT gene

In an EMS mutant screen in Arabidopsis, Liljegren *et al.* (2004) identified an allelic series of mutants that were named *indehiscent* (*ind*) due to their failure to open the fruit. As for the *shp1 shp2* double mutant, the phenotypic defect of *ind* mutants is confined to the valve margin region. In strong alleles of *ind*, cells in both the lignified layer and the separation layer of the valve margin fail to differentiate (Plate 11). However, for weak alleles, lignification occurred in the apical part of the fruit (Liljegren *et al.*, 2004).

The mutations responsible for the indehiscent phenotype were identified by map-based cloning and found to be in a gene encoding a basic Helix-Loop-Helix (bHLH)-type transcription factor. An enhancer trap line, GT140, with the T-DNA inserted approximately 3kb upstream of the translational start site, conferred specific valve margin expression (Plate 10; Ferrándiz *et al.*, 2000; Liljegren *et al.*, 2004). This was also the case for transgenic lines harboring a construct having this 3kb region upstream of the GUS reporter gene (Østergaard, unpublished).

7.5.4 The ALCATRAZ gene

In a different genetic screen, Rajani and Sundaresan (2001) identified a genetrap transposon line producing indehiscent fruit. This mutant was named *alcatraz* (*alc*), and although *alc* mutant fruits produce lignified cells at the valve margin, separation layer cells fail to develop (Plate 11). The transposal insertion event in *alc* has occurred in a gene encoding another bHLH-type transcription factor. Expression of the *ALC* gene during fruit development as detected by GUS activity was observed broadly in the fruit including valves and valve margins at early stages, but later in development expression becomes restricted to the valve margin (Plate 10). An additional allele of *alc* was also identified in the EMS screen carried out by Liljegren *et al.* (2004) and shown to have a similar phenotype.

7.5.5 *The REPLUMLESS gene*

A key regulator of replum formation was discovered by Roeder *et al.* (2003). Since the replum is difficult to see without using a microscope, these authors took advantage of the enlarged replum in the *ful* mutant to do a suppressor screen of this phenotype and isolated a mutant line with no replum development and named this mutant *replumless* (*rpl*). After crossing out the *ful* mutation from the *rpl ful* double mutant, inspection of the replum region revealed a defect in replum development, with valves encroaching from both sides (Plate 11). The mutation was identified by map-based cloning, and expression of the affected gene was detected in the replum (Plate 10; Roeder *et al.*, 2003) as well as in other tissues of the plant. The *RPL* gene encodes a homeodomain box transcription factor of the BELL1 family. This gene has also been observed to have functions in meristem formation and phyllotaxis control under the names of *BELLRINGER* (Byrne *et al.*, 2003), *PENNYWISE* (Smith and Hake, 2003) and *VAAMANA* (Bhatt *et al.*, 2004).

One of the obvious differences between Arabidopsis fruits and those of *Brassica* species is the replum development. Although all the same cell types can be identified in siliques from these species, the outer replum of the *Brassica* silique is much reduced, with a morphology resembling fruits from weak mutant alleles of Arabidopsis *rpl* (Plate 11). This difference is especially interesting since a *Brassica* genomic sequence entry is present in the database encoding a protein with 85% sequence identity at the amino acid level (Østergaard, unpublished). Such an identity percentage suggests orthologous function of the two proteins, although a subfunctionalization between the species may have occurred. Several explanations for the reduced replum in *Brassica* are possible at this point, including insufficient expression of the homologous gene in the correct tissue, or a difference in affinity between the proteins toward promoters of downstream targets.

7.6 Genetic interactions during Arabidopsis fruit development

The different phenotypes of the Arabidopsis mutants and gain-of-function transgenics suggest that a regulatory genetic network underlies the developmental processes that lead to Arabidopsis fruit morphology. Studying combinations of these genotypes by crossing has indeed provided interesting information about how the key regulators interact to establish the sharp borders and to differentiate the distinct cell types that make up the mature fruit (reviewed in Dinneny and Yanofsky, 2005).

The epistatic relationship between *SHP* and *IND* in defining the valve margin was established when *IND* expression was found to be compromised in *shp1 shp2* mutant fruits (Liljegren *et al.*, 2000), and it could therefore be concluded that the *SHP* genes are positive regulators of *IND*. However, valve margin development in *shp1 shp2 ind* triple mutant fruit is even more affected than in the individual *shp1 shp2* and *ind* mutants (Liljegren *et al.*, 2004), indicating that the *SHP* genes and the *IND* gene have functions that are independent of each other.

The cells that develop in the valves of *ful* mutant fruit are small and lignified and therefore have characteristics shared with valve margin cells. In accordance with this, Ferrándiz *et al.* (2000) found that expression of the valve margin identity genes expand to include the valves of *ful* mutant fruit. Furthermore, valves of *ful shp1 shp2* triple mutants are partially rescued compared to *ful* valves, with the reappearance of stomatal cells interspersed with the valve cells (Ferrándiz *et al.*, 2000). The dramatic mutant phenotype of *ful* fruit is even further rescued in the *ind* mutant background (Liljegren *et al.*, 2004). Besides developing stomata, the *ful ind* double mutant partially rescues the short-fruit phenotype of *ful* fruit. These results therefore show that *FUL* specifies valve tissue at least partially by inhibiting the expression of valve margin identity genes in the valves.

Fruit length is progressively restored by crossing mutants in valve margin identity genes into the *ful* background such that in the quintuple *ful shp1 shp2 ind alc* mutant the fruit length is almost completely restored to that of the wild type; however, these fruit are still completely indehiscent. In contrast to a strictly linear relationship between the valve margin identity factors, these experiments therefore show that SHP1, SHP2 and ALC have activities that are independent of IND. In addition, the quintuple mutant also revealed a redundant function of these five genes in specifying the inner lignified en*b* layer of the valves, since this layer is absent from quintuple mutant fruit (Liljegren *et al.*, 2004).

Similar to the situation in valves, *SHP* and *IND* were found to be ectopically expressed in the reduced replum of *rpl* mutants and mutations in the *SHP* genes were able to rescue the *rpl* phenotype in *rpl shp1 shp2* mutant fruit (Roeder *et al.*, 2003). Therefore, *RPL* functions similar to *FUL* to inhibit valve margin gene expression in the replum, thereby restricting their expression to the narrow file of cells where the valve margin forms.

It has been reported that *FUL* expression is slightly inhibited in valve cells that are close to the valve margin of *shp1 shp2* and *ind* mutants and significantly inhibited in the *shp1 shp2 ind alc* quadruple mutant (Ferrándiz *et al.*, 2000; Liljegren *et al.*, 2004). This observation correlates with an inhibition of en*b* lignification close to the margins in the same quadruple mutant (Plate 11). Liljegren *et al.* (2004) suggest that the valve margin identity factors block a signal from the replum, which otherwise expands to the valve and inhibit valve specification.

The genetic network that has been unraveled so far regarding fruit development in Arabidopsis clearly shows how expression of the valve margin identity genes is restricted to the few cell rows where valve margin and dehiscence zone formation will take place. However, it still poses the very important question of what early events set up this pattern. Which factors act upstream of this pathway? For example, we know that *FUL* acts as a translational repressor to exclude the expression of valve margin identity genes in the valves, but it is unclear what excludes *FUL* expression from the other fruit tissues where it is initially expressed, so that it becomes restricted to the valve tissue.

Recent work by Dinneny *et al.* (2005) sheds light on parts of these questions. These authors identified redundant actions of the *JAGGED* (*JAG*), *FILAMENTOUS FLOWER* (*FIL*) and *YABBY3* (*YAB3*) genes in activating *FUL*, *SHP1/2* and *IND* expressions in the presumptive valve and valve margin tissues in unfertilized gynoecia. Moreover, the expression of these upstream activators was detected in valve and valve margin primordial cells (Plate 10), but it appeared to be negatively regulated in the replum by the action of RPL (Plate 12). In fact, the ubiquitous expression of *JAG* as in the *jag-D* mutant allele gives rise to a phenotype similar to *rpl* mutant fruits. *JAG*, *FIL* and *YAB3* were previously characterized as being important for the development of lateral organs (Bowman, 2000; Dinneny *et al.*, 2004), and whereas *FIL* and *YAB3* were shown to establish polarity in organ development (Bowman, 2000), *JAG* has previously been found to promote growth during organ development (Dinneny *et al.*, 2004). It is therefore likely that multiple developmental pathways are merging to activate *FUL* and *SHP* expression in the fruit as speculated by Dinneny *et al.* (2005).

The genetic interactions depicted in Plate 12 may or may not be direct. All these genes, which are known to be important for proper fruit development, encode transcription factors that are expected to regulate expression of their target genes through binding to regulatory elements in promoter regions. However, no direct binding to a promoter region of a specific gene has been shown. In fact, still very little is known about the downstream cascade of gene activity leading to cell separation at the dehiscence zone.

The oilseed rape cDNA clone SAC29 may be derived from a gene involved in such a signal transduction pathway. SAC29 represents a 0.6kb transcript, which is abundantly expressed in the *B. napus* pod dehiscence zone 40 days after anthesis (DAA). SAC29 encodes a putative protein of 136 amino acids, which shows homology to prokaryotic and eukaryotic response regulator proteins (Whitelaw *et al.*, 1999). Response regulator proteins are part of the bacterial two-component systems, which also consists of a sensor protein (Parkinson and Kofoid, 1992). The sensor protein contains an input domain and a histidine kinase domain while the response regulator protein typically consists of a receiver domain linked to an output domain. These proteins are modular and can exist in a variety of configurations. By comparison to bacterial two-component-like proteins, SAC29 encodes the receiver domain of the response regulator protein and possesses the conserved amino acid residues required for phosphorylation of the receiver domain. It has been shown that the two Arabidopsis receptor proteins ETR1 and ERS, which are related to bacterial two-component histidine kinases (Hall *et al.*, 2000), interact directly with a kinase in the ethylene response pathway (Clark *et al.*, 1998). By analogy, a potential role for SAC29 could be to initiate a MAP kinase cascade. This could lead to changes in the expression of genes involved in pod dehiscence, such as up-regulation of genes encoding cell wall hydrolytic enzymes like PG and glucanases and down-regulation of genes encoding proteins involved in cell wall biosynthesis. SAC29 is highly expressed in the dehiscence zone of *B. napus* pods in the days leading to shatter, and could itself be regulated by phytohormones such as ethylene or auxin both of which peak at around 40 DAA. Further work is required to elucidate the true function of SAC29 and may contribute to the study of signal transduction mechanisms in plants.

7.7 Anther dehiscence

In most flowering plants, the anther wall is composed of four cell layers: epidermis, endothecium, middle layer and tapetum. Prior to dehiscence, during the period between microspore release and pollen maturation, the tapetum degenerates and the endothecium enlarges and develops thick secondary walls. Anther dehiscence is a complex process, which includes the lytic opening of a longitudinal line of weakness in the epidermis, known as the stomium, and retraction of the anther wall to widen the stomium and permit pollen release. Rupture of the stomium, the functional equivalent to the fruit dehiscence zone, and retraction of the anther wall are caused by differential shrinkage of thickened and unthickened regions of the endothecium. Studies with transgenic tobacco plants have shown that a functional stomium is required for anther dehiscence, since targeted ablation of the stomium using a cytotoxic *barnase* gene driven by a cell-specific promoter generates anthers that do not dehisce (Beals and Goldberg, 1997). This indicates that dehiscence is not simply a mechanical process, but involves specific, timed, cellular events. In self-pollinating species, anther developmental processes have to occur in a timely controlled manner to ensure that the stigma is receptive for pollination when anther dehiscence occurs.

Screening of mutagenized Arabidopsis lines has identified a number of male-sterile mutants with defects in anther dehiscence. These mutants can be classified as either nondehiscing or late-dehiscing mutants. In the *nondehiscence1* mutant line, cells within the anther undergo a striking cell-death program late in anther development, in which the endothecium and connective degenerate completely resulting in a bilocular anther filled with pollen grains surrounded by a thin epidermis in which breakage of the stomium region does not occur (Sanders *et al.*, 1999). In the nondehiscent mutant *myb26*, a MYB transcription factor mutant (and allelic mutant of *ms35* (Dawson *et al.*, 1999)), the endothecium cells do not expand and they lack lignified cellulosic secondary wall thickenings (Steiner-Lange *et al.*, 2003). While breakage of the stomium takes place in this mutant, retraction of the anther wall does not occur, inhibiting pollen release. Thus, endothecium secondary wall strengthening is a prerequisite for anther dehiscence. The lack of this in the mutant suggests a role for AtMYB26 in the differentiation of this cell layer (Steiner-Lange *et al.*, 2003).

Several dehiscence mutants with a delayed phenotype have been isolated and characterized. In the *delayed dehiscence1* (*dde1*)/*opr3* mutant, anther development is similar to that of wild-type anthers but stomium breakage and pollen release occur later than in wild-type anthers at a time when the stigma is no longer receptive to pollen (Sanders *et al.*, 2000; Stintzi and Browse, 2000). The dehiscence program in delayed dehiscence2 (*dde2*) anthers is similar to that observed in *dde1* (Sanders *et al.*, 1999). Other delayed dehiscence mutants include *coi1* identified by root insensitivity to jasmonic acid (JA) (Feys *et al.*, 1994), the triple mutant fad3–2, fad7-2, fad8 (McConn and Browse, 1996) and *dad1* (Ishiguro *et al.*, 2001). The dehiscence mutants *dde1*, *dde2*, *dad1* and the *fad* triple mutant have been shown to be mutated in genes encoding enzymes in the JA biosynthesis pathway (Sanders *et al.*, 2000; Stintzi and Browse, 2000; Ishiguro *et al.*, 2001; Park *et al.*, 2002; Von Malek *et al.*, 2002), while the mutant *coi1* is defective in JA perception (Xie *et al.*, 1998; Deveto *et al.*, 2005). It is therefore likely that JA performs a signaling function in anther development, to control the proper timing of stomium breakdown and rupture.

Anther dehiscence is also influenced by other hormones such as gibberellin (GA) and ethylene. Overexpression of the *HvGAMYB* gene in barley, a GA-induced transcriptional regulator (Murray *et al.*, 2003), perhaps acting in cooperation with a micro-RNA (Achard *et al.*, 2004), leads to anthers that fail to dehisce. The correct timing of anther dehiscence in tobacco also appears to be controlled by ethylene (Rieu *et al.*, 2003). In tobacco plants rendered insensitive for ethylene reception, either by expression of the mutant *etr1-1* ethylene receptor allele from Arabidopsis, or chemically by treatment with 1-methylcyclopropane, pollen development occurred normally but degeneration of the stomium cells, as well as dehydration, was delayed. This led to the suggestion that ethylene acts directly on anthers to promote dehiscence as mentioned above.

Screening of T-DNA and EMS-mutagenized Arabidopsis lines has thus far not identified mutants with defects in the breakdown of the stomium, although enzymatic activities for cell wall degrading enzymes potentially involved in this process occurs in anthers (del Campillo and Lewis, 1992; Neelam and Sexton, 1995). That stomium breakage depends on cell wall hydrolytic enzymes is inferred from analysis of the endo-PG involved in pod dehiscence in Arabidopsis and oilseed rape. Transgenic plants expressing GUS under the control of the endo-PG promoter showed GUS staining in the anther stomium in addition to the pod dehiscence zone, while transgenic plants downregulated in endo-PG expression were male sterile (Jenkins *et al.*, 1999; Sander *et al.*, 2001).

7.8 Prospects for engineering shatter resistance

Engineering shatter resistance into *Brassica* species, especially oilseed rape, has been attempted since the early 1990s. In soybean and sesame, nontransgenic approaches not dealt with here are dominant. Two main approaches have been pursued in oilseed rape: interference with cell separation in the dehiscence zone was the original focus, while manipulating the differentiation of the dehiscence zone has taken center stage more recently. Table 7.1 reflects the current state of patent applications with respect to engineering shatter resistance. The newest and the oldest applications (listed first and last as #1 and #11) in the table both attempt to embrace the engineering of shatter resistance in quite general terms. The application (#11) by Isaac *et al.* (WO9423043) proposed to interfere with any conceivable aspect of dehiscence zone development and function using antisense constructs, gene products that interfere with hormonal control as well as cell ablation through the expression

Table 7.1

#	Title	Assignee	Inventors	Target	PCT	Issued patents	Reference
1	Reducing seed shattering in a plant, preferably a Brassicaceae plant by creating a population of transgenic lines of the plant, where the transgenic lines of the population exhibit variation in pod shatter resistance	Bayer Bioscience and Univ. California	Vancanneyt G.; Yanofsky M.; Kempin S.	ALC, IND, SHP1, SHP2 (At5g67110, At4g00120, At3g58780, AtAt2g42830)	WO2004113542		Rajani and Sundaresan, 2001; Liljegren *et al.*, 2000
2	Isolated nucleic acids for modulating fruit dehiscence in plants, comprises an indehiscent 1 (IND1) polynucleotide sequence	Univ. California	Liljegren S.; Yanofsky M.F.	Transcription factor IND1 (At4g00120)	WO200179517		Liljegren *et al.*, 2004
3	Modulating cell death, growth and stress resistance in eukaryotes, specifically plants, used, e.g to impart fungus or nematode resistance	Bayer Bioscience *et al.*	Babiychuk E.; Kushnir S.; De Block M.; Block M.D.	Programmed cell death by poly (ADP-ribose) polymerases (orthologous to At2g31320, At4g02390)	WO200004173	US6693185	De Block *et al.*, 2005
4	A nucleic acid encoding a signal transduction protein involved in plant dehiscence, useful for producing shatter resistant male-sterile plants	Biogemma UK Ltd *et al.*	Wyatt P.; Roberts J.A.; Whitelaw C.; Paul W.	Response regulator SAC29 (homologous to At3g04280)	WO9949046		Whitelaw *et al.*, 1999

5	Control of pod dehiscence or shatter	Biogemma UK Ltd and Nickerson Biochem Ltd	Paul W.; Roberts J.A.; Whitelaw C.	DZ15, putative electron transfer flavoprotein homologous to At5g43430	WO9915681		
6	New *Brassica napus* nucleic acid and protein useful for regulating pod dehiscence and/or abscission by producing transgenic plants and propagating material	Biogemma UK Ltd and Nickerson Biochem Ltd	Paul W.; Roberts J.A.; Whitelaw C.	ORS7(9), a xyloglucan endo-transglycosylase[a]	WO9915680		
7	Generation of male-sterile plants by controlling anther dehiscence	Biogemma UK Ltd and Nickerson Biochem Ltd	Roberts J.A; Paul W.; Craze M.	ADPG1 promoter sequences (At3g57510)	WO9913089		Jenkins *et al.*, 1999
8	Seed plants characterized by delayed seed dispersal	Univ. California	Yanofsky M.F.; Ferrándiz C.	AGL8-like gene product (At5g60910)	WO9900502	US6198024	Mandel and Yanofsky, 1995
9	Transgenic plant containing dehiscence zone selective chimeric gene has modified dehiscence properties, especially delayed pod dehiscence	Bayer Crop Science	Ulvskov P.; Child R.; Van Onckelen H.; Prinsen E.; Borkhardt B.; Sander L.; Petersen M.; Bundgaard Poulsen G.; Botternan J.	RDPG1 Endopolygalacturonase and promoter, orthologous to At3g57510	WO9713865	US6420628 US6797861	Petersen *et al.*, 1996; Sander *et al.*, 2001

(*Continued*)

Table 7.1 (*Continued*)

#	Title	Assignee	Inventors	Target	PCT	Issued patents	Reference
10	Control of seed pod dehiscence – using polygalacturonase or nucleic acid sequences derived from polygalacturonase gene	Biogemma UK Ltd and Nickerson Biochem Ltd	Roberts J.A.; Coupe S.A.; Jenkins E.S.	SAC66 oilseed rape polygalacturonase, orthologous to At3g57510	WO9630529	US6096946	Jenkins *et al.*, 1996
11	Recombinant and isolated nucleic acids – encode enzymes and proteins in plant abscission or dehiscence	Biogemma UK Ltd and Nickerson Biochem Ltd	Isaac P.G., Roberts J.A., Coupe S.A.	SAC51, secreted proline-rich protein[a]	WO9423043	US5907081	Coupe *et al.*, 1993

[a] Similar to more than one gene in Arabidopsis – no obvious ortholog can be suggested.

of toxic gene products. The application disclosed SAC51, a gene that is specifically expressed in the dehiscence zone and probably encodes a dehydration-induced extensin-like structural cell-wall protein (similar to At1g62510, which is expressed in both seeds and siliques and which features the Pfam protease inhibitor/seed storage/lipid transfer protein family protein domain, making functional assignment based on the sequence alone impossible).

The application WO2004113542 (#1) by Vancanneyt *et al.* tries to embrace interference of dehiscence zone formation using pin-loop-induced silencing of *any* gene with a role in silique opening. Their broadest claim thus appears to define the class of genes relevant to the invention by the definition of the invention itself: if silencing of any gene x leads to shatter resistance, then x is covered by the invention. The more restricted claims build in part on applications/patents #2 and #8 and focus on the transcription factors ALC, IND and SHP, and further emphasize the use of weak promoters or pin-loop constructs with some level of mismatch with the target gene. The weak promoter or target mismatch has been introduced following observations concerning threshability. When dehiscence zone formation is compromised altogether, a considerable amount of force has to be applied in the combine harvester to open the siliques. Such force will damage the seeds impairing protein and oil yields. The optimal product will therefore need a strongly adherent dehiscence zone rather than lacking it completely. Partial suppression of *IND* or *SHP* expression is likely to have the desired effect.

Promoters that direct expression specifically in the dehiscence zone either during dehiscence zone development or during cell separation are important tools for engineering shatter resistance, in particular, once it was realized that the endo-PG was also involved in anther dehiscence. The xylogucan endo-transglycosylase disclosed in #6 is another example of a cell wall modifying enzyme that is involved in dehiscence but not exclusively as it is also expressed in abscission zones. Jeremy Roberts' laboratory has therefore applied various differential techniques to the identification of promoters that direct expression specifically to the dehiscence zone. SAC25 (Coupe *et al.*, 1994), and the applications #11 and #5 exemplify early results of these efforts and the most recent outcome, respectively. No cell wall modifying enzyme solely committed to cell separation in dehiscence has been identified so far, and if identification of the missing ß-glucanase or other polysaccharide hydrolases expressed in the dehiscence zone does not change this, then it becomes relevant to consider systemic spread of gene silencing. Will silencing of a polysaccharide hydrolase in the silique dehiscence zone spread to, e.g., the stomium of younger flowers and hence render them male sterile, although the silencing construct was driven by a promoter exclusively expressed in the dehiscence zone? Systemic transmission of silencing signals is rather poorly understood presently. Only short-distance ($\sim$15 cells) transmission of silencing has been demonstrated for endogenous genes (Himber *et al.*, 2003), while systemic transmission is well documented in model systems where a transgene is targeted by the silencing construct see Mlotshwa *et al.* (2002) for a discussion. Proper address of this question and cloning of the ß-glucanase would strengthen the set of tools available for engineering shatter resistance by interfering with cell separation. If the use of transcription factor silencing constructs with mismatches or driven by weak promoters mentioned above yields traits that are stable under field conditions, then this approach is also viable. Patents/applications #3 and #4 regard signaling leading to, e.g., programmed cell death (#3), and may define an additional class of technologies for engineering shatter resistance. In conclusion, it appears plausible that both Bayer Crop Science and the Biogemma/Nickerson Seed/Group Limagrain consortium are in a position to make and market shatter-resistant oilseed varieties and will be able to do so without infringing the other party's patent portfolio.

7.9 Development of model-to-crop approaches

Progress in the genetics of fruit development in Arabidopsis has been and will continue to be facilitated by the generation of functional genomics tools, which have been developed over the past 5–10 years. These include sequencing and annotation of the whole Arabidopsis genome and establishment of numerous near-saturation mutant collections and publicly available expression data from genome-wide microarray experiments.

It will be important in the nearest future to develop strategies to exploit the knowledge that we have acquired so far in Arabidopsis in agronomical efforts for crop improvement. *Brassica* species including oilseed rape are closely related to Arabidopsis and represent ideal candidates for model-to-crop approaches since they include important crop plants. The similar morphology of fruit from these two species suggests that strategies to modify pod shatter in Arabidopsis may be readily transferable into various *Brassica* species. For example, we know from a recent work that ectopic expression of the Arabidopsis *FUL* gene under control of the CaMV 35S promoter in *B. juncea* results in pod shatter-resistant *Brassica* fruits that lack both the lignified valve margin layer as well as separation layer cells. Moreover, the *B. juncea SHP* genes were inactivated in these transgenic lines similar to what was observed in Arabidopsis *35S::FUL* plants (Ferrándiz *et al.*, 2000; Østergaard *et al.*, 2006). This shows that at least part of the genetic pathway leading to valve margin specification is conserved between Arabidopsis and *Brassicas*.

Although much less genomics tools are presently available for *Brassica* research compared to Arabidopsis, the international effort to sequence the *B. rapa* genome will be extremely valuable for future attempts to control developmental processes in oilseed rape. As more sequence information becomes available, we will be able to isolate orthologs of the fruit development genes in *Brassica* and determine their chromosomal localization. Besides providing strategies for transgenic approaches to reduce pod shatter in oilseed rape, this information will be essential for marker-assisted breeding combined with a candidate gene approach.

Acknowledgments

Robin Child is acknowledged for additional light- and electron micrographs and Dr. Peter M. Ray for reading and critically commenting on the manuscript. B.B. and P.U. are supported by The Danish

National Research Foundation and the EU-project PectiCoat. L.Ø. is supported by the Biotechnology and Biological Sciences Research Council, United Kingdom.

References

Achard, P., Herr, A., Baulcombe, D.C. and Harberd, N.P. (2004) Modulation of floral development by a gibberellin-regulated microRNA. *Development* **131**, 3357–3365.

Agrawal, A.P., Basarkar, P.W., Salimath, P.M. and Patil, S.A. (2002) Role of cell wall-degrading enzymes in pod-shattering process of soybean. *Glycine max* (L.) Merrill. *Current Science* **82** (1), 58–61.

Beals, T.P. and Goldberg, R.B. (1997) A novel cell ablation strategy blocks tobacco anther dehiscence. *Plant Cell* **9**, 1527–1545.

Beitz, E. (2000) TeXshade: shading and labeling of multiple sequence alignments using LaTeX2e. *Bioinformatics* **16**, 135–139.

Bhatt, A.M., Etchells, J.P., Canales, C., Lagodienko, A. and Dickinson, H. (2004) VAAMANA – a BEL1-like homeodomain protein, interacts with KNOX proteins BP and STM and regulates inflorescence stem growth in Arabidopsis. *General* **328**, 103–111.

Bowman, J.L. (2000) The YABBY gene family and abaxial cell fate. *Current Opinion in Plant Biology* **3**, 17–22.

Byrne, M.E., Groover, A.T., Fontana, J.R. and Martienssen, R. (2003) Phyllotactic pattern and stem cell fate are determined by the Arabidopsis homeobox gene *BELLRINGER*. *Development* **130**, 3941–3950.

Campillo, E. del and Lewis, L.N. (1992) Occurrence of 9.5 cellulase and other hydrolases in flower reproductive organs undergoing major cell wall disruption. *Plant Physiology* **99**, 1015–1020.

Chandler, J. Corbesier, L. Spielmann, P. Dettendorfer, J. Stahl, D. Apel, K. and Melzer, S. (2005) Modulating flowering time and preventing pod shatter in oilseed rape. *Molecular Breeding* **15**, 87–94.

Chauvaux, N., Child, R., John, K., Ulvskov, P., Borkhardt, B., Prinsen, E., van Onckelen, H.A. (1997) The role of auxin in cell separation in the dehiscence zone of oilseed rape pods. *Journal of Experimental Botany* **48** (312), 1423–1429.

Child, R.D., Chauvaux, N., John, K., Ulvskov, P. and Van Onckelen, H.A. (1998) Ethylene biosynthesis in oilseed rape pods in relation to pod shatter. *Journal of Experimental Botany* **49**, 829–838.

Child, R.D., Summers, J.E., Babij, J., Farrent, J.W. and Bruce, D.M. (2003) Increased resistance to pod shatter is associated with changes in the vascular structure in pods of a resynthesize *Brassica napus* line. *Journal of Experimental Botany* **54** (389), 1919–1930.

Christiansen, L.C., Dal Degan, F., Ulvskov, P. and Borkhardt, B. (2002) Examination of the dehiscence zone in soybean pods and isolation of a dehiscence-related endopolygalacturonase gene. *Plant Cell and Environment* **25**, 479–490.

Clark, K.L., Larsen, P.B., Wang, X. and Chang, C. (1998) Association of the Arabidopsis CTR1 Raf-like kinase with the ETR1 and ERS ethylene receptors. *Proceedings of the National Academy of Sciences of the United States of America* **95** (9), 5401–5406.

Coupe, S.A., Taylor, J.E., Isaac, P.G. and Roberts, J.A. (1993) Identification and characterization of a proline-rich mRNA that accumulates during pod development in oilseed rape (*Brassica napus* L.). *Plant Molecular Biology* **23**, 1223–1232.

Coupe, S.A., Taylor, J.E., Isaac, P.G. and Roberts, J.A. (1994). Characterization of a mRNA that accumulates during development of oilseed rape pods. *Plant Molecular Biology* **24**, 223–227.

Dal Degan, F., Child, R., Svendsen, I. and Ulvskov, P. (2001) The cleavable N-terminal domain of plant endo-polygalacturonases from clade B may be involved in a regulated secretion mechanism. *Journal of Biological Chemistry* **176**, 35297–35304.

Dawson, J., Sözen, E., Vizir, I., Waeyenberge, S.V., Wilson, Z.A. and Mulligan, B.J. (1999) Characterization and genetic mapping of a mutation (ms35) which prevents anther dehiscence in

Arabidopsis thaliana by affecting secondary wall thickening in the endothecium. *New Phytologist* **144**, 213–222.

De Block, M., Verduyn, C., De Brouwer, D. and Cornelissen, M. (2005) Poly(ADP-ribose) polymerase in plants affects energy homeostasis, cell death and stress tolerance. *Plant Journal* **41** (1), 95–106.

Devoto, A., Ellis, C., Magusin, A., Chang, H.S., Chilcott, C., Zhu, T. and Turner, J.G. (2005) Expression profiling reveals COI1 to be a key regulator of genes involved in wound- and methyl jasmonate-induced secondary metabolism, defence, and hormone interactions. *Plant Molecular Biology* **58**, 497–513.

Dinneny, J.R., Yadegari, R., Fisher, R.L., Yanofsky, M.F. and Weigel, D. (2004) The role of *JAGGED* in shaping lateral organs. *Development* **131**, 1101–1110.

Dinneny, J.R., Weigel, D. and Yanofsky, M.F. (2005) A genetic framework for fruit patterning in *Arabidopsis thaliana. Development* **132**, 4687–4696.

Dinneny, J.R. and Yanofsky, M.F. (2005) Drawing lines and borders: how the dehiscent fruit of Arabidopsis is patterned. *BIOESSAYS* **27** (1), 42–49.

Edgar, R.C. (2004) MUSCLE: a multiple sequence alignment method with reduced time and space complexity. *BMC Bioinformatics* **5** (1), 113.

Esau, K. (1977) *Anatomy of Seed Plants.* John Wiley, New York.

Ferrándiz, C., Liljegren, S.J. and Yanofsky, M.F. (2000) Negative regulation of the *SHATTERPROOF* genes by FRUITFULL during Arabidopsis fruit development. *Science* **289**, 436–438.

Ferrándiz, C., Pelaz, S. and Yanofsky, M.F. (1999) Control of carpel and fruit development in Arabidopsis. *Annual. Review. Biochemistry* **68**, 321–354.

Feys, B.J., Benedetti, C.E., Penfold, C.N. and Turner, J.G. (1994) Arabidopsis mutants selected for resistance to the phytotoxin coronatine are male sterile, insensitive to methyl jasmonate, and resistant to a bacterial pathogen. *Plant Cell* **6**, 751–759.

Goldberg, R.B., Beals, T.P. and Sanders, P.M. (1993) Anther development: basic principles and practical applications. *Plant Cell* **5**, 1217–1229.

Gu, Q., Ferrándiz, C., Yanofsky, M.F. and Martienssen, R. (1998) The *FRUITFULL* MADS-box gene mediates cell differentiation during Arabidopsis fruit development. *Development* **125**, 1509–1517.

Hall, A.E., Findell, J.L., Schaller, G.E., Sisler, E.C. and Bleecker, A.B. (2000) Ethylene perception by the ERS1 protein in Arabidopsis. *Plant Physiology* **123**, 1449–1458.

Himber, C., Dunover, P., Moissiad, G., Ritzenthaler C. and Voinnet, O. (2003) Transitivity-dependent and–independent cell-to-cell movement of RNA silencing. *EMBO Journal* **22** (17), 4523–4533.

Ishiguro, S., Kawai-Oda, A., Ueda, J., Nishida, I. and Okada, K. (2001) The DEFECTIVE IN ANTHER DEHISCHENCE1 gene encodes a novel phospholipase A1 catalyzing the initial step of jasmonic acid biosynthesis, which synchronizes pollen maturation, anther dehiscence, and flow opening in Arabidopsis. *Plant Cell* **13**, 2191–2209.

Iwai, H., Ishii, T. and Satoh, S. (2001) Absence of arabinan in the side chains of the pectic polysaccharides strongly associated with cell walls of Nicotiana plumbaginifolia non-organogenic callus with loosely attached constituent cells. *Planta* **213**, 907–915.

Jarvis, M.C., Briggs, S.P.H. and Knox, J.P. (2003) Intercellular adhesion and cell separation in plants. *Plant, Cell and Environment* **26**, 977–989.

Jenkins, E.S., Paul, W., Coupe, S.A., Bell, S.J., Davies, E.C. and Roberts, J.A. (1996) Characterization of an mRNA encoding a polygalacturonase expressed during pod development in oilseed rape (*Brassica napus* L.). *Journal of Experimental Botany* **47**, 111–115.

Jenkins, E.S., Paul, W., Craze, M., Whitelaw, C.A., Weigand, A. and Roberts, J.A. (1999) Dehiscence-related expression of an *Arabidopsis thaliana* gene encoding a polygalacturonase in transgenic plants of *Brassica napus. Plant, Cell and Environment* **22**, 159–167.

Johnson-Flanagan, A.M. and Spencer, M.S. (1994) Ethylene production during development of mustard (*Brassica juncea*) and canola (*Brassica napus*) seed. *Plant Physiology* **106**, 601–606.

Liljegren, S.J., Ditta, G.S., Eshed, Y., Savidge, B., Bowman, J.L. and Yanofsky, M.F. (2000) *SHATTERPROOF* MADS-box genes control seed dispersal in Arabidopsis. *Nature* **13**, 766–770.

Liljegren, S.J., Roeder, A.H., Kempin, S.A., Gremski, K., Østergaard, L., Guimil, S., Reyes, D.K. and Yanofsky, M.F. (2004) Control of fruit patterning in Arabidopsis by *INDEHISCENT*. *Cell* **16**, 843–853.

Kondra, Z.P. and Steffansson, B.R. (1970) Inheritance of the major glucosinolates of rapeseed (*Brassica napus*) meal. *Canadian Journal of Plant Science* **50**, 643–647.

Lord, E.M. (2003) Adhesion and guidance in compatible pollination. *Journal of Experimental Botany* **54**, 47–54.

Mandel, M.A. and Yanofsky, M.F. (1995) The Arabidopsis *AGL8* MADS box gene is expressed in inflorescence meristems and is negatively regulated by APETALA1. *Plant Cell* **7**, 1763–1771.

McConn, M. and Browse, J. (1996) The critical requirement for linolenic acid is pollen development, not photosynthesis, in an Arabidopsis mutant. *Plant Cell* **8**, 403–416.

Meakin, P.J. and Roberts, J.A. (1990a) Dehiscence of fruit in oilseed rape (*Brassica napus* L.). *Journal of Experimental Botany* **41**, 995–1002.

Meakin, P.J. and Roberts, J.A. (1990b) Dehiscence of fruit in oilseed rape. The role of cell wall degrading enzymes. *Journal of Experimental Botany* **41**, 1003–1011.

Mlotshwa, S., Voinnet, O., Mette, M.F., Matzke, M., Vaucheret, H., Ding, S.W., Pruss, G. and Vance, V.B. (2002) RNA silencing and the mobile silencing signal. *Plant Cell* **14** (Suppl. S), S289–S301.

Morgan, C.L., Bavage, A., Bancroft, I., Bruce, D., Child, R., Chinoy, C., Summers, J. and Arthur, E. (2003) Using novel variation in *Brassica* species to reduce agricultural inputs and improve agronomy of oilseed rape – a case study in pod shatter resistance. *Plant Genetic Resources* **1** (1) 59–65.

Morgan, C.L., Bruce, D.M., Child, R., Ladbrooke, Z.L. and Arthur, A.E. (1998) Genetic variation for pod shatter resistance among lines of oilseed rape developed from synthetic *B. napus*. *Field Crop Research* **58**, 153–165.

Morris, D.A. (1993) The role of auxin in the apical regulation of leaf abscission in cotton (Gossypium hirsutum L). *Journal of Experimental Botany* **44** (261), 807–814.

Murray, F., Kalla, R., Jacobsen, J. and Gubler, F. (2003) A role for HvGMYB in anther development. *Plant Journal* **33**, 481–491.

Neelam, A. and Sexton, R. (1995) Cellulase (endo b-1,4 glucanase) and cell wall breakdown during anther development in sweet pea (Lathyrus odoratus L.): isolation and characterization of partial cDNA clones. *Journal of Plant Physiologyl* **146**, 622–628.

Østergaard, L. and Yanofsky, M.F. (2004) Establishing gene function by mutagenesis in *Arabidopsis thaliana*. *Plant Journal* **39**, 682–696.

Østergaard, L., Kempin, S.A., Bies, D., Klee, H.J. and Yanofsky, M.F. (2006) Pod shatter-resistant fruit produced by ectopic expression of the *FRUITFULL* gene. *Plant Biotechnology Journal* **4**, 45–51.

Park, J.-H., Halitschike, R., Kim, H.B., Baldwin, I.T., Feldmann, K.A. and Feyereisen, R. (2002) A knock-out mutation in allele oxidase synthase results in male sterility and defective wound signal transduction in Arabidopsis due to a block in jasmonic acid biosynthesis. *Plant Journal* **31**, 1–12.

Parkinson, J.S. and Kofoid, E.C. (1992) Communication modules in bacterial signaling proteins. *Annual Review of Genetics* **26**, 71–112.

Petersen, M., Sander, L., Child, R., Van Onckelen, H., Ulvskov, P. and Borkhardt, B. (1996) Isolation and characterization of a pod dehiscence zone-specific polygalacturonase from *Brassica napus*. *Plant Molecular Biology* **31**, 517–527.

Pinyopich, A., Ditta, G.S., Savidge, B., Liljegren, S.J., Baumann, E., Wisman, E. and Yanofsky, M.F. (2003) Assessing the redundancy of MADS-box genes during carpel and ovule development. *Nature* **424**, 85–88.

Rajani, S. and Sundaresan, V. (2001) The Arabidopsis myc/bHLH gene *ALCATRAZ* enables cell separation in fruit dehiscence. *Current Biology* **11** (24), 1914–1922.

Rieu, I., Wolters-Arts, M., Derksen, J., Mariani, C. and Weterings, K. (2003) Ethylene regulates the timing of anther dehiscence in tobacco. *Planta* **217** (1), 131–137.

Roberts, J.A., Elliot, K.A. and Gonzalez-Carranza, Z.H. (2002) Abscission, dehiscence and other cell separation processes. *Annual Review of Plant Biology* **53**, 131–158.

Rodriguez-Gacio, M.D. and Matilla, A.J. (2001) The last step of the ethylene biosynthesis in turnip tops (*Brassica rapa*) seeds: alterations related to development and germination and its inhibition during desiccation. *Physiologia Plantarum* **112** (2), 273–279.

Roeder, A.H., Ferrándiz, C. and Yanofsky, M.F. (2003) The role of the REPLUMLESS homeodomain protein in patterning the Arabidopsis fruit. *Current Biology* **13**, 1630–1635.

Sander, L., Child, R., Ulvskov, P., Albrechtsen, M. and Borkhardt, B. (2001) Analysis of a dehiscence zone endo-polygalacturonase in oilseed rape (*Brassica napus*) and *Arabidopsis thaliana*: evidence for roles in cell separation in dehiscence and abscission zones, and in stylar tissues during pollen tube growth. *Plant Molecular Biology* **46** (4), 469–479.

Sanders, P.M., Bui, A.Q., Weterings, K., McIntire, K.N., Hsu, Y.C., Lee, P.Y., Truong, M.T., Beals, T.P. and Goldberg, R.B. (1999) Anther developmental defects in *Arabidopsis thaliana* male-sterile mutants. *Sexual Plant Reproduction* **11**, 297–322.

Sanders, P.M., Lee, P.Y., Biegsen, C., Boone, J.D., Beals, T.P., Weiler, E.W. and Goldberg, R.B. (2000) The Arabidopsis DELAYED DEHISCENCE1 gene encodes an enzyme in the jasmonic acid synthesis pathway. *Plant Cell* **12**, 1041–1062.

Savidge, B., Rounsley, S. and Yanofsky, M.F. (1995) Temporal relationship between the transcription of two Arabidopsis MADS box genes and the floral identity genes. *Plant Cell* **7**, 721–733.

Seyis, F., Friedt, W., Voss, A. and Lühs, W. (2004) Identification of individual *Brassica oleracea* plants with low erucic acid content. *Asian Journal of Plant Sciences* **3** (5), 593–596.

Smith, H.M. and Hake, S. (2003) The interaction of two homeobox genes, *BREVIPEDICELLUS* and *PENNYWISE*, regulates internode patterning in the Arabidopsis inflorescence, *Plant Cell* **15**, 1717–1727.

Smyth, D.R., Bowman, J.L. and Meyerowitz, E.M. (1990) Early flower development in Arabidopsis. *Plant Cell* **2**, 755–767.

Spence, J., Vercher, Y., Gates, P. and Harris, N. (1996) 'Pod shatter' in *Arabidopsis thaliana*, *Brassica napus* and *B. juncea*. *Journal of Microscopy* **181**, 195–203.

Steiner-Lange, S., Unte, U.S., Eckstein, L., Yang, C., Wilson, Z.A., Schmeizer, E., Dekker, K. and Saedler, H. (2003) Disruption of *Arabidopsis thaliana* MYB26 results in male sterility due to nondehiscent anthers. *Plant Journal* **34**, 519–528.

Stintzi, A. and Browse, J. (2000) The Arabidopsis male sterile mutant, opr3, lacks the 12-oxophytodienoic acid reductase required for jasmonate biosynthesis. *Proceedings of the National Academy of Sciences of the United States of America* **97**, 10625–10630.

Tiwari, S. and Bhatia, V.S. (1995) Characters of pod anatomy associated with resistance to pod-shattering in soybean. *Annals of Botany* **76**, 483–485.

Tiwari, S. and Bhatnagar, P. (1991) Pod shattering as related to other agronomic attributes in soybean. *Tropical Agriculture (Trinidad)* **68**, 102–103.

Wang, M., Hoeckstra, S., van Bergen, S., Lamers, G.E.M., Oppedijk, B.J., van der Heijden, M.W., de Priester, W. and Schilperoort, R.A. (1999) Apoptosis in developing anthers and the role of ABA in this process during androgenesis in *Hordeum vulgare* L. *Plant Molecular Biology* **39**, 489–501.

Whitelaw, C.A., Paul, W., Jenkins, E.S., Taylor, V.M., Roberts, J.A. (1999) An mRNA encoding a response regulator protein from *Brassica napus* is up-regulated during pod development. *Journal of Experimental Botany* **50** (332), 335–341.

Willats, W.G.T., Orfila, C., Limberg, G., Buchholt, H.C., van Alebeek, G.W.M., Voragen, A.G.J., Marcus, S.E., Christensen, T., Mikkelsen, J.D., Murray, B.S. and Knox, J.P. (2001) Modulation of the degree and pattern of methyl-esterification of pectic homogalacturonan in plant cell walls. Implications for pectin methyl esterase action, matrix properties, and cell adhesion. *The Journal of Biological Chemistry* **276** (22), 19404–19413.

Von Malek, B., van der Graaff, E., Schneitz, K. and Keller, B. (2002) The Arabidopsis male-sterile mutant dde2–2 is defective in the ALENE OXIDASE SYNTHASE gene encoding one of the key enzymes of the jasmonic acid biosynthesis pathway. *Planta* **216**, 187–192.

Xie, D.X., Feys, B.F., James, S., Nieto-Rostro, M. and Turner, J.G. (1998) COI1: an Arabidopsis gene required for jasmonate-regulated defense and fertility. *Science* **280**, 1091–1094.

Zimmermann, P., Hirsch-Hoffmann, M., Hennig, L. and Gruissem, W. (2004) GENEVESTIGATOR. Arabidopsis microarray database and analysis toolbox. *Plant Physiology* **136**, 2621–2632.

8 Fruit ripening

Catherine Martel and James Giovannoni

8.1 Introduction

The ripening process, by which physiological and biochemical changes render a fruit attractive for consumption, has always been a major area of interest for plant biologists due to the uniqueness of this developmental mechanism but, perhaps more significantly, because of the importance of fruit in the human diet. The advent of molecular biology and the emergence of high-throughput genomic technologies have generated a new set of tools, approaches and opportunities to dissect, understand and manipulate fruit biology.

8.2 Fruits

8.2.1 Fruit physiology

Fruits are botanically defined as seed receptacles that develop from ovaries (Seymour, 1993). This definition accurately describes fruits like tomato, melon and stone fruit. Extension of this classical definition to fruits derived from extracarpellary (i.e. non-ovary derived) tissues is necessary to include fruits such as apple, strawberry and pineapple (Giovannoni, 2001). Fruits perform the dual functions of protecting seeds during their maturation and assisting in their dispersion through a range of different strategies from simple shattering to attraction of predators. Fruits are classified as being either dry or fleshy depending on the developmental outcome of fruit development in a given species. *Arabidopsis thaliana*, for example, produces dry pod-like siliques, whereas tomato produces fleshy expanded fruits. Fleshy fruits are believed to have evolved as a mean to promote seed dispersal by animal vectors in regions where abiotic vectors, such as wind, were less reliable (Seymour,1993; Giovannoni, 2001).

8.2.2 Fruit development

Following pollination, fleshy fruit development can be subdivided into four broad stages. The first stage involves intensive cell division of the immature fruit-forming tissue. This increase in cell number is followed by a phase of rapid cell expansion that generates a full or near full-sized fruit typically containing enlarge polyploid cells. Finally, the ripening phase, which is responsible for converting the fleshy seed receptacle into a palatable edible tissue attractive to seed-dispersing organisms, includes different biochemical and structural changes resulting in the characteristic aroma, colour, texture and composition of the mature fruit. Finally, most fleshy fruit undergo a stage of over-ripening and eventual senescence characterized by enhanced susceptibility to opportunistic pathogens (if they are not consumed by seed-dispersing organisms).

Most of the basic knowledge regarding factors involved in regulation of early fruit development and maturation arises from studies on the model organism Arabidopsis. The availability of numerous mutant lines, as a result of large-scale mutagenesis programs, has been exploited to identify molecular factors that affect flower, carpel and fruit development. One notable feature of these studies is the predominant role played by the MADS box family of transcription factors in these phenomena. Precise spatio-temporal expression of different members of this gene family has been shown to be essential for the correct initiation and completion of these developmental programs. Basic floral and fruit development mechanisms are likely to be shared between Arabidopsis and other fruit species although regulatory nuances are likely to be manifested to account for the array of morphologies and developmental programs displayed in fruits of different species. For example, the tomato AGAMOUS protein (TAG) has been shown to have the same role in carpel determinancy as its Arabidopsis orthologue (Pnueli *et al.*, 1994). Species-specific expansion and neo-functionalization of certain gene family members are likely to have contributed to the evolution and establishment of fleshy-fruit-specific processes. It is therefore not surprising that a MADS box gene has been shown to be involved in the regulation of fleshy fruit ripening (see below).

8.2.3 Fleshy fruit ripening

The ripening process observed in fleshy fruit was first believed to be a senescence-driven event in which programmed degradative processes were ultimately responsible for the loss of integrity of fruit tissue. However, the subsequent demonstration of the requirement of *de novo* protein and RNA synthesis demonstrated that ripening is an anabolic process requiring energy and ongoing transcription to occur (Seymour, 1993). Although the exact nature of the fruit modifications associated with ripening varies, depending on the species examined, it generally includes modification of cell wall ultrastructure, conversion of starch to sugars, increase in susceptibility to post-harvest pathogens, changes in the accumulation and biosynthesis of pigments and a rise in the production of aroma- and flavour-associated volatile compounds (Seymour, 1993; Giovannoni, 2001).

Analyses of ripening mechanisms in different fleshy fruit species have led to further subdivision of these fruits into two classes. Climacteric fruits, such as tomato, cucurbits, apple, avocado, banana and stone fruits, are characterized by an increase in respiration levels, principally due to increased flux through the glycolytic pathway, often accompanied by a burst in ethylene biosynthesis (Seymour, 1993; Lelievre *et al.*, 1997). The gaseous hormone, produced by climacteric fruit, is essential for the initiation and completion of the ripening program and is hypothesized to favour the synchronization of ripening in an individual or bunch of fruit. Physiologically defined non-climacteric fruits such as strawberry, grape and citrus do not show a ripening-associated increase in respiration and neither produce nor require ethylene to initiate and maintain their ripening program. Climacteric and non-climacteric fruits have therefore evolved different strategies to co-ordinate the ripening process. In some species, the existence of both climacteric and non-climacteric varieties (e.g. melon) suggests that non-climacterism might be the result of an alteration in ethylene synthesis or signalling. Whether this observation can be generalized to all non-climacteric fruits remains to be determined. Although a dependency towards ethylene seems to have diverged during the evolution of different species, it is believed that the regulatory switches involved in triggering the ripening program may be shared between these two classes of fleshy fruit and may possibly have common characteristics with the developmental programs of dry fruits.

8.2.4 Tomato as a fleshy fruit model for ripening

Although undeniably useful in many studies of plant biology, the Arabidopsis model has limited value for the study of ripening due to the dehiscent nature of its dry fruits. Among the numerous species producing fleshy fruits, tomato has emerged as one of the best model systems to dissect the molecular mechanisms underlying ripening. Tomato (*Solanum lycopersicum*), a member of the Solanaceae species, is native to South America (Seymour, 1993). Today, tomato represents the 11th largest crop in the US and is an essential part of human diet all over the world. This role is but one of numerous characteristics that renders the species attractive as a model organism for fruit-related studies. Technical advantages of using tomato includes its short generation time, relatively small genome (950MB), abundance of expressed sequence tags (ESTs) and ongoing genome sequencing project, ease of sexual hybridization, year-round growing potential in greenhouses and efficient transformation allowing ready generation of transgenic plants for functional analyses (Fei *et al.*, 2004; Mueller *et al.*, 2005). The long domestication and breeding history of tomato have resulted in an impressive germplasm collection, further completed by several mutagenized populations and introgressed lines that can be used to dissect molecular mechanisms such as ripening (www.sgn.cornell.edu).

8.3 Regulation of ripening

8.3.1 Ethylene-regulated ripening

The characteristic burst in ethylene biosynthesis, seen at the onset of climacteric ripening, has fostered many studies to understand its effects on fruit tissue at the physiological and molecular levels. Studies in tomato and Arabidopsis have lead to the elucidation of the signalling pathway and cellular responses involved in ethylene action during climacteric fruit ripening.

8.3.2 *Biosynthesis*

Ethylene is synthesized from the methionine-derived compound S-adenosyl-methionine (SAM) through the sequential action of two enzymes (Yang *et al.*, 1985). SAM is first converted to 1-aminocyclopropane-1-carboxylic acid (ACC) by the action of ACC synthase (ACS) before being metabolized to ethylene by ACC oxidase. Although both steps are considered to be rate limiting, the high level of ACO activity found in most plant tissue suggests that the regulation of ACS levels primarily controls the rate of ethylene synthesis. Multiple members of ACS and ACO gene families have been found in tomato, some of which are expressed in a fruit-specific manner (Rottmann *et al.*, 1991; Barry *et al.*, 2000). Among the four ACS genes expressed in tomato fruit, LeACS1 and LeACS4 are developmentally up-regulated at the onset of ripening and are responsible for the initiation of the ethylene-regulated ripening program. Two ACO genes have also been showed to be up-regulated by ethylene during tomato fruit ripening (Barry *et al.*, 2000).

8.3.3 *Signalling*

Extensive characterization of the ethylene signal transduction pathway has been performed in Arabidopsis (Bleecker and Kende, 2000; Stepanova and Ecker, 2000) and numerous studies have shown that the basic signalling components are conserved in tomato and other plants, although inter-specific variations in family composition and regulation are observed (Wilkinson *et al.*, 1997; Adams-Phillips *et al.*, 2004).

The gaseous hormone is first bound by ethylene-specific receptors (ETR), which are similar to bacterial two-component histidine kinase receptors (Bleecker, 1999; Chang and Stadler, 2001). Whereas five ETRs have been found in Arabidopsis, six have been identified in tomato although only five of these have, so far, been shown to bind ethylene. The tomato NR and LeETR4 receptors are largely responsible for ethylene responses of fruit ripening (Tieman and Klee, 1999). Upon binding to ethylene, these receptors interact with the MAPKKK protein CTR1 (Clark *et al.*, 1998; Gao *et al.*, 2003), a negative regulator of the ethylene response signal. Knockout mutations of either the ETR receptors or CTR kinase produce a constitutive ethylene response phenotype (Hua and Meyerowitz, 1998; Hall and Bleecker, 2003; Huang *et al.*, 2003; Wang *et al.*, 2003). Several members of the tomato CTR-like (LeCTR) family are expressed during ripening, but only LeCTR1 was found to be up-regulated during this process (Leclercq *et al.*, 2002; Adams-Phillips *et al.*, 2004). Activation of the kinase cascade ultimately leads to the post-translational up-regulation of an ethylene-responsive transcription factor, EIN3 (Guo and Ecker, 2003; Potuschak *et al.*, 2003), which binds to and activates the transcription of the *ethylene-response factor* (*ERF1*) gene. ERF1 is a transcription factor that regulates the expression of numerous ethylene responsive genes. Four EIN3-like (EIL) and four ERF orthologues have been identified in tomato and one member of each family (LeEIL4 and LeERF2) shows a ripening-associated expression of the gene pattern (Tieman *et al.*, 2001; Tournier *et al.*, 2003; Yokotani *et al.*, 2003).

8.3.4 Developmentally regulated ripening

Ethylene represents a major regulator of climacteric fruit ripening; however, it is itself insufficient to trigger this program as demonstrated by the inability of an immature fruit to ripen in the presence of exogenous ethylene. The initial cue responsible for timely induction of ethylene synthesis (through increased expression of *ACS* and *ACO* genes) and up-regulation of other signalling components (such as the NR ethylene receptor) has until recently been a mystery and remains an area of active investigation. A developmental ripening signal must necessarily be involved in the acquisition, by unripe mature fruit, of a competency to respond to ethylene. Ripening-impaired tomato mutants that fail to ripen in the presence of exogenous ethylene, yet possessing fully functional ethylene signalling networks (Giovannoni, 2004), have been identified. These mutants are believed to be impaired in ripening competency acquisition and therefore represent valuable tools for the elucidation of the developmental signalling network required for ethylene competency acquisition and subsequent climacteric ripening (Figure 8.1).

8.3.5 Developmental mutants

The *ripening-inhibitor* (*rin*) (Davies and Hobson, 1981) tomato mutant produces non-climacteric fruits that remain green for a prolonged period. A yellowish colour eventually develops but the fruit never becomes red, presumably due to a deficiency in carotenoid biosynthesis or accumulation. The non-ripening (*nor*) (Davies and Hobson, 1981) mutation produces phenotype similar to *rin*, but it is associated with a different locus. Fruits of the tomato *nor* mutant show the following characteristic features: absence of softening during maturation, cracking resistance and abnormal carotenoid biosynthesis resulting in an orange pericarp and locular jelly. Map-based cloning of the *RIN* and *NOR* genes has revealed that they both encode transcription factors (Vrebalov *et al.*, 2002); Vrebalov and Giovannoni, unpublished). It was therefore hypothesized that these transcription factors could act as master regulators of ripening competency acquisition by regulating the expression of other ripening genes.

RIN is a member of the MADS box family of transcriptional regulators whose essential roles in development have been described in numerous organisms including plants, animals and fungi. In plants, MADS box proteins have been demonstrated to play essential roles in processes such as control of vegetative growth and flowering time, and formation of flower and reproductive structures (Figure 8.1). A full characterization of *NOR* is in progress.

The identification of these two transcription factors paves the way for numerous studies that will lead to a better understanding of the developmental ripening signalling network responsible for acquisition of ripening competency. Consequently, a detailed characterization of RIN and NOR activities and regulation represents an essential step towards the elucidation of the ripening process.

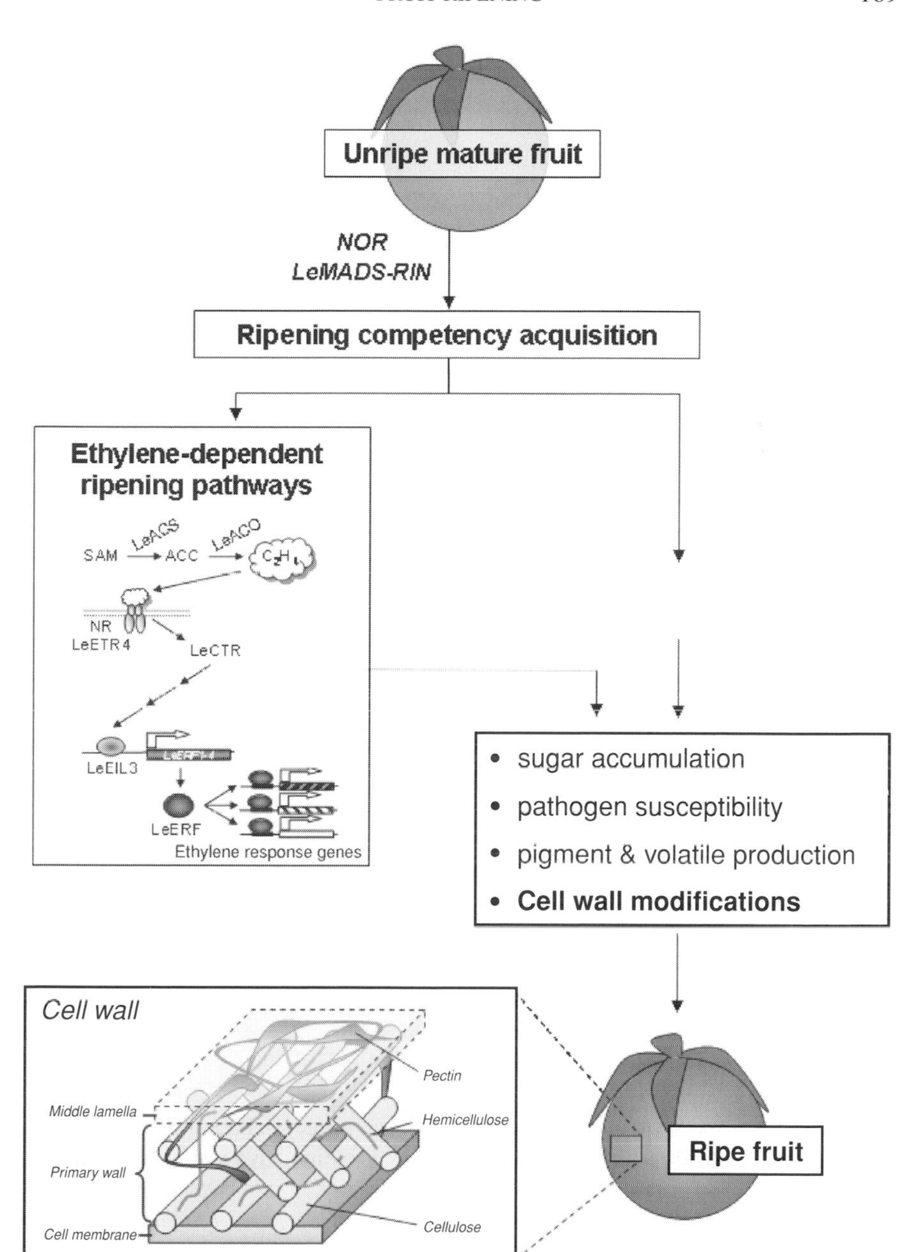

Figure 8.1 Regulation and physiology of tomato ripening. The unripe fruit establishes a competency to respond to the ethylene produced in part as a result of normal action of the RIN and NOR genes. These regulators influence ripening via both ethylene and non-ethylene pathways with the former shown to be comprised of similar signalling components as defined in *Arabidopsis*. The activity of ripening regulators including RIN, NOR and ethylene leads to modification of biochemical pathways influencing carbohydrate composition, pigment accumulation, susceptibility to opportunistic pathogens and cell wall modifications, which together confer the ripe phenotype. The complexity of the cell wall and the array of associated metabolic activities summarized in the text represents the most complex set of biochemical activities associated with gross ripening changes.

8.4 Cell wall metabolism during fruit ripening

Among the numerous physical and biochemical modifications occurring during the ripening process, fruit softening has received particular attention from scientists due to its influence on post-harvest fruit quality – a critical aspect of commercial fruit production. Indeed, fruit shelf life and quality of fruit-derived products (juice, paste, sauces) are directly influenced by fruit-softening characteristics. A better understanding of the molecular mechanisms underlying this process could allow for the development of a more efficient strategy for designing fruits with increased shelf life. Moreover, targeted modulation of fruit softening is less likely to affect other desirable ripening-associated modifications, such as flavour and nutritional content. The general dissatisfaction of consumers with fresh market production varieties likely reflects that genetic and cultural practices, in addition to post-harvest handling, are currently geared towards methods of ripening inhibition as an approach for promoting firmer fruit texture.

8.4.1 Fruit cell wall composition

The fruit pericarp is composed mainly of unlignified parenchymous cells surrounded by a type I primary cell wall (Carpita and Gibeaut, 1993) composed of rigid cellulose microfibrils embedded in a network of glycans, glycoproteins and pectins (Carpita and Gibeaut, 1993; McCann, 1997; Brummell and Harpster, 2001). The pectin portion of the cell wall is more significant in fruits than in many other plant tissues, accounting for 40–60% of the typical fruit cell wall (Redgwell and Fischer, 2001). It should be noted that due to the high level of variability seen in the composition of fruit cell walls and their concomitant ripening-associated modifications, the phenomena described in model fruit species might not apply to all other species.

8.4.2 Ripening-induced cell wall modifications

Fruit texture develops as a result of controlled disassembly and modification of the fruit cell wall. A distinction should be made between cell wall disassembly, as observed during fruit ripening and abscission, and cell wall loosening, which is associated with cell elongation. Indeed, whereas the latter involves transient and reversible modifications of the wall as well as extensive cell wall biosynthesis, the former reflects a more irreversible, yet tightly regulated process leading to dissolution of cell wall components. Although the outcomes differ, both processes are likely to share some common mechanisms (Rose and Bennett, 1999). Moreover, cell wall modifications during ripening are not exclusively catabolic as new components are also incorporated into the cell wall throughout the ripening process (Greve and Labavitch, 1991). Fruit texture changes are brought about by the combined effects of non-physiological dehydration of the fruit tissue (i.e. loss of cell turgor), starch degradation and modification of cell wall properties. Cell wall modifications are the primary factors affecting fruit softening and texture properties of most species, although the two former mechanisms might play a more or less important role among different fruit species (Seymour *et al.*, 1993).

Loss of fruit firmness results from disassembly of primary cell walls and dissolution of the pectin-rich middle lamella region of contact between neighbouring cells. This takes place through an increase in hydration (i.e. swelling), and an alteration in pectin gel properties results in the cohesiveness between weakened neighbouring cells. The extent of these modifications directly influences the texture of each fruit species (Jarvis, 1984). For example, in apples, cell walls show limited swelling producing a characteristic crispier texture, whereas in strawberry, a softer texture is the result of a higher degree of swelling (Redgwell *et al.*, 1997). In tomato, depolymerization and solubilization of pectins has been shown to increase cell wall porosity (Huber and O'Donoghue, 1993; Brummell *et al.*, 1997; Chun and Huber, 1998). Other modifications of the cell wall observed in numerous fruit species include a loss of neutral sugars, mainly galactose (Seymour *et al.*, 1993), a modification of ion concentrations and a decrease in pH (Almeida and Huber, 1999). These changes are believed to play an important role in regulating the activity of cell-wall-localized enzymes (Figure 8.1), which in turn have an impact on cell wall metabolism and texture (Chun and Huber, 1998; Almeida and Huber, 1999).

In addition to the nature of the modifications taking place during ripening, the time at which they take place also greatly influences the final texture of the mature fruit. Softness of a ripe fruit reflects the degree of cell–cell adhesion in the tissue as a result of the time at which ripening-regulated cell wall enzymes start dissolving the pectin-rich middle lamella that holds cells together (for a further discussion, see Chapter 9). In tomato, de-esterification of pectins in the middle lamella first starts at the mature green (MG) stage in tomato before extending to the entire cell wall during subsequent ripening stages (Roy *et al.*, 1992; Blumer *et al.*, 2000). The same phenomenon happens latter in development of apples (Ben-Arie *et al.*, 1979).

8.5 Cell wall modifying enzymes

As mentioned above, modification of the cell wall during fruit ripening involves the highly regulated and co-ordinated action of a spectrum of enzymes. Although the identity of those involved may be common between different types of fruit, differences in their relative level of expression, time of action and activity result in a high degree of variability in cell wall ultrastructure and fruit texture.

8.5.1 Pectin and pectinases

Considering the pectin-rich nature of fruit cell wall, it is not surprising that many enzymes affecting pectin solubilization are involved in ripening-associated fruit softening.

Other factors such as pH and ion concentration in the apoplast can also play a critical role in regulating pectin degradation through their influence on wall porosity and associated substrate accessibility.

8.5.1.1 Polygalacturonase

Polygalacturonase (PG) catalyses the hydrolysis of the α-1,4-D-galacturonan backbone of pectin. PG activity was shown to be highly up-regulated during ripening in numerous species, including avocado, peach and tomato (Hobson, 1962; Gross, 1979; Huber, 1983; Redgwell and Fischer, 2001). In tomato, *PG* mRNA levels typically increase more than 2000 times at the onset of ripening, representing up to 2% of total mRNAs (DellaPenna *et al.*, 1986, 1987; Fischer and Bennett, 1991). This elevation in *PG* mRNA level and activity is however not observed in all fruits; strawberry, apple and melon do not display such an increase in PG mRNA levels during their ripening. PG up-regulation is not observed in *rin, nor* and *Nr* tomato mutants consistent with its transcriptional regulation by ethylene (Sitrit *et al.*, 1999).

The involvement of PG in ripening-associated fruit softening has been examined in tomato. Surprisingly, down-regulation (Sheehy, 1988, Smith *et al.*, 1988) or complete inhibition (Cooley and Yoder, 1998) of PG activity has no effect on ripening-associated softening. Furthermore, overexpression of a *PG* transgene in the *rin* background fails to promote softening (Giovannoni *et al.*, 1989), although normal pectin solubilization levels were observed in transgenic fruits (Giovannoni *et al.*, 1989; DellaPenna *et al.*, 1990). These results suggest that PG activity and pectin solubilization are neither sufficient nor necessary for fruit softening in tomato. It should be noted that antisense-*PG* fruits show a marginal increase in post-harvest shelf life (Schuch *et al.*, 1991; Kramer *et al.*, 1992; Langley *et al.*, 1994). Although not a major player in tomato fruit softening by itself, PG might play a more important role in the softening other fruits. For instance, in peach and pepper, *PG* genes have been closely linked to major fruit softening loci (Lester *et al.*, 1994; Lester *et al.*, 1996; Rao and Paran, 2003).

8.5.1.2 Pectin methyl esterases

Another pectin modification that is associated with ripening is de-esterification. During cell wall synthesis, pectin is incorporated in the cell wall in a methyl esterified form and is then demethylated by cell-wall localized enzymes (Kauss and Hassid, 1967; Lau *et al.*, 1985). Demethylated pectin molecules possess charged residues that allow them to bind ions, such as Ca^{2+}, contributing to the formation of a denser pectate gel (Grant *et al.*, 1973; Jarvis, 1984; Carpita and Gibeaut, 1993; Almeida and Huber, 1999). Modification of the degree of methylation of pectin in a cell wall consequently affects its overall structure, which in turn influences the accessibility and activity of other cell wall modifying enzymes (Koch and Nevins, 1989, Wakabayashi, 2000).

The enzymes responsible for pectin de-esterification are the pectin methyl esterases (PMEs), which include over 80 members in *A. thaliana*. In the fruit-ripening model organism tomato, only four PME genes have been identified to date (Ray *et al.*, 1988; Hall *et al.*, 1994; Gaffe *et al.*, 1997). PME activity is detected early during fruit development and shows a marked increased at the onset of tomato ripening (Tucker *et al.*, 1982; Fischer and Bennett, 1991). *rin/rin* fruits show a strong reduction in PME expression at the onset of ripening. Knockdown of the major ripening-associated

PME isoform produces fruits with a net increase in methylesterified pectin (due to a 15–40% reduction in de-esterification level), but this modification does not correlate with any significant loss of softening or effect on fruit ripening (Tieman *et al.*, 1992; Tieman and Handa, 1994). Post-harvest fruits, however, show defects in tissue integrity, a phenotype that could be due to the inability of the cell wall to bind divalent cations required for efficient cross-linking of pectins (Tieman and Handa, 1994), or by affecting the activity of other cell wall enzymes (such as PG).

8.5.1.3 Pectate lyase

Pectate lyase (PL) is another class of pectin-modifying enzyme that has been shown to be expressed in a ripening-dependent way in many fruit species such as banana, grape and strawberry (Dominguez-Puigjaner *et al.*, 1997; Nunan *et al.*, 2001; Pua *et al.*, 2001) where it is believed to play an important role in the softening process (Medina-Escobar *et al.*, 1997; Jimenez-Bermudez *et al.*, 2002). Although initially reported as being absent from tomato fruit (Besford and Hobson, 1972), a PL EST was recently detected in a tomato fruit library (Marin-Rodriguez *et al.*, 2002). Major studies focusing on the role of PL in fruit softening are lacking in part because the best correlations of activity and softening are found in species such as banana, which are difficult to study at the molecular level. As a consequence, the exact function of PL during ripening remains unclear.

8.5.1.4 β-galactosidase/arabinosidases

β-Galactosidases and arabinosidases are pectin side chain modifying enzymes. Another major change occurring in the cell wall during fruit ripening is the breaking down of polymeric galactose resulting in free galactose accumulation (Gross and Wallner, 1979; Gross, 1984). Although up to 70% of pectin's galactose side chains are lost during fruit ripening (Gross, 1984; Redgwell *et al.*, 1997), no direct correlation seems to exist between the extent of galactose loss and the degree of softening among different fruit species (Redgwell and Fischer, 2001). *rin* and *nor* fruits display a reduction in free galactose accumulation levels (Gross, 1984). A high, yet constant level of β-galactosidase expression has been reported during tomato fruit ripening (Wallner and Walker, 1975; Carey *et al.*, 1995). A closer examination of the seven different β-galactosidases isozymes (Smith and Gross, 2000) revealed specific temporal expression patterns of certain members of the family during the different stages of ripening (Pressey, 1983; Carey *et al.*, 1995; Carrington and Pressey, 1996; Brummell and Harpster, 2001). This observation suggests that some members might have a ripening-specific activity that is not shared by other members of the family. Generation of tomato lines with reduced expression of ripening-specific β-galactosidases further supports this hypothesis (Carey *et al.*, 2001; Smith *et al.*, 2002). Knockdown experiments revealed that reduction in β-galactosidase activity during the early stages of ripening reduces degradation of galactose polymers and decreases softening by 40% (Smith *et al.*, 2002). Interestingly, down-regulation of the β-galactosidase activity latter during ripening fails to prevent softening even though galactose polymer degradation is effectively prevented (Brummell and Harpster, 2001). This observation supports an indirect role for β-galactosidase in promoting softening. Cell wall models suggest that it might increase cell wall porosity, therefore allowing other softening-related enzymes to access their substrate more efficiently.

Although increases in arabinase levels during ripening of Japanese pear, tomato and carambola fruits have been reported (Tateishi *et al.*, 1996; Chin *et al.*, 1999; Sozzi *et al.*, 2002), the role of these pectin side chain modifying enzymes in fruit softening remains largely unknown.

8.5.2 Cellulose and cellulose-modifying enzymes

For a long time, cellulose structure was believed to remain largely unaffected during the ripening process and cellulose-modifying enzymes were therefore overlooked as major fruit softening contributors (Gross and Wallner, 1979; Gross *et al.*, 1986; Maclachlan and Brady, 1994; Nunan *et al.*, 1998). More recently, analyses of the cellulose state of raspberry and avocado fruits during ripening using more accurate analytical techniques seem to suggest that cellulose microfibrils undergo modification during the ripening process (O'Donoghue *et al.*, 1994; Stewart *et al.*, 2001). Cellulose-modifying enzymes might therefore play a more important role in fruit softening then previously thought.

8.5.2.1 Endo-1,4-β-glucanases

Based largely on *in vitro* studies, endo-1,4-β-glucanases (EGases) are able to catalyse the hydrolysis of cellulose microfibrils, xyloglucan (hemicellulose) and possibly glucomannans (Brummell and Harpster, 2001; Levy *et al.*, 2002). In *Arabidopsis thaliana*, 27 genes are predicted to be members of the EGase family (Henrissat *et al.*, 2001). The expression of EGases has been shown to increase during the ripening of numerous fruits, including avocado (Cass *et al.*, 1990), peach (Bonghi *et al.*, 1998), strawberry (Llop-Tous *et al.*, 1999), pepper (Harpster *et al.*, 1997) and tomato (Lashbrook *et al.*, 1994; Kalaitzis *et al.*, 1999). EGases however show wide variation of activity across these different fruits. In tomato, EGase is expressed early in an ethylene-dependent fashion and declines in ripe fruits (Hobson, 1962; Lashbrook *et al.*, 1994). The involvement of EGase in mediating fruit softening seems marginal since EGase knockdown tomato and strawberry lines do not show any specific ripening or softening phenotypes (Lashbrook *et al.*, 1998; Brummell *et al.*, 1999a; Woolley *et al.*, 2001). Redundancy between different EGase isoforms has been postulated to account for the lack of effect, but the observation that overexpression of a pepper EGase in tomato fails to increase softening of the fruit (Brummell and Harpster, 2001) further strengthens the idea that EGases are not major players in fruit-softening processes.

8.5.3 Expansins

Expansins (Exp) represent a novel type of cell wall modifying protein that promotes loosening of the wall without apparent hydrolysis of any of its constituents. Although specific mechanisms of action and substrates remains unclear, expansin activity is thought to result from disruption of hydrogen bonding between cellulose and hemicelluloses (McQueen-Mason *et al.*, 1992; McQueen-Mason and Cosgrove, 1995; Cosgrove, 2000; Whitney *et al.*, 2000; Brummell and Harpster, 2001). Exp have been detected in many fruits including strawberry (Civello *et al.*, 1999; Harrison *et al.*, 2001), peach (Hayama *et al.*, 2000) and pear (Hiwasa *et al.*, 2003). In tomato, LeExp1 was shown to be specifically up-regulated, in an ethylene-dependent fashion at the onset of ripening (Rose *et al.*, 1997, 2000; Brummell *et al.*, 1999b,c). This up-regulation seems to be important for ripening-induced softening as fruit produced by LeExp1-suppressed lines show a 20% increase in firmness. Moreover, overexpression of LeExp1 during ripening accelerates fruit softening (Brummell *et al.*, 1999b).

8.5.4 Hemicellulose and hemicellulases

Limited changes in hemicellulose structure have been detected during ripening although these changes seems to make an important contribution to fruit softening (Gross *et al.*, 1986; Maclachlan and Brady, 1994; Rose *et al.*, 1998; Brummell and Harpster, 2001). Actual data indicate an increase in xyloglucan endotransglucosylase–hydrolase (XTH) (Maclachlan and Brady, 1994; Nunan *et al.*, 2001), β-mannanase (Pressey, 1989; Bewley *et al.*, 2000; Bourgault *et al.*, 2001, Carrington *et al.*, 2002) and xylanases (Yamaki and Kakiuchi, 1979; Paull and Chen, 1983) activities during fruit ripening but their specific role in the process remains unknown. Further studies are needed to assess more clearly the role of these hemicellulases during fruit ripening and their respective impacts on softening.

8.6 Colourless non-ripening mutant: a model to study fruit ripening cell wall modifications

More information regarding the important players involved in ripening-induced fruit softening will be identified from the study of the newly identified *Colourless non-ripening* (*Cnr*) tomato-ripening mutant (Thompson *et al.*, 1999). In addition to a lack of pigmentation, fruit of this mutant shows an important reduction in cell-to-cell adhesion resulting from abnormal pericarp cell wall structure (Orfila *et al.*, 2001). A closer analysis indicates that ripe *Cnr* fruits have larger intracellular spaces and thinner cell walls than wt fruits. Reduction in Ca+ binding capacity of *Cnr* cell walls was also reported and attributed to modification of the pectic polysaccharide structure, including a reduction in the amount of de-esterified pectin, reduce arabinan deposition and decreased solubility. The pleiotropic phenotype of *Cnr* fruit suggests a lesion in an upstream regulator of the expression of cell wall modifying enzymes. RT-PCR analyses have been performed to assess the level of some of the best characterized ripening-associated cell wall enzymes and this has revealed a significant reduction in the level of PG, PE, β-galactosidase and XET enzymes seen during fruit-ripening stages (Eriksson *et al.*, 2004). Lack of normal up-regulation of these genes further supports the notion that *Cnr* encodes a major upstream regulator of cell wall modifying enzymes, probably in the context of a more global ripening regulation similar to that seen in other pleiotropic ripening mutants such as *rin* and *nor*. The rapid expansion of genomics technologies, specifically related to tomato and including the ongoing sequencing of the tomato genome (www.sgn.cornell.edu), will certainly facilitate isolation of regulatory loci such as that encoded by *Cnr* and provide a more complete repertoire of cell wall modifying enzymes active during fruit development and ripening.

8.7 Conclusions

In this chapter, we have described some of the events that contribute to the process of fruit ripening. Whilst in a fleshy fruit it is clear that, during softening, major changes in the adhesion between cells take place, the precise biochemical and molecular changes that bring these about have yet to be determined. Application of the new 'omic' technologies to the study of fruit ripening in a range of species should help to elucidate how cell separation takes place and identify the mechanisms that co-ordinate the process. Such information will prove to be of value in our quest to manipulate phenomena such as shelf life and the nutritional properties of a major component of our diet.

References

Adams-Phillips, L., Barry, C., Kannan, P., Leclercq, J., Bouzayen, M. and Giovannoni, J. (2004) Evidence that CTR1-mediated ethylene signal transduction in tomato is encoded by a multigene family whose members display distinct regulatory features. *Plant Molecular Biology* **54**, 387–404.

Almeida, D.P.F. and Huber, D.J. (1999) Apoplastic pH and inorganic ion levels in tomato fruit: a potential means for regulating cell wall metabolism during ripening. *Physiologia Plantarum* **105**, 506–512.

Barry, C.S., Llop-Tous, M.I. and Grierson, D. (2000) The regulation of 1-aminocyclopropane-1-carboxylic acid synthase gene expression during the transition from system-1 to system-2 ethylene synthesis in tomato. *Plant Physiology* **123**, 979–986.

Ben-Arie, R., Kislev, N. and Frenkel, C. (1979) Ultrastructural changes in the cell walls of ripening apple and pear fruit. *Plant Physiology* **64**, 197–202.

Besford, R.T. and Hobson, G.E. (1972) Pectic enzymes associated with softening of tomato fruit. *Phytochemistry* **11**, 873–881.

Bewley, J.D., Banik, M., Bourgault, R., Feurtado, J.A., Toorop, P. and Hilhorst, H.W. (2000) Endo-beta-mannase activity increases in the skin and outer pericarp of tomato fruits during ripening. *Journal of Experimental Botany* **51**, 529–538.

Bleecker, A.B. (1999) Ethylene perception and signaling: an evolutionary perspective. *Trends in Plant Science* **4**, 269–274.

Bleecker, A.B. and Kende, H. (2000) Ethylene: a gaseous signal molecule in plants. *Annual Review Cell and Developmental Biology* **16**, 1–18.

Blumer, J.M., Clay, R.P., Bergmann, C.W., Albersheim, P. and Darvill, A.G. (2000) Characterization of changes in pectin methylesterase expression and pectin esterification during tomato fruit ripening. *Canadian Journal of Botany* **78**, 607–618.

Bonghi, C., Ferrarese, L., Ruperti, B., Tonutti, P. and Ramina, A. (1998) Endo-b-1,4-glucanases are involved in peach fruit growth and ripening, and are regulated by ethylene. *Physiologia Plantarum* **102**, 346–352.

Bourgault, R., Bewley, J.D., Alberici, A. and Decker, D. (2001) Endo-b-mannanase activity in tomato and other ripening fruits. *Horticultural Science* **36**, 72–75.

Brummell, D.A., Bird, C.R., Schuch, W. and Bennett, A.B. (1997) An endo-1,4-beta-glucanase expressed at high levels in rapidly expanding tissues. *Plant Molecular Biology* **33**, 87–95.

Brummell, D.A., Hall, B.D. and Bennett, A.B. (1999a) Antisense suppression of tomato endo-1,4-beta-glucanase Cel2 mRNA accumulation increases the force required to break fruit abscission zones but does not affect fruit softening. *Plant Molecular Biology* **40**, 615–622.

Brummell, D.A. and Harpster, M.H. (2001) Cell wall metabolism in fruit softening and quality and its manipulation in transgenic plants. *Plant Molecular Biology* **47**, 311–340.

Brummell, D.A., Harpster, M.H., Civello, P.M., Palys, J.M., Bennett, A.B. and Dunsmuir, P. (1999b) Modification of expansin protein abundance in tomato fruit alters softening and cell wall polymer metabolism during ripening. *Plant Cell* **11**, 2203–2216.

Brummell, D.A., Harpster, M.H. and Dunsmuir, P. (1999c) Differential expression of expansin gene family members during growth and ripening of tomato fruit. *Plant Molecular Biology* **39**, 161–169.

Carey, A.T., Holt, K., Picard, S., Wilde, R., Tucker, G.A., Bird, C.R., Schuch, W. and Seymour, G.B. (1995) Tomato exo-(1→4)-beta-D-galactanase. Isolation, changes during ripening in normal and mutant tomato fruit, and characterization of a related cDNA clone. *Plant Physiology* **108**, 1099–1097.

Carey, A.T., Smith, D.L., Harrison, E., Bird, C.R., Gross, K.C., Seymour, G.B. and Tucker, G.A. (2001) Down-regulation of a ripening-related beta-galactosidase gene (TBG1) in transgenic tomato fruits. *Journal of Experimental Botany* **52**, 663–668.

Carpita, N.C. and Gibeaut, D.M. (1993) Structural models of primary cell walls in flowering plants: consistency of molecular structure with the physical properties of the walls during growth. *Plant Journal* **3**, 1–30.

Carrington, C.M.S. and Pressey, R. (1996) b-galactosidase II activity in relation to changes in cell wall galactosyl composition during tomato fruit ripening. *Journal of American Social Horticulture Science* **121**, 132–136.

Carrington, C.M.S., Vendrell, M. and Dominguez-Puigjaner, E. (2002) Characterisation of an endo-(1,4)-b-mannanase (LeMAN4) expressed in ripening tomato fruit. *Plant Science* **163**, 599–606.

Cass, L.G., Kirven, K.A. and Christoffersen, R.E. (1990) Isolation and characterization of a cellulase gene family member expressed during avocado fruit ripening. *Molecular and General Genetics* **223**, 76–86.

Chang, C. and Stadler, R. (2001)Ethylene hormone receptor action in Arabidopsis. *Bioessays* **23**, 619–627.

Chin, L.H., Ali, Z.M. and Lazan, H. (1999) Cell wall modifications, degrading enzymes and softening of carambola fruit during ripening. *Journal of Experimental Botany* **50**, 767–775.

Chun, J.P. and Huber, D.J. (1998) Polygalacturonase-mediated solubilization and depolymerization of pectic polymers in tomato fruit cell walls. Regulation by pH and ionic conditions. *Plant Physiology* **117**, 1293–1299.

Civello, P.M., Powell, A.L., Sabehat, A. and Bennett, A.B. (1999) An expansin gene expressed in ripening strawberry fruit. *Plant Physiology* **121**, 1273–1280.

Clark, K.L., Larsen, P.B., Wang, X. and Chang, C. (1998) Association of the Arabidopsis CTR1 Raf-like kinase with the ETR1 and ERS ethylene receptors. *Proceedings of the National Academy of Sciences of the United States of America* **95**, 5401–5406.

Cooley, M. and Yoder, J.I. (1998) Insertional inactivation of the tomato polygalacturonase gene. *Plant Molecular Biology* **38**, 521–530.

Cosgrove, D.J. (2000) Loosening of plant cell walls by expansins. *Nature* **407**, 321–326.

Davies, J.N. and Hobson, G.E. (1981) The constituents of tomato fruit—the influence of environment, nutrition, and genotype. *Critical Reviews in Food Science and Nutrition* **15**, 205–280.

DellaPenna, D., Kates, D.S. and Bennett, A.B. (1986) Molecular cloning of tomato fruit polygalacturonase : analysis of polygalacturonase mRNA levels during ripening. *Proceedings of the National Academy of Sciences of the United States of America* **83**, 6420–6424.

DellaPenna, D., Kates, D.S. and Bennett, A.B. (1987) Polygalacturonase gene expression in Rutgers, rin, nor and Nr tomato fruits. *Plant Physiology* **85**, 502–507.

DellaPenna, D., Lashbrook, C.C., Toenjes, K., Giovannoni, J.J., Fischer, R.L. and Bennett, A.B. (1990) Polygalacturonase isozymes and pectin depolymerization in transgenic rin tomato fruit. *Plant Physiology* **94**, 1882–1886.

Dominguez-Puigjaner, E., LLop, I., Vendrell, M. and Prat, S. (1997) A cDNA clone highly expressed in ripe banana fruit shows homology to pectate lyases. *Plant Physiology* **114**, 1071–1076.

Eriksson, E.M., Bovy, A., Manning, K., Harrison, L., Andrews, J., De Silva, J., Tucker, G.A. and Seymour, G.B. (2004) Effect of the colorless non-ripening mutation on cell wall biochemistry and gene expression during tomato fruit development and ripening. *Plant Physiology* **136**, 4184–4197.

Fei, Z., Tang, X., Alba, R.M., White, J.A., Ronning, C.M., Martin, G.B., Tanksley, S.D. and Giovannoni, J.J. (2004) Comprehensive EST analysis of tomato and comparative genomics of fruit ripening. *Plant Journal* **40**, 47–59.

Fischer, R.L. and Bennett, A.B. (1991) Role of cell wall hydrolases in fruit ripening. *Annual Review of Plant Physiology and Plant Molecular Biology* **42**, 675–703.

Gaffe, J., Tiznado, M.E. and Handa, A.K. (1997) Characterization and functional expression of a ubiquitously expressed tomato pectin methylesterase. *Plant Physiology* **114**, 1547–1556.

Gao, Z., Chen, Y.F., Randlett, M.D., Zhao, X.C., Findell, J.L., Kieber, J.J. and Schaller, G.E. (2003) Localization of the Raf-like kinase CTR1 to the endoplasmic reticulum of Arabidopsis through participation in ethylene receptor signaling complexes. *Journal of Biological Chemistry* **278**, 34725–34732.

Giovannoni, J.J. (2001) Molecular biology of fruit maturation and ripening. *Annual Review of Plant Physiology and Plant Molecular Biology* **52**, 725–749.

Giovannoni, J.J. (2004) Genetic regulation of fruit development and ripening. *Plant Cell* **16** (Suppl.), S170–S180.

Giovannoni, J.J., DellaPenna, D., Bennett, A.B. and Fischer, R.L. (1989) Expression of a chimeric polygalacturonase gene in transgenic rin (ripening inhibitor) tomato fruit results in polyuronide degradation but not fruit softening. *Plant Cell* **1**, 53–63.

Grant, G.T., Morris, E.R., Rees, D.A., Smith, P.J.C. and Thom, D. (1973) Biological interactions between polysaccharides and divalent cations: the egg box model. *FEBS Letters* **32**, 534–543.

Greve, L.C. and Labavitch, J.M. (1991) Cell wall metabolism in ripening fruit. *Plant Physiology* **97**, 1456–1461.

Gross, F. (1979) Constraints of drug regulation on the development of new drugs. *Archives of Toxicology* **43**, 9–17.

Gross, K.C. (1984) Fractionation and partial characterization of cell walls from normal and non-ripening tomato fruit. *Physiologia Plantarum* **62**, 25–32.

Gross, K.C. and Wallner, S.J. (1979) Degradation of cell wall polysaccharides during tomato fruit ripening. *Plant Physiology* **63**, 117–120.

Gross, K.C., Watada, K.E., Kang, M.S., Kim, S.D. and Lee, S.W. (1986) Biochemical changes associated with the ripening of hot pepper fruit. *Physiologia Plantarum* **66**, 31–36.

Guo, H. and Ecker, J.R. (2003) Plant responses to ethylene gas are mediated by SCF(EBF1/EBF2)-dependent proteolysis of EIN3 transcription factor. *Cell* **115**, 667–677.

Hall, A.E. and Bleecker, A.B. (2003) Analysis of combinatorial loss-of-function mutants in the Arabidopsis ethylene receptors reveals that the ers1 etr1 double mutant has severe developmental defects that are EIN2 dependent. *Plant Cell* **15**, 2032–2041.

Hall, L.N., Bird, C.R., Picton, S., Tucker, G.A., Seymour, G.B. and Grierson, D. (1994) Molecular characterisation of cDNA clones representing pectin esterase isozymes from tomato. *Plant Molecular Biology* **25**, 313–318.

Harpster, M.H., Lee, K.Y. and Dunsmuir, P. (1997) Isolation and characterization of a gene encoding endo-beta-1,4-glucanase from pepper (Capsicum annuum L.). *Plant Molecular Biology* **33**, 47–59.

Harrison, E.P., McQueen-Mason, S.J. and Manning, K. (2001) Expression of six expansin genes in relation to extension activity in developing strawberry fruit. *Journal of Experimental Botany* **52**, 1437–1446.

Hayama, H., Shimada, T., Haji, T., Ito, A., Kashimura, Y. and Yoshioka, H. (2000) Molecular cloning of a ripening-related expansin cDNA in peach: evidence for no relationship between expansin accumulation and change in fruit firmness during storage. *Journal of Plant Physiology* **157**, 567–573.

Henrissat, B., Coutinho, P.M. and Davies, G.J. (2001) A census of carbohydrate-active enzymes in the genome of Arabidopsis thaliana. *Plant Molecular Biology* **47**, 55–72.

Hiwasa, K., Rose, J.K., Nakano, R., Inaba, A. and Kubo, Y. (2003) Differential expression of seven alpha-expansin genes during growth and ripening of pear fruit. *Physiologia Plantarum* **117**, 564–572.

Hobson, G.E. (1962) Determination of polygalacturonase in fruits. *Nature* **195**, 804–805.

Hua, J. and Meyerowitz, E.M. (1998) Ethylene responses are negatively regulated by a receptor gene family in Arabidopsis thaliana. *Cell* **94**, 261–271.

Huang, Y., Li, H., Hutchison, C. E., Laskey, J. and Kieber, J.J. (2003) Biochemical and functional analysis of CTR1, a protein kinase that negatively regulates ethylene signaling in Arabidopsis. *Plant Journal* **33**, 221–233.

Huber, D.J. (1983) The role of cell wall hydrolases in fruit softening. *Horticultural Review* **5**, 169–219.

Huber, D.J. and O'Donoghue, E.M. (1993) Polyuronides in avocado (persea americana) and tomato (Lycopersicon esculentum) fruits exhibit markedly different patterns of molecular weight downshifts during ripening. *Plant Physiology* **102**, 473–480.

Jarvis, M.C. (1984) Structure and properties of pectin gels in plant cell walls. *Plant, Cell and Environment* **7**, 153–164.

Jimenez-Bermudez, S., Redondo-Nevado, J., Munoz-Blanco, J., Caballero, J.L., Lopez-Aranda, J.M., Valpuesta, V., Pliego-Alfaro, F., Quesada, M.A. and Mercado, J.A. (2002) Manipulation of strawberry fruit softening by antisense expression of a pectate lyase gene. *Plant Physiology* **128**, 751–759.

Kalaitzis, P., Hong, S.B., Solomos, T. and Tucker, M.L. (1999) Molecular characterization of a tomato endo-beta-1,4-glucanase gene expressed in mature pistils, abscission zones and fruit. *Plant Cell Physiology* **40**, 905–908.

Kauss, H. and Hassid, W.Z. (1967) Enzymatic introduction of the methyl ester groups of pectin. *Journal of Biological Chemistry* **242**, 3449–3453.

Koch, J.L. and Nevins, D.J. (1989) Tomato fruit cell wall I. Use of purified tomato polygalacturonase and pectin methylesterase to identify developmental changes in pectins. *Plant Physiology* **91**, 816–822.

Kramer, M., Sanders, R., Bolkan, H., Waters, C., Sheehy, R.E. and Hiatt, W.R. (1992) Postharvest evaluation of transgenic tomatoes with reduced levels of polygalacturonase: processing, firmness and disease resistance. *Postharvest Biology and Technology* **1**, 241–255.

Langley, K.R., Martin, A. and Stenning, R. (1994) Mechanical and optical assessment of the ripening of tomato fruit with reduced polygalacturonase activity. *Journal of the Science of Food and Agriculture* **66**, 547–554.

Lashbrook, C.C., Giovannoni, J.J., Hall, B.D., Fischer, R.L. and Bennett, A.B. (1998) Transgenic analysis of tomato endo-b-1,4-glucanase gene function. Role of cell in floral abscission. *Plant Journal* **13**, 303–310.

Lashbrook, C.C., Gonzalez-Bosch, C. and Bennett, A.B. (1994) Two divergent endo-beta-1,4-glucanase genes exhibit overlapping expression in ripening fruit and abscising flowers. *Plant Cell* **6**, 1485–1493.

Lau, J.M., McNeil, M., Darvill, A.G. and Albersheim, P. (1985) Structure of backbone of rhamnogalacturonan I, a pectic polysaccharide in the primary cell walls of plants. *Carbohydrate Research* **137**, 111–125.

Leclercq, J., Adams-Phillips, L.C., Zegzouti, H., Jones, B., Latche, A., Giovannoni, J.J., Pech, J.C. and Bouzayen, M. (2002) LeCTR1, a tomato CTR1-like gene, demonstrates ethylene signaling ability in Arabidopsis and novel expression patterns in tomato. *Plant Physiology* **130**, 1132–1142.

Lelievre, J.M., Tichit, L., Dao, P., Fillion, L., Nam, Y.W., Pech, J.C. and Latche, A. (1997) Effects of chilling on the expression of ethylene biosynthetic genes in Passe-Crassane pear (Pyrus communis L.) fruits. *Plant Molecular Biology* **33**, 847–855.

Lester, D.R., Sherman, W.B. and Atwell, B.J. (1996) Endopolygalacturonase and the melting flesh (M) locus in peach. *Journal of the American Society for Horticultural Science* **121**, 231–234.

Lester, D.R., Speirs, J., Orr, G. and Brady, C.J. (1994) Peach (Prunus persica) endopolygalacturonase cDNA isolation and mRNA analysis in melting and non-melting peach cultivars. *Plant Physiology* **105**, 225–231.

Levy, I., Shani, Z. and Shoseyov, O. (2002) Modification of polysaccharides and plant cell wall by endo-1,4-beta-glucanase and cellulose-binding domains. *Biomolecular Engineering* **19**, 17–30.

Llop-Tous, I., Dominguez-Puigjaner, E., Palomer, X. and Vendrell, M. (1999) Characterization of two divergent endo-beta-1,4-glucanase cDNA clones highly expressed in the nonclimacteric strawberry fruit. *Plant Physiology* **119**, 1415–1422.

Maclachlan, G. and Brady, C. (1994) Endo-1,4-[beta]-glucanase, Xyloglucanase, and Xyloglucan Endo-Transglycosylase activities versus potential substrates in ripening tomatoes. *Plant Physiology* **105**, 965–974.

Marin-Rodriguez, M.C., Orchard, J. and Seymour, G.B. (2002) Pectate lyases, cell wall degradation and fruit softening. *Journal of Experimental Botany* **53**, 2115–2119.

McCann, M. (1997) Tracheary element formation: building up to a dead end. *Trends in Plant Science* **2**, 333–338.

McQueen-Mason, S.J. and Cosgrove, D.J. (1995) Expansin mode of action on cell walls. Analysis of wall hydrolysis, stress relaxation, and binding. *Plant Physiology* **107**, 87–100.

McQueen-Mason, S.J., Durachko, D.M. and Cosgrove, D.J. (1992) Two endogenous proteins that induce cell wall extension in plants. *Plant Cell* **4**, 1425–1433.

Medina-Escobar, N., Cardenas, J., Moyano, E., Caballero, J.L. and Munoz-Blanco, J. (1997) Cloning, molecular characterization and expression pattern of a strawberry ripening-specific cDNA with sequence homology to pectate lyase from higher plants. *Plant Molecular Biology* **34**, 867–877.

Mueller, L.A., Solow, T.H., Taylor, N., Skwarecki, B., Buels, R., Binns, J., Lin, C., Wright, M.H., Ahrens, R., Wang, Y., Herbst, E.V., Keyder, E.R., Menda, N., Zamir, D. and Tanksley, S.D. (2005) Genomics Network: a comparative resource for Solanaceae biology and beyond. *Plant Physiology* **138**, 1310–1317.

Nunan, K.J., Davies, C., Robinson, S.P. and Fincher, G.B. (2001) Expression patterns of cell wall-modifying enzymes during grape berry development. *Planta* **214**, 257–264.

Nunan, K.J., Sims, I.M., Bacic, A., Robinson, S.P. and Fincher, G.B.(1998) Changes in cell wall composition during ripening of grape berries. *Plant Physiology* **118**, 783–792.

O'Donoghue, E.M., Huber, D.J., Timpa, J.D., Erdos, G.W. and Brecht, J.K. (1994) Influence of avocado (Persea americana) Cx-cellulase on the structural features of avocado cellulose. *Planta* **194**, 573–584.

Orfila, C., Seymour, G.B., Willats, W.G., Huxham, I.M., Jarvis, M.C., Dover, C.J., Thompson, A.J. and Knox, J.P. (2001) Altered middle lamella homogalacturonan and disrupted deposition of (1→5)-alpha-L-arabinan in the pericarp of Cnr, a ripening mutant of tomato. *Plant Physiology* **126**, 210–221.

Paull, R.E. and Chen, N.J. (1983) Postharvest variation in cell wall-degrading enzymes of papaya (Caria papaya) during fruit ripening. *Plant Physiology* **72**, 382–385.

Pnueli, L., Hareven, D., Rounsley, S.D., Yanofsky, M.F. and Lifschitz, E. (1994) Isolation of the tomato AGAMOUS gene TAG1 and analysis of its homeotic role in transgenic plants. *Plant Cell* **6**, 163–173.

Potuschak, T., Lechner, E., Parmentier, Y., Yanagisawa, S., Grava, S., Koncz, C. and Genschik, P. (2003) EIN3-dependent regulation of plant ethylene hormone signaling by two Arabidopsis F box proteins: EBF1 and EBF2. *Cell* **115**, 679–689.

Pressey, R. (1983) b-galactosidases in ripening tomatoes. *Plant Physiology* **71**, 32–135.

Pressey, R. (1989) Endo-b-mannanase in ripening tomatoes. *Pythochemistry* **28**, 3277–3280.

Pua, E.-C., Ong, C.K., Liu, P. and Liu, J.-Z. (2001) Isolation and expression of two pectate lyase genes during fruit ripening of banana (Musca acuminata). *Physiologia Plantarum* **113**, 92–99.

Rao, G.U. and Paran, I. (2003) Polygalacturonase: a candidate gene for the soft flesh and deciduous fruit mutation in Capsicum. *Plant Molecular Biology* **51**, 135–141.

Ray, J., Knapp, J., Grierson, D., Bird, C. and Schuch, W. (1988) Identification and sequence determination of a cDNA clone for tomato pectin esterase. *European Journal of Biochemistry* **174**, 119–124.

Redgwell, R.J. and Fischer, M. (2001) Fruit texture, cell wall metabolism and consumer perceptions. In: *Fruit Quality and its Biological Basis* (ed. Knee, M.). CRC Press LLC, Boca Raton, FL.

Redgwell, R.J., MacRae, E., Hallet, I., Fischer, M., Perry, J. and Harker, R. (1997) *In vivo* and in vitro swelling of cell walls during fruit ripening. *Planta* **203**, 162–173.

Rose, J.K. and Bennett, A.B. (1999) Cooperative disassembly of the cellulose-xyloglucan network of plant cell walls: parallels between cell expansion and fruit ripening. *Trends in Plant Science* **4**, 176–183.

Rose, J.K., Cosgrove, D.J., Albersheim, P., Darvill, A.G. and Bennett, A.B. (2000) Detection of expansin proteins and activity during tomato fruit ontogeny. *Plant Physiology* **123**, 1583–1592.

Rose, J.K., Hadfield, K.A., Labavitch, J.M. and Bennett, A.B. (1998) Temporal sequence of cell wall disassembly in rapidly ripening melon fruit. *Plant Physiology* **117**, 345–361.

Rose, J.K., Lee, H.H. and Bennett, A.B. (1997) Expression of a divergent expansin gene is fruit-specific and ripening-regulated. *Proceedings of the National Academy of Sciences of the United States of America* **94**, 5955–5960.

Rottmann, W.H., Peter, G.F., Oeller, P.W., Keller, J.A., Shen, N.F., Nagy, B.P., Taylor, L.P., Campbell, A.D. and Theologis, A. (1991) 1-Aminocyclopropane-1-carboxylate synthase in tomato is encoded by a multigene family whose transcription is induced during fruit and floral senescence. *Journal of Molecular Biology* **222**, 937–961.

Roy, S., Vian, B. and Roland, J.-C. (1992) Immunocytochemical study of the deesterification pattern during cell wall autolysis in the ripening of cherry tomato. *Plant Physiology and Biochemistry* **30**, 139–146.

Schuch, W., Kanczler, J. and Robertson, D. (1991) Fruit quality characteristics of transgenic tomato fruit with altered polygalacturonase activity. *Horticultural Science* **26**, 1517–1520.

Seymour, G., Taylor, J.E. and Tucker, G.A. (1993) *Biochemistry of Fruit Ripening*. Chapman and Hall, London.

Sheehy, J.L. (1988) Cholesteatoma surgery: canal wall down procedures. *Annals of Otology, Rhinology and Laryngology* **97**, 30–35.

Sitrit, Y., Hadfield, K.A., Bennett, A.B., Bradford, K.J. and Downie, A.B. (1999) Expression of a polygalacturonase associated with tomato seed germination. *Plant Physiology* **121**, 419–428.

Smith, C.J.S., Watson, C.F. and Ray, J. (1988) Antisense RNA inhibition of polygalacturonase gene expression in transgenic tomatoes. *Nature* **334**, 724–726.

Smith, D.L., Abbott, J.A. and Gross, K.C. (2002) Down-regulation of tomato beta-galactosidase 4 results in decreased fruit softening. *Plant Physiology* **129**, 1755–1762.

Smith, D.L. and Gross, K.C. (2000) A family of at least seven beta-galactosidase genes is expressed during tomato fruit development. *Plant Physiology* **123**, 1173–1183.

Sozzi, G.O., Greve, L.C., Prody, G.A. and Labavitch, J.M. (2002) Gibberellic acid, synthetic auxins, and ethylene differentially modulate alpha-L-Arabinofuranosidase activities in antisense 1-aminocyclopropane-1-carboxylic acid synthase tomato pericarp discs. *Plant Physiology* **129**, 1330–1340.

Stepanova, A.N. and Ecker, J.R. (2000) Ethylene signaling: from mutants to molecules. *Current Opinion in Plant Biology* **3**, 353–360.

Stewart, D., Iannetta, P.P. and Davies, H.V. (2001) Ripening-related changes in raspberry cell wall composition and structure. *Phytochemistry* **56**, 423–428.

Tateishi, A., Kanayama, Y. and Yamaki, S. (1996) a-L-arabinofuranosidase from cell walls of Japanese pear fruits. *Phytochemistry* **42**, 295–299.

Thompson, A.J., Tor, M., Barry, C.S., Vrebalov, J., Orfila, C., Jarvis, M.C., Giovannoni, J.J., Grierson, D. and Seymour, G.B. (1999) Molecular and genetic characterization of a novel pleiotropic tomato-ripening mutant. *Plant Physiology* **120**, 383–390.

Tieman, D.M., Ciardi, J.A., Taylor, M.G. and Klee, H.J. (2001) Members of the tomato LeEIL (EIN3-like) gene family are functionally redundant and regulate ethylene responses throughout plant development. *Plant Journal* **26**, 47–58.

Tieman, D.M. and Handa, A.K. (1994) Reduction in pectin methylesterase activity modifies tissue integrity and cation levels in ripening tomato (lycopersicon esculentum mill.) fruits. *Plant Physiology* **106**, 429–436.

Tieman, D.M., Harriman, R.W., Ramamohan, G. and Handa, A.K. (1992) An antisense pectin methylesterase gene alters pectin chemistry and soluble solids in tomato fruit. *Plant Cell* **4**, 667–679.

Tieman, D.M. and Klee, H.J. (1999) Differential expression of two novel members of the tomato ethylene-receptor family. *Plant Physiology* **120**, 165–172.

Tournier, B., Sanchez-Ballesta, M.T., Jones, B., Pesquet, E., Regad, F., Latche, A., Pech, J.C. and Bouzayen, M. (2003) New members of the tomato ERF family show specific expression pattern and diverse DNA-binding capacity to the GCC box element. *FEBS Letters* **550**, 149–154.

Tucker, G.A., Robertson, N.G. and Grierson, D. (1982) Purification and changes in activities of tomato pectinesterases enzymes. *Journal of the Science of Food and Agriculture* **33**, 396–400.

Vrebalov, J., Ruezinsky, D., Padmanabhan, V., White, R., Medrano, D., Drake, R., Schuch, W. and Giovannoni, J. (2002) A MADS-box gene necessary for fruit ripening at the tomato ripening-inhibitor (rin) locus. *Science* **296**, 343–346.

Wakabayashi, K. (2000) Changes in cell wall polysaccharides during fruit ripening. *Journal of Plant Research* **113**, 231–237.

Wallner, S.J. and Walker, J.E. (1975) Glycosidases in cell wall-degrading extracts of ripening tomato fruits. *Plant Physiology* **55**, 94–98.

Wang, W., Hall, A.E., O'Malley, R. and Bleecker, A.B. (2003) Canonical histidine kinase activity of the transmitter domain of the ETR1 ethylene receptor from Arabidopsis is not required for signal transmission. *Proceedings of the National Academy of Sciences of the United States of America* **100**, 352–357.

Whitney, S.E., Gidley, M.J. and McQueen-Mason, S.J. (2000) Probing expansin action using cellulose/hemicellulose composites. *Plant Journal* **22**, 327–334.

Wilkinson, J.Q., Lanahan, M.B., Clark, D.G., Bleecker, A.B., Chang, C., Meyerowitz, E.M. and Klee, H.J. (1997) A dominant mutant receptor from Arabidopsis confers ethylene insensitivity in heterologous plants. *Nature Biotechnology* **15**, 444–447.

Woolley, L.C., James, D.J. and Manning, K. (2001) Purification and properties of an endo-beta-1,4-glucanase from strawberry and down-regulation of the corresponding gene, cel1. *Planta* **214**, 11–21.

Yamaki, S. and Kakiuchi, N. (1979) Changes in hemicellulose-degrading enzymes during development and ripening of Japanese pear fruit. *Plant Cell Physiology* **20**, 301–309.

Yang, D.J., Richmond, C.D., Teets, V.J., Brown, P.I. and Rankin, G.O. (1985) Effect of succinimide ring modification on N-(3,5-dichlorophenyl)succinimide-induced nephrotoxicity in Sprague-Dawley and Fischer 344 rats. *Toxicology* **37**, 65–77.

Yokotani, N., Tamura, S., Nakano, R., Inaba, A. and Kubo, Y. (2003) Characterization of a novel tomato EIN3-like gene (LeEIL4). *Journal of Experimental Botany* **54**, 2775–2776.

9 The role of polymer cross-linking in intercellular adhesion

Keith W. Waldron and Christopher T. Brett

9.1 Introduction

The morphology of a plant is a manifestation of a hierarchy of interrelated structures (Figure 9.1), all of which are dynamic at some time during plant growth and development, involving co-ordinated and controlled expansion of adjacent but attached cells. The adhesion of plant cells to one another is fundamental to the formation and maintenance of plant structure. Consequently, the nature, control and regulation of cell adhesion relates to numerous important plant functions throughout the life cycle.

In most supporting tissues, cell adhesion maintains the contact of plant cells under different loads and stresses. However, some plant structures exhibit controlled cell separation events. These include flower and leaf abscission (see Chapter 6) (Roberts *et al.*, 2000, 2002), dehiscence of seed pods and associated dynamic seed dispersal (see Chapter 7), pollen formation and fruit ripening (see Chapter 8) (Brett and Waldron, 1996). External biological agents may seek to compromise cell adhesion. Invasion of plant tissues by pathogens often involves the separation of cells, and the plant may respond to this through modification of the cell wall chemistry associated with adhesion (Waldron *et al.*, 1997).

In addition to the maintenance of natural physiological function, the role of cell adhesion on the mechanical properties of plant tissues has important socio-economic impacts at different stages along the food chain (Figure 9.2). The texture of plant-based foods is dependent to a large extent on the nature of tissue fracture during mastication. If cell adhesion is maintained, fracture involves cell wall rupture (Figure 9.3). In the case of vegetables and fruits, this is associated with a crisp texture and associated release of flavours and nutrients (Waldron *et al.*, 1997). Ripening related softening of fruits can involve the weakening of the cell walls, and a reduction in the strength of cell adhesion. If the adhesion is sufficiently weak, the cells will separate during eating. The encapsulation of the juices and flavours creates a mealy texture, often considered to be a low quality characteristic (Waldron *et al.*, 2003). Consequently, there is significant interest in developing means to prevent such deterioration, and to extend shelf life. Edible fruits and particularly vegetables are often thermally softened (cooked) to make them acceptable for ingestion. Vegetables include leaf-based organs (e.g. cabbage, onion), stem-based organs (broccoli, cauliflower, asparagus), roots (carrot, potato) and seeds (beans). Thermal softening involves an increase in the ease of cell separation (Waldron *et al.*, 2003) and will be strongly influenced by the nature of the chemical cross-links

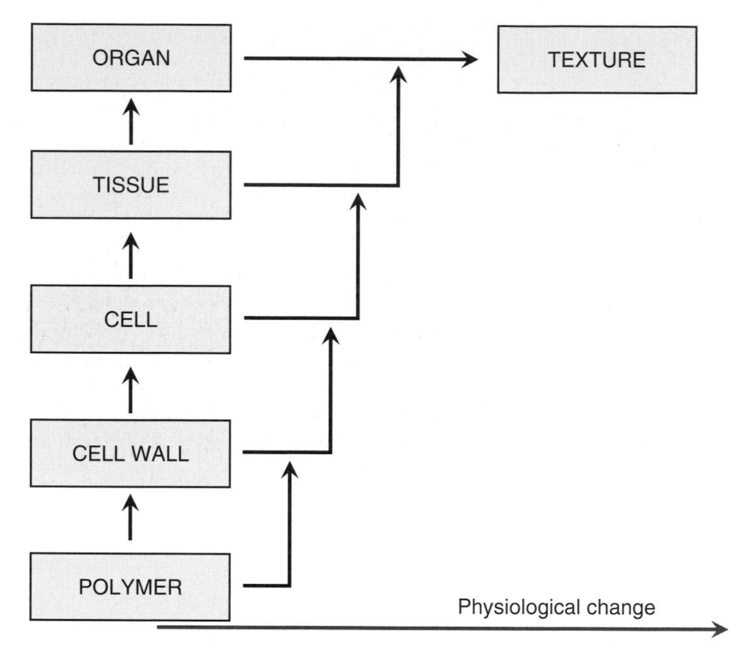

Figure 9.1 The hierarchy of structures, which underlie the morphology of plants.

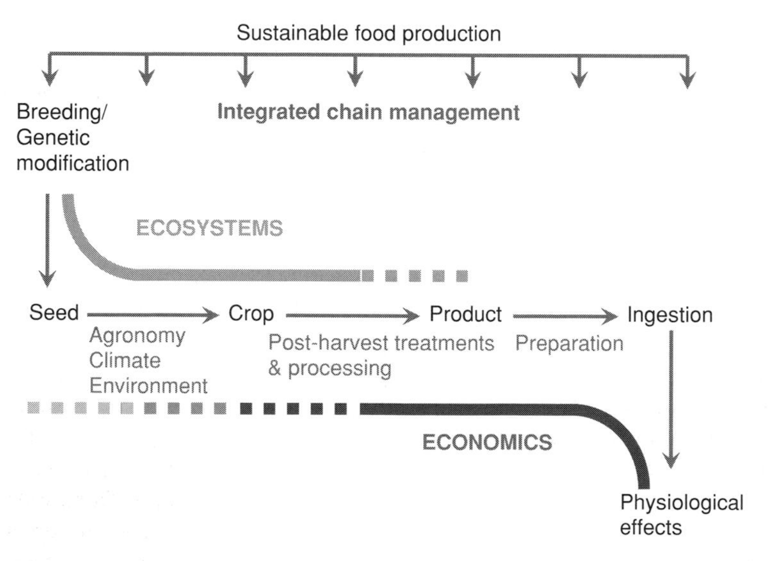

Figure 9.2 The food chain.

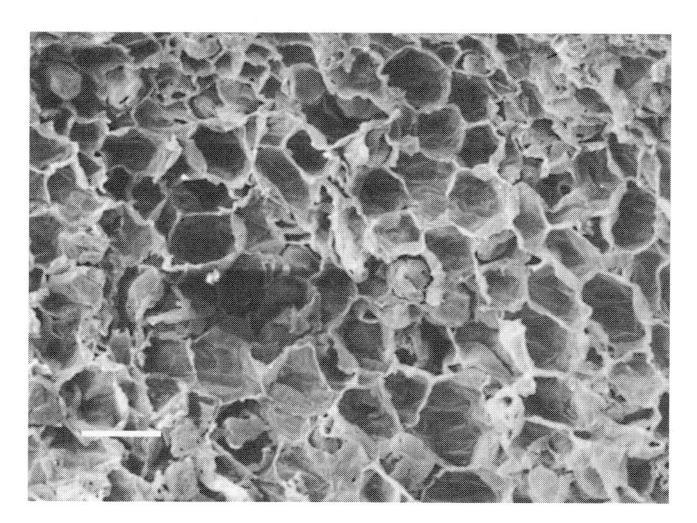

Figure 9.3 Cell wall rupture in unripe olive parenchyma.

involved in cell adhesion. That, in turn, will be dependent on the physiological nature of the tissues, food-chain-related changes in the tissues during plant growth and development (Figure 9.2), the influence of post-harvest activities such as storage and the nature and extent of the food processing activities. Such criteria are not only important to the food chains serving economically wealthy countries, but are of considerable importance in developing countries where staple foods may be compromised by deterioration in textural quality. For example, the hard-to-cook defect and the associated hard-shell defect in grain legumes can reduce the acceptability of high-protein dry beans to an extent that it may impinge on nutritional well-being (see Section 9.3.3; Liu, 1995).

Hence, an understanding of the chemistry and biochemistry of cell adhesion is of considerable importance for the future control and exploitation of plants used for both food and non-food uses. In this respect, it is often assumed that by understanding the physics, chemistry, biochemistry and molecular biology underpinning cell adhesion it should be possible to manipulate plant structure and, therefore, cell-wall-related quality characteristics such as texture. Whilst there is no doubt such knowledge will contribute to an understanding of how to control quality, it should be noted that 'quality' characteristics of edible fruits and vegetables should be considered from a food-chain perspective in which different parts of the food chain required dictate different aspects of quality, with the consumer interests dominating (Bech *et al.*, 2001; Waldron *et al.*, 2003).

9.2 The location of intercellular cross-links within the wall

Plant cell walls typically exhibit layered structures as visualised by light and electron microscopy (Brett and Waldron, 1996). In immature and fleshy tissues, the cell walls are made up of a primary cell wall from each cell separated by the middle lamella and intercellular spaces. Within the cell wall, the main components (cellulose, hemicelluloses, pectins, proteins, phenolics) are arranged in a complex but ordered manner (for examples of wall models, see Brett and Waldron 1996 and references therein). During cell extension, the polymers that constitute the cell wall and middle lamella are synthesised by the cells on either side of the wall and secreted into the space between them. Intercellular adhesion requires the formation of cross-links, either covalent or non-covalent, between polymers that lie between the two primary walls. In this context, cross-links are defined as bonds that link molecules derived from each of the two cells. Without such cross-linking, intercellular adhesion cannot occur. Hence, the formation of such cross-links and its control are key aspects of intercellular adhesion.

Cell separation can occur by degrading either these cross-links or the bonds within and between the polymers derived from the individual cells. Hence, the focus of this chapter is on cell adhesion rather than cell separation. Nevertheless, the nature of the linkages involved in cell adhesion may be elucidated tentatively by investigations that have concentrated on the induction of cell separation. These include physical and (bio)chemical approaches and also molecular genetic modification (see, e.g. Jarvis *et al.*, 2003). This chapter considers the nature of these cross-links, and also how intercellular adhesion may be controlled by altering the cross-linking process.

Most plant cells retain the same neighbours for life. A new cell wall forms between the daughter cells at cytokinesis, and the two cells remain linked by the same wall until cell separation or cell death occurs (Verma, 2001). The middle lamella forms the boundary between the two cells, as most or all of the material in each primary wall is thought to be contributed only by the cell on that side of the middle lamella. Hence, the cross-links responsible for intercellular adhesion are likely to be in the middle lamella. Because the middle lamella is between the adjacent cells, it was previously assumed that the middle lamella is generally the area of cell adhesion, and that cell adhesion occurs across the entirety of the middle lamella. However, research during the last 20 years has provided sufficient evidence to indicate that cell adhesion is probably dependent on cross-links present predominantly at the edges of cell faces, also known as tricellular junctions. The evidence for this is as follows.

9.2.1 *Cell adhesion at tricellular junctions*

9.2.1.1 *Intercellular space formation*

The formation of new tricellular junctions is a particularly critical point in the generation of intercellular adhesion. In studies on germinating pea seeds, Kolloffel and Linssen (1984) explored the later stages of cell division. They observed that a highly localised region of the parent cell is controllably degraded to facilitate connection of the newly formed middle lamella between the daughter cells, to the middle lamella surrounding the parent cell. Intercellular space formation initiates at the central point of contact between the two daughter cells and the adjacent cell, spreading along the middle lamella until it terminates at what they observed as pre-existing

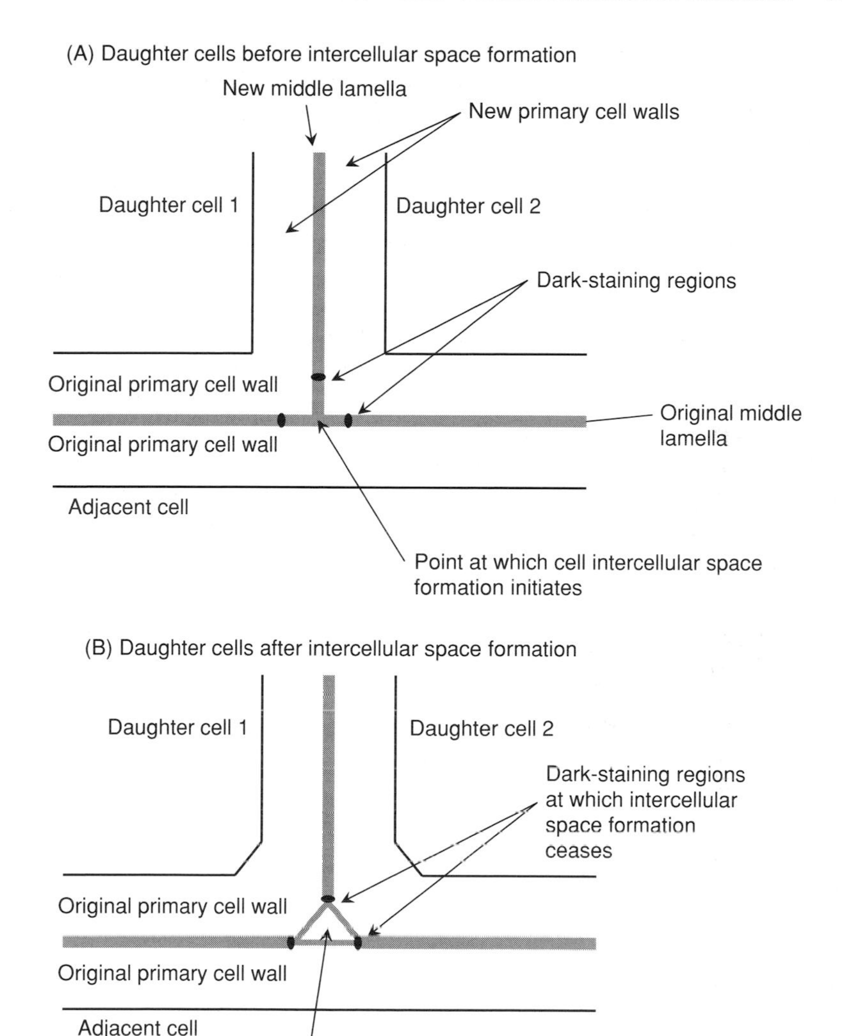

Figure 9.4 Intercellular space formation. (A) Formation of electron-dense intra-wall structures that (B) enable intercellular space formation without full cell separation.

electron-dense intra-wall structures (Figure 9.4). These are then located at the corners of the pre-determined intracellular spaces and act as a 'zipstop' in that if intercellular space formation was allowed to continue, the cells would eventually separate.

9.2.1.2 Studies of the cell surface after cell separation

Parker *et al.* (2001) have carried out in-depth studies of the surface of potato cells after having induced cell separation by mild chemical or thermal treatments. The

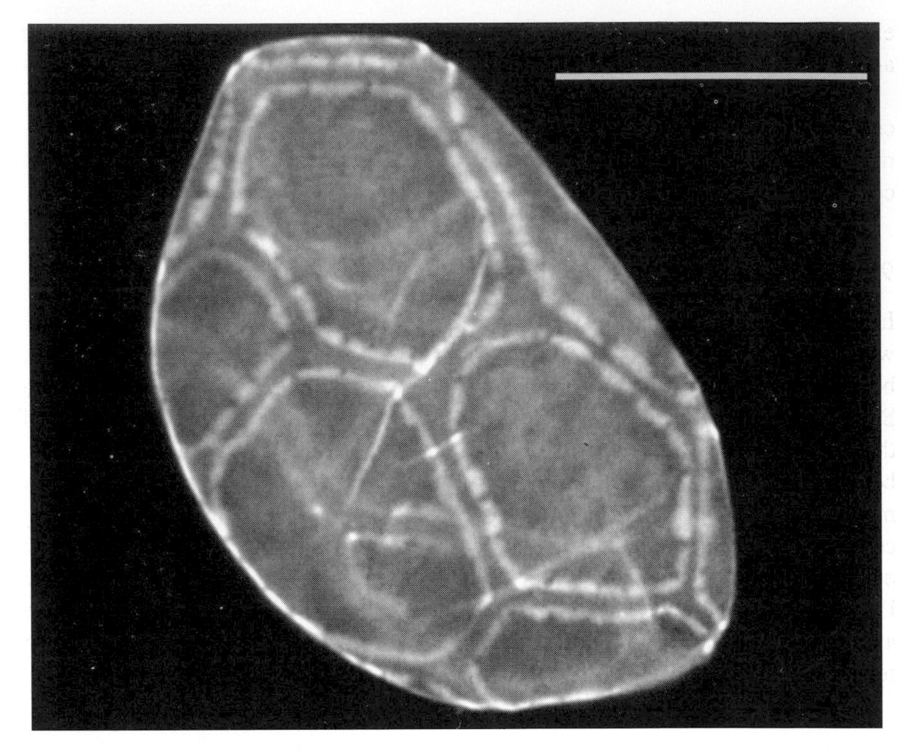

Figure 9.5 Fluorescence micrograph of single cell of Chinese water chestnut showing location of ferulic acid components on cell surface.

results from both approaches were similar and demonstrated clearly that the middle lamella can be viewed as a separate entity, which is very poorly cross-linked to the outer surface area of the plant cell, except at the edges of the cell faces where it is strongly adhered.

9.2.1.3 Chemical nature of components at the edge of the cell face

A number of studies involving monoclonal antibodies specific to cell wall polysaccharides (e.g. Willats *et al.*, 2001a,b) and fluorescence microscopy of a range of cell wall components (Figure 9.5) (Parker and Waldron, 1995; Waldron *et al.*, 1997) have demonstrated that the chemistry of the edges of the cell faces differs considerably from that of the rest of the primary cell wall and middle lamella. Furthermore, moieties located at the edges of the faces are typical of polymers that can form cross-links, which have been associated with cell adhesion. These will be elaborated below.

9.2.1.4 Mathematical modelling

Since most plant cells are polyhedral rather than spherical, the greatest stress arising from turgor pressure is concentrated at the cell corners, i.e. at the edge of each cell face (Jarvis, 1998). So, it is the middle lamella at these cell corners (tricellular junctions) that is probably most important in this respect, precisely where electron-dense material (Kollofel and Linssen, 1984) and cross-linking polymers and moieties have been located.

It should also be noted that the cell wall normally undergoes stretching during cell growth, and hence new cell wall material is deposited by intussusception to maintain wall thickness. Hence, maintenance of intercellular adhesion requires the continued formation of new cross-links as the wall stretches.

9.2.2 *Cell adhesion in locations other than the tricellular junction*

In addition to adhesion at the edges of the cell faces, it is also clear that the primary walls of adjacent cells may be firmly fixed to each other via pit fields. This has been shown to be of relevance to textural and eating qualities in pears and olives. In Spanish pears (Martin-Cabrejas *et al.*, 1994), ripening results in general swelling of the cell walls and cell separation except at the pit fields. These act like 'spot welds' between the cells, and provide points of wall rupture in very ripe fruits. This property may be important in preventing the mealy character from developing by ensuring cell rupture occurs and juice is released during eating. A similar situation has been observed in onions and potatoes during thermal processing (Ng and Waldron, 1997a; LeCain *et al.*, 1999) and olives (Figure 9.6). In potatoes, such a character may have a role in the release of starch in cooked potato. In olives, this property may be of relevance to the degree to which oil may be extracted during first pressing.

9.2.3 *Role of secondary thickening in cell adhesion*

Although not the focus of this chapter, changes in cell wall chemistry during cell differentiation and secondary thickening may also affect adhesion of primary-walled cells. This has been highlighted in the case of sclereid development in pears

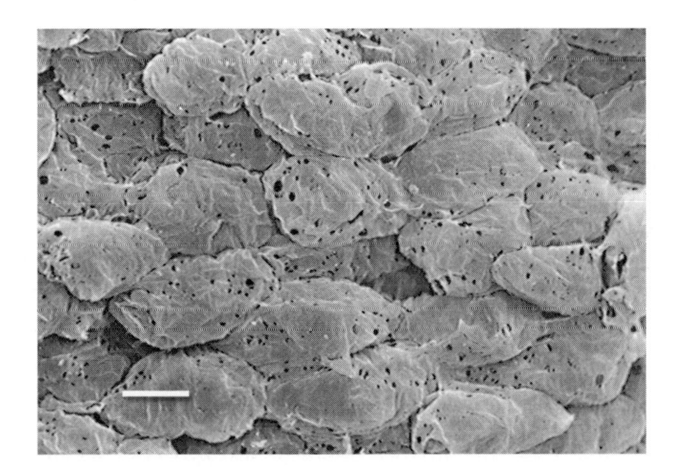

Figure 9.6 Holes in separated cells from ripe olives, which appear to result from localised tearing due to pit-field wall adhesion.

(Martin-Cabrejas *et al.*, 1994). Sclereid lignification can result in cross-linking, which affects cell walls of adjacent parenchyma cells. This can enhance cell adhesion between the sclereid cluster and adjacent parenchyma cell, preventing ripening-related cell separation and ensuring rupture with associated juice release during mastication.

9.3 The nature of intercellular polymers and their cross-links

Investigations into the possible nature of polymers and interpolymeric cross-links involved in cell adhesion can be broadly divided into two categories. First, a number of studies have sought to perturb cell adhesion through the use of chemical, biochemical, physical and genetic approaches (Van Buren, 1979; Parker and Waldron, 1995; Shevell *et al.*, 2000; Orfilla *et al.*, 2001; Atkinson *et al.*, 2002; Eriksson *et al.*, 2004). From the changes (thought to be) made in the cell wall, which result in modifications to the nature and extent of cell adhesion, the identity and character of polymers involved in cell adhesion might be inferred. Second, histochemical studies have been carried out on regions of the cell wall in which adhesion is thought to occur (e.g. Willats *et al.*, 2001a). Investigations involving the use of antibodies specific to cell wall polymer epitopes have been particularly important. Many studies involve a combination of these approaches.

9.3.1 Role of pectic polysaccharides and their cross-links in intercellular adhesion

Pectic polymers consist of polysaccharides rich in galacturonic acid (GalA) often with significant amounts of rhamnose (Rha), arabinose (Ara) and galactose (Gal). They are generally ubiquitous to primary cell walls of both monocots and dicots, and probably form a gel-like matrix within the cellulose–hemicellulose network. The three main polymer types are homogalacturonan (HG), rhamnogalacturonan I (RG I) and rhamnogalacturonan II (RG II) (Brett and Waldron, 1996).

Pectic polysaccharides, especially HGs, are the major constituents of the middle lamella, and are thus the strongest candidates. HGs can be ionically cross-linked by calcium ions binding to galacturonate residues on each of two neighbouring HG molecules. The strength of such ionic cross-bridges is greatly increased when they act synergistically, which occurs if several ionised galacturonate residues are adjacent to one another on each of the two homogalacturonate chains (Morris *et al.*, 1982). Early evidence for the role of pectic polysaccharides in cell adhesion came from studies on chemical and thermal softening of fruits and vegetables, and studies on fruit ripening.

9.3.1.1 Evidence for the involvement of pectic polymers in cell adhesion

Thermally- and chemically-induced cell separation

Thermal softening of edible, immature plant tissues generally results from an increase in the ease of cell separation, and is accompanied by the dissolution of pectic polymers, mainly as a result of β-elimination (van Buren, 1979). Numerous studies have been carried out to compare and correlate the two, often involving mathematical modelling of the texture (Tijskens *et al.*, 1997a,b).

In many immature tissues, cell separation can also be induced by extracting the tissues in chelating agents such as EDTA and CDTA (Brett and Waldron, 1996). Conversely, perfusing tissues with calcium salts (Ng and Waldron, 1997a) or increasing the levels in the tissues relative to monovalent cations through agronomic intervention (Van Buren and Peck, 1982) increases the resistance of tissues to thermal softening. This indicates that divalent cations such as calcium, through the cross-linking of pectic polysaccharides, play an important role in cell adhesion.

Since HGs are secreted in a relatively highly esterified form, intercellular adhesion generated by this mechanism probably requires the prior action of pectin methylesterases (PMEs) *in muro* (Micheli, 2001). PMEs can act in a number of different ways (Willats *et al.*, 2001b). Some remove methyl groups contiguously, resulting in a stretch of fully demethylated HG. Others remove them at random, resulting in a HG with a random distribution of methyl groups. The overall degree of methylation resulting from their action also varies. PMEs form a large gene family, and plants contain both the types (Micheli, 2001). PMEs are thought to be required for 'pre-cooking'-enhanced texture. If tissues such as carrot root or bean pod are subjected to thermal treatment at 50°C for 30 min, their subsequent thermal softening at 100°C will be considerably reduced in rate and extent. This is thought to be due to the thermal stimulation of wall-bound PMEs, which de-esterify the pectic polymers involved in cell adhesion, thereby increasing the propensity for calcium cross-linking (indeed, added calcium then has a larger impact after precooking; Ng and Waldron, 1997a,b), and reducing the thermal depolymerisation of the pectic polysaccharides themselves. The latter effect is due to the decrease in susceptibility for β-eliminative degradation of pectic polymers, which is greatest if they are methyl-esterified (Van Buren, 1979; Brett and Waldron, 1996).

Fruit ripening and the role of cell wall degrading enzymes

Studies on ripening of fruits have also provided a large body of evidence for the involvement of pectic polymers in cell adhesion. In particular, the changes in wall-degrading enzymes and the accompanying dissolution of pectic polymers during cell separation is well documented. Aspects of ripening are covered in Chapter 8. Related and important evidence for the role of pectic polymers in cell adhesion has also come from studies in which cell separation is induced by exogenously applied polygalacturonase (PG; Brett and Waldron, 1996).

9.3.1.2 *Nature of pectic polymers and their cross-links*

Histochemistry of pectic polymers and their cross-links

A number of monoclonal antibodies are now available, specific for particular pectin epitopes (Willats *et al.*, 2001a). Some of these antibodies bind to pectic HG with specific patterns of methyl esterification of the GalA carboxyl groups. In view of the possible roles of calcium cross-links between neighbouring HG chains in intercellular adhesion, it is of particular interest that the regions of the cell wall most strongly involved adhesion contain specific patterns of these epitopes.

JIM5 and JIM7 are monoclonal antibodies, which bind to HG with different degrees of methylation. The JIM5 epitope contains from 0 to 50% methylation, while JIM7 binds to 35–90% methylated HG (Knox *et al.*, 1990). Studies using transmission electron microscopy indicate that the JIM7 epitope is evenly distributed in all parts of the primary cell wall. However, JIM5 epitopes, which are more likely to form calcium cross-links, were concentrated at the cell corners (Knox *et al.*, 1990; Liners and Van Cutsem, 1992; Roy *et al.*, 1994). This has been confirmed by examining the surfaces of separated potato parenchyma cells by scanning electron microscopy (SEM) (Parker *et al.*, 2001). These SEM studies have indicated that the JIM5 epitope, but not the JIM7 epitope, is selectively concentrated at the edges of the cell faces, which are equivalent in three dimensions of the cell corners seen in two-dimensional TEM. In the same SEM study, it was also seen that the middle lamella was most strongly attached to the primary wall at the edges of the cell faces (see above). This was observed by bringing about a partial separation, either by storage of the tissue in methanol or by thermal treatment.

These results complement studies using secondary ion mass spectroscopy (SIMS). This technique has indicated that calcium ions are selectively concentrated at the edges of the cell face (Roy *et al.*, 1995). Since calcium ions are thought to act as ionic cross-linking agents for de-esterified pectin (see above), their presence at the cell corners further strengthens the hypothesis that such cross-linking is important in intercellular adhesion.

More recently, additional monoclonal antibodies binding to partially esterified HG have become available. Epitopes for two of these, LM7 and PAM1, are located specifically at the cell corners (Willats *et al.*, 2001b). LM7 binds to pectin with a random pattern of esterification; this epitope is found most strongly at the corners of intercellular spaces in all dicot tissues examined, at the point where the maximum cell separation stress is concentrated in turgid cells. To a lesser extent, the LM7 epitope also occurs along the adjacent cell wall layers lining the intercellular spaces. The PAM1 epitope contains long blocks of contiguously de-esterified pectin; it also is found in the cell walls lining the intercellular spaces, but the LM7 and PAM1 epitopes do not overlap in the wall, even though they are very closely aligned. *In vitro* studies with model pectins containing substantial amounts of the LM7 epitope indicate that they can form calcium-cross-linked gels, and that these gels have distinctive properties. The PAM1 epitope is also able to participate in calcium cross-linking.

As discussed above, there is good evidence for a role for intermolecular calcium cross-links between partially de-esterified HG molecules in intercellular adhesion. However, it is unlikely that this is the only mechanism involved. Extraction of dicot cell walls with chelating agents solubilises only part of the pectin present, and seldom achieves full cell separation (Jarvis *et al.*, 2003). So other bonds, probably of a covalent nature, are likely to be involved as well.

Alkali-labile bonds involving pectin

Extraction of cell walls with mild alkali after extraction with chelating agents further solubilises considerable amounts of pectin, and is often needed to bring about cell separation (McCartney and Knox, 2002). The strength of alkali involved is sufficient to break ester or amide bonds. Ester bonds could involve the C-6 carboxylic acid groups of GalA, linked to a hydroxyl group on another polysaccharide or a protein. Alternatively, the carboxyl group could be provided by another molecule, e.g. a protein, linked to a pectin hydroxyl group. Amide linkages could involve an amino group on a protein or a polyamine.

In support of these ideas, measurements of the total pectin carboxyl groups involved in alkali-labile linkages suggest that this number significantly exceeds the amount of methanol released by mild alkali (Kim and Carpita, 1992; Mackinnon *et al.*, 2002). The presumption is that non-methyl esters or amides contribute to the remainder. There is no evidence for any other low-molecular weight compounds being released, so it seems likely that these non-methyl esters or amides link pectin to other macromolecules in the cell wall. Possible candidates include borate cross-links and phenolic cross-links.

Borate cross-linking of rhamnogalacturonan II

RG II is a minor component of pectin, but its structure is highly conserved in all dicots, indicating an important function. It contains a backbone of α-(1,4)-galacturonan, to which four distinct oligosaccharide side chains are attached (Ridley *et al.*, 2001; O'Neill *et al.*, 2004). One of these contains an apiose residue, which is normally cross-linked via a borate diester to an apiose residue in another RG II domain (O'Neill *et al.*, 1996; Ishii *et al.*, 1999). Mutants in which the borate diester cross-links fail to form show a severe cell adhesion defect, and callus cultures of the mutant are non-organogenic (Iwai *et al.*, 2002). Hence, one role for RG II may be in intercellular adhesion. However, immunocytochemical evidence suggests that RG II is present in the primary wall rather than the middle lamella (Williams *et al.*, 1996; Matoh *et al.*, 1998), and so this role is assumed to be indirect, rather than the borate diester being involved physically in holding the cells together.

Phenolics

There is now very good evidence that cinnamic acid derivatives play an important role in cell adhesion in both dicotyledonous and monocotyledonous plants through the cross-linking of both pectic and hemicellulosic polymers. This is presented in Section 9.4.

9.3.2 *Cell adhesion and pectin biosynthesis*

The immunocytochemical evidence shows that the location of pectins in the cell wall is very finely controlled, especially at the cell corners. It also shows that the structures found there could have a role as cross-linking polymers. Two important questions need to be considered: first, how might such pectic polymers be formed? and second, how might they come to be located at particular points in the wall with such precision?

9.3.2.1 *Pectin biosynthesis*

Pectin is formed in the Golgi apparatus. The polysaccharide synthases, which form the pectic backbone (HG and the alternating GalA–Rha backbone of RG I), are located there, as are the glycosyl transferases, which add the side chains (Brett and Waldron, 1996). It is possible that the process is completed in the Golgi vesicles during exocytosis, but no further polymerisation is thought to occur after the pectin is deposited in the wall. Likewise, pectin methyltransferase (PMT) is localised in the Golgi, so the newly synthesised pectin that is exported to the cell wall has a relatively high degree of methylation. PMT adds methyl groups to form a random pattern of esterification, so it is possible that some cell wall pectins, e.g. those which contain the LM7 epitope, are synthesised in the Golgi apparatus with their correct, final structure and are not further modified during secretion to the wall or during their subsequent lifetime in the wall.

9.3.2.2 Pectin modification in muro

It is highly likely, although, that most pectin in the wall is modified *in muro* by the extensive array of PME isoforms found in the cell wall (Micheli, 2001). Such modification may indeed begin earlier, since PMEs are secreted via the Golgi apparatus and Golgi vesicles and could begin to demethylate pectin there. It has been suggested that the PME pro-region might act as an inhibitor of PME activity prior to its removal during post-translational processing, since it has significant sequence homology with known PME inhibitors. However, there is no experimental evidence available on this point as yet.

Whether or not PMEs may begin de-esterification prior to exocytosis, the de-esterification of pectins by PME in the cell wall is thought to be a normal part of wall maturation, since the degree of esterification decreases as cells age. The best-studied plant PMEs are the type-1 PMEs, which generally produce blockwise demethylation. In addition, the type-2 PMEs are also known to be present in plants, and they tend to produce random demethylation of pectin. However, the activity and mode of de-esterification of PMEs are influenced by factors such as pH and cation concentration, so it is possible that the detailed pattern of PME activity in the wall may depend on local conditions *in muro* (Denes *et al.*, 2000; Micheli, 2001).

9.3.2.3 Possible mechanisms for the generation of specific patterns of pectin methylation at the edges of cell faces

(1) One possible mechanism is that the specific pectin structures might be synthesised as such in the Golgi apparatus and secreted specifically to the correct point at the cell surface. Such targeted exocytosis of specific polysaccharides to particular points in the wall has not been reported. However, some targeting to more general regions of the cell wall must occur; for instance, the bulk of polysaccharide secretion in extending cells in the root or shoot must be to the sidewalls, which are extending, rather than to the end walls, which are not.

(2) Formation of particular pectin epitopes could occur by targeted secretion of particular PMEs, with specific patterns of de-esterification activity, to particular points in the cell wall, e.g. the edges of the cell faces. There, they could act on non-specifically secreted pectins to produce the specific epitopes found at the cell corners. Such targeting of particular proteins to particular regions of the cell surface has been observed, e.g. for auxin transporters. However, it would need to be very precisely targeted to produce the epitope pattern observed.

(3) The observed epitope distribution could be generated by the action of PMEs on HG, neither of which is secreted in a targeted fashion, but directed by the presence of particular conditions of, e.g., pH or ionic concentration at the cell corners.
(4) Self-assembly mechanisms might cause non-specifically secreted pectins to order themselves into precise structures after exocytosis. The very precise pattern of LM7 and PAM1 epitopes at cell corners suggests that self-assembly might be operating, at least over the short distances involved here.
(5) At present there is little evidence to suggest which of these mechanisms might operate. Of course, they are not mutually exclusive, and a combination of several mechanisms may be needed to achieve the observed patterns.

9.3.2.4 Other possible linkages

The middle lamella contains proteins (Smallwood *et al.*, 1994), which could be involved in a range of interactions affecting intercellular adhesion. These might include specific non-covalent interactions with pectins or with each other. They might also include covalent cross-links, e.g. including amides or esters or isodityrosine linkages.

Other linkages, which may be needed for intercellular adhesion, are those that attach the middle lamella to the components of the primary cell wall (at the edges of the cell faces – see Section 9.2). Some pectin molecules are large enough to cross the cell wall from the primary wall on one side, through the middle lamella, into the primary wall on the other side (Round, 1999; Morris *et al.*, 2003). If this occurs, its contribution to intercellular adhesion will depend on the covalent bonds within the pectin molecule and the interaction between the pectin and the primary wall components on either side. Recently, it has become clear that at least some pectin molecules are covalently linked to xyloglucan (Femenia *et al.*, 1999; Thompson and Fry, 2000; Brett *et al.*, 2005; Cumming *et al.*, 2005). This allows the possibility of binding of these pectin–xyloglucan complexes to the primary wall on each side of the middle lamella, since xyloglucan component of the complex would hydrogen-bond readily with the cellulose in the primary wall microfibrils. It has also been reported that pectin may bind to some arabinogalactan proteins (AGPs) (Baldwin *et al.*, 1993; Carpita and Gibeaut, 1993). Since some AGPs are anchored to cell membrane by phosphotidylinositide tails, this in principle allows bonding from one plasma membrane across the entire wall to the plasma membrane of the neighbouring cell. Another indication that AGPs might be involved in intercellular adhesion is that some contain fasciclin-like domains. Fasciclins are cell adhesion proteins found in Drosophila and Volvox (Gaspar *et al.*, 2001).

9.3.3 Phenolics in the primary wall and middle lamella

Some important families of angiosperms, especially the Poaceae and Chenopodeaceae, contain significant amounts of phenolic compounds in the primary wall and middle lamella (Waldron *et al.*, 1997). In these plants, phenolics play an important role in intercellular adhesion. The main phenolic compound involved is ferulic acid (FA), which forms ester linkages with cell wall sugars, mainly arabinose. In the Chenopodeaceae, the FA is esterified to the side chains of RG I, mainly to arabinose as well as, in lesser amounts, to galactose. In the Poaceae, which contain only small amounts of pectin, FA is linked to the single arabinose residues present as side chains on the xylan backbone of arabinoxylan. In both families, the FA moieties can be cross-linked to each other by the action of peroxidase and hydrogen peroxide (Thibault, 1986; Ng *et al.*, 1997). This results in the formation of ferulate dehydrodimers, of which six different types have been found (Ralph *et al.*, 1994; Waldron *et al.*, 1997). Subsequent cross-linking can occur, giving rise to trimers and higher oligomers of ferulate (Funk *et al.*, 2005; Rouau *et al.*, 2003). It is thought that such oligomers can act as nucleation sites for lignin biosynthesis (Bunzel *et al.*, 2004).

9.3.3.1 Studies on Chinese water chestnut

Ferulate dehydrodimers have been shown to play a key role in intercellular adhesion in those plants in which they are abundant. Chinese water chestnut (CWC) is prized for maintaining a crisp texture after heating, which is due to a very high thermal stability of cell adhesion. The parenchyma cells in this tissue are not separated by thermal treatments, which leave the ferulate cross-links intact. Nevertheless, cell separation can be induced by several different approaches, which affect different cell wall components, including hot and cold alkali treatments, which would be expected to release ferulic and diferulic acids (DiFAs); dilute, hot acid treatment, which would hydrolyse arabinofuranoside substituents on arabinoxylans; and pure endoxylanase, which would break down cell wall xylans. This has led to the conclusion that cell adhesion in CWC is thermally stable due to DiFA cross-linking of arabinoxylans (Figure 9.7) (Parker and Waldron, 1995). The role of ferulates has been supported by the observation that they are concentrated at the edges of the cell faces, as can readily be shown by fluorescence microscopy of cells separated in hot, dilute alkali (see Section 9.2). In more recent work, Parker *et al.* (2003) have investigated which particular dehydrodiferulates may be most important in intercellular adhesion in this tissue. Strips of CWC parenchyma tissue were incubated in progressively higher concentrations of alkali, and five out of the six dehydrodiferulates could be removed from tissue without reducing tissue strength. At the highest concentration of alkali, the final dehydrodiferulate, the 8,8′-DiFA (aryltetralyn form), was removed, and this coincided with the loss of intercellular adhesion. Hence the 8,8′-DiFA (aryltetralyn form) may be specifically responsible for intercellular adhesion in this tissue. The role of ferulic acid in cell adhesion has been similarly reported in Chufa (Parker *et al.*, 2000).

9.3.3.2 Asparagus

Spears of this vegetable contain a range of tissues, each of which contributes to the overall mechanical and textural properties (Waldron and Selvendran, 1990a). Thermal processing results in cell separation in apical parts of the stem, and associated tissue softening. Asparagus spears undergo post-harvest toughening, particularly if

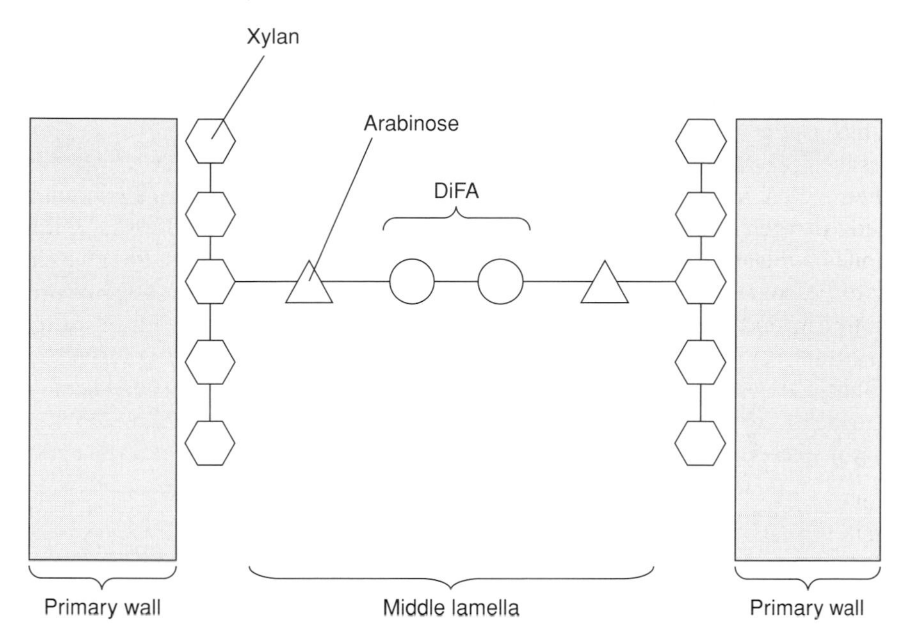

Figure 9.7 Schematic diagram of intercellular cross-linking of arabinoxylans via dehydrodiferulic acid moieties.

stored at ambient temperature (Waldron and Selvendran, 1990b). Early studies on asparagus spears (Waldron and Selvendran, 1990a,b, 1992) strongly indicated that such storage-related toughening is closely associated with significant changes in cell wall composition. Some components decreased, particularly galactan side chains of pectic polysaccharides (Waldron and Selvendran, 1990b). Others increased and included the formation of polymer complexes involving pectic polysaccharides, xylans and, at the time, unidentified phenolic components (Waldron and Selvendran, 1992). It is now clear that the cell walls contain considerable quantities of ferulic acid and its dehydrodimers (Rodriguez-Arcos *et al.*, 2002, 2004a,b). These increase appreciably during maturation (basipetally) and extensively during post-harvest storage and are probably responsible for the increase in cooked textural firmness. The increases in this cross-linking occurs in all tissues, including the inner parenchyma, which show little change in lignification but an increase in thermal stability of texture consistent with enhanced cell adhesion. Hence, these results are consistent with a role of ferulic acid intercellular cross-linking in the post-harvest toughening of asparagus spears. Additionally, the observation that the maturation-related basipetal increase in cell wall ferulic and DiFA moieties in asparagus is much less than the increase induced by post-harvest storage suggests that the storage-related increase in toughness is initiated by the process of harvesting, and may be due to a wound response rather than a continued maturation effect. This could provide researchers with additional biochemical pathways and mechanisms for controlling cell adhesion (Waldron *et al.*, 1997).

9.3.3.3 Sugar beet

The importance of ferulic-acid-substituted pectic polysaccharides in sugar beet was highlighted originally through research by Thibault (1986) who demonstrated that soluble sugar beet pectin could be made to gel by cross-linking the polymer peroxidatively. The rationale was that DiFA would be formed and thus create a covalently cross-linked, hydrated network. Further, from the work on Chinese water chestnut, Ng *et al.* (1998) investigated whether or not the ferulic acid in the walls of sugar beet could have an impact on thermal stability of cell adhesion. Initial studies involved comparisons between sugar beet, which was observed to have a longer softening time than commercial beetroot, which softened in about 30 min at 100°C. The characteristics of the intercellular cross-links responsible for cell–cell adhesion were obtained by subjecting sugar beet and beetroot tissues to a range of chemical and biochemical treatments designed to cleave cell-wall chemical bonds selectively. They observed that neither forms of *Beta vulgaris* would undergo vortex-induced cell separation (VICS) if treated with chelating agents or weak base (0.05 M Na_2CO_3). However, VICS could be induced after extraction in dilute, cold alkali (0.05–0.1 M KOH) and was accompanied by loss of wall autofluorescence associated with cinnamic acids. Interestingly, in sugar beet, over 20% of the FA was in dehydrodiferulic acid form. In beetroot, however, the value was only 10%. The main FA dimers were 8–0-4′DiFA and 8,5′DiFA (benzofuran form). The results indicated that the degree of thermal stability of cell–cell adhesion and, therefore, texture in *Beta vulgaris* tissues could be related to the degree of FA cross linking between pectic polysaccharides.

This hypothesis was tested further by exploiting the indigenous cell wall peroxidase activity in beetroot cell walls to see whether it was possible to peroxidatively cross-link polymers involved in cell adhesion, and enhance thermal stability of texture. Beet root (*Beta vulgaris*, var. Detroit 2 crimson) tissue was incubated in the presence or absence of hydrogen peroxide (H_2O_2) for 18 h after which it was subjected to thermal treatment. Incubation in H_2O_2 dramatically increased the time taken for tissues to soften at 100°C from an average of 130–650 min and resulted in a significantly higher tensile strength in heat-treated tissues. This was accompanied by a large decrease in esterified *cis*- and *trans*-ferulic acid and an increase in 5,5′-, 8-O-4′- and 5,8′-(benzofuran form)-DiFA moieties. In addition, the yield of hot-water-soluble wall polymers was much lower, consistent with increased cross-linking of the pectic polysaccharides. Interestingly, the carbohydrate composition and degree of uronide methylesterification of cell walls from H_2O_2-treated tissues was similar to that of fresh tissue. In contrast, incubation of control samples in the absence of H_2O_2 had no effect on thermal softening or phenolic chemistry of the walls. These results provided very good evidence to support the hypothesis that the H_2O_2-mediated changes in mechanical properties of the walls and the rate of thermal softening resulted from enhanced phenolic cross-linking of pectic polymers involved in cell–cell adhesion, due to oxidative coupling of ferulic acid moieties. Although not assessed in beetroot, cell wall peroxidase has been observed in the corners of tricellular junctions in several fruits and vegetables (Brett and Waldron, 1996; Ingham *et al.*, 1998) precisely where cross-linking would be expected to enhance cell adhesion.

Unlike CWC, beetroot and sugar beet, even after enhanced phenolic cross-linking, eventually undergo VICS under prolonged thermal treatment. This is because the intercellular cross-linking involves methyl-esterified pectic polysaccharides, which will undergo β-eliminative cleavage.

9.3.4 *Cell adhesion and the biosynthesis of feruloylated wall polymers*

Ferulic acid is thought to be added to pectin and arabinoxylan by the action of a feruloyltransferase in the Golgi apparatus (Brett *et al.*, 2000). The most likely feruloyl donor is feruloyl-CoA, which is a known intermediate in the synthesis of lignin. The feruloylated polymers are then transported to the cell surface in Golgi-derived vesicles, and incorporated into the wall by exocytosis (Waldron *et al.*, 1997). Dimerisation of the polysaccharide-linked ferulic acid then occurs in the wall, by the action of hydrogen peroxide and peroxidase, as discussed above. There is also evidence that some dimerisation may occur intracellularly (Fry *et al.*, 2000); such pre-formed dimers might then participate in further cross-linking to form trimers and higher oligomers of ferulate.

The observation (discussed above) that wall-bound ferulates are concentrated at the edges of cell faces raises questions about the control mechanisms that bring this about. Since feruloylation occurs intracellularly, the observed distribution is likely to be brought about by targeted exocytosis of feruloylated polymers to the edges of the cell faces. How this is achieved is not known.

The ratios of the six different dehydrodiferulates are different in different tissues (Wende *et al.*, 2000). There is evidence that one particular dehydrodiferulate, the 8,8′-DiFA (aryltetralyn form), may be specifically responsible for intercellular adhesion (Parker *et al.*, 2003). This raises the important question of what controls the ratios of the different types of dehydrodimers formed. *In vitro* studies indicate that hydrogen peroxide concentration may be one important factor (Baydoun *et al.*, 2004). In recent work, we have studied the effects of pectin structural variation on cross-linking *in vitro*, by pre-treating the pectins from sugar beet with specific polysaccharidases and glycosidase prior to cross-linking with hydrogen peroxide and peroxidase. These studies have indicated that the structure of the polysaccharide has a major effect on the degree and nature of the cross-links formed (Baydoun, Abdel-Massih, Waldron and Brett, unpublished). It is likely that the micro-structure of the pectin in the vicinity of the ferulate linkage influences the nature of the coupling reaction, either by interactions between the pectin and the peroxidase enzyme, or by affecting the mutual orientation of the interacting ferulate radicals.

9.3.5 *The Hard-To-Cook defect*

The Hard-To-Cook (HTC) defect in legume seeds occurs during storage in conditions of high temperature and high humidity prevalent in tropical conditions. This condition prevents cell separation from occurring in cotyledon tissues during cooking, resulting in increased cooking times. There are a number of hypotheses to explain the HTC phenomenon. The earliest (Mattson, 1946) involves a storage-induced decrease in intracellular phytic acid, which would normally chelate calcium. In this mechanism, phytase degrades the phytate thereby releasing calcium, which diffuses into the cell walls on seed imbibition. The free calcium would therefore displace monovalent cations, resulting in pectin insolubilisation and strengthening of cell adhesion. This potential mechanism was supported by studies that demonstrated that imbibing beans in solutions of Mg^{2+} and Ca^{2+} increased thermal softening times considerably. Furthermore, soaking beans in chelators could reverse the defect. However, no correlation has been found between cooking time and the level of phyate content (Liu, 1995). Other hypotheses to explain the HTC defect include lignification-like reactions, roles for tannins (Stanley, 1992) and possibly cinnamic acid derivatives (Garcia *et al.*, 1998). However, as for many other 'bulk tissue' approaches to understanding cell adhesion issues, these mechanisms are not robust.

However, one microscopic study carried out by Shomer *et al.* (1990) showed that development of the HTC defect was accompanied by changes in the cell walls at the corners of the tricellular junctions, precisely where cell adhesion is controlled. There appears to be little further research surrounding this observation.

9.4 Conclusions

A number of possible cross-linking mechanisms might be important in generating intercellular adhesion. In order to clarify which mechanisms are most important, and to understand them better, it will be important to make imaginative use of the techniques now becoming available. The use of synchronised cell suspension cultures (Menges and Murray, 2002; Menges *et al.*, 2003) should permit a detailed study of the events occurring during the formation of tricellular junctions and intercellular spaces at cytokinesis. Immunoaffinity chromatography using LM7 and other monoclonal antibodies offers the possibility of isolating and purifying the polymers containing the epitopes found at cell corners. HPLC and other analytical techniques may be used to analyse the polysaccharides to which the 8,8′-DiFA (aryltetralyn form) is attached in Chinese water chestnut and other model systems such as sugar beet. As more pectin synthesis mutants of Arabidopsis become available, it will be possible to investigate which parts of the pectin molecules are important for intercellular adhesion.

Acknowledgements

The Authors thank Dr M.L. Parker of the Institute of Food Research for the micrographs.

References

Atkinson, R.G., Schroder, R., Hallett, I.C., Cohen, D. and MacRae, E. (2002) Overexpression of polygalacturonase in transgenic apple trees leads to a range of novel phenotypes involving changes in cell adhesion. *Plant Physiology* **129**, 122–133.

Baldwin, T.C., McCann, M.C. and Roberts, K. (1993) A novel hydroxyproline-deficient AGP secreted by suspension-cultured cells of Daucus carota. Purification and partial characterisation. *Plant Physiology* **103** (1), 115–123.

Baydoun, E.A.-H., Pavlencheva, N., Cumming, C.M., Waldron, K.W. and Brett, C.T. (2004) Control of dehydrodiferulate cross-linking in pectins from sugar-beet tissues. *Phytochemistry* **65**, 1107–1115.

Bech, A.C., Gruinert, K.G., Bredahl, L., Juhl, H.L. and Poulsen, C.S. (2001) Consumer's quality perception. In: *Food, People and Society: A European Perspective of Consumer's Food Choices* (eds Frewer, L.J., Risvik, E. and Schifferstein, H.), pp. 97–113. Springer, Berlin, Heidelberg, Germany.

Brett, C.T., Baydoun, E.A.H. and Abdel-Massih, R.M. (2005) Pectin-xyloglucan linkages in type I primary cell walls of plants. *Plant Biosystems* **139**, 54–59.

Brett, C.T. and Waldron, K. (1996) *Physiology and Biochemistry of Plant Cell Walls*, 2nd edn., Chapman & Hall, London.

Brett, C.T., Wende, G., Smith, A.C. and Waldron, K.W. (2000) Biosynthesis of cell-wall ferulate and diferulates. *Journal of the Science of Food and Agriculture* **79**, 421–424.

Bunzel, M., Ralph, J., Lu, F., Hatfield, R.D. and Steinhart, H., (2004) Lignins and ferulate-coniferyl alcohol cross-coupling products in cereal grains. *Journal of Agricultural and Food Chemistry* **52**, 6496–6502.

Carpita, N. and Gibeaut D.M. (1993) Structural models of primary cell walls in flowering plants-consistancy of molecular structure with the physical properties of the walls during growth. *The Plant Journal* **3**, 1–3.

Cumming, C.M., Rizkallah, H.D., McKendrick, K.A., Abdel-Massih, R.M., Baydoun, E.A.H. and Brett, C.T. (2005) Biosynthesis and cell-wall deposition of a pectin-xyloglucan complex in pea. *Planta*.

Denes, J.-M., Baron, A., Renard, C.M.C.G., Pean, C. and Drilleau, J.-F. (2000) Different action Pattern for apple pectin methylesterase at pH 7.0 and 4.5. *Carbohydrate Research* **327**, 385–393.

Eriksson, E.M., Bovy, A., Manning, K., Harrison, L., Andrews, J., De Silva, J., Tucker, G.A. and Seymour, G.B. (2004) Effect of the colorless non-ripening mutation on cell wall biochemistry and gene expression during tomato fruit development and ripening. *Plant Physiology* **136**, 4184–4197.

Femenia, A., Rigby, N.M., Selvendran, R.R. and Waldron, K.W. (1999) Investigation of the occurrence of pectic-xylan-xyloglucan complexes in cell walls of cauliflower stem tissues. *Carbohydrate Polymers* **39**, 151–164.

Fry, S.C., Willis, S.C. and Paterson, A.E.J. (2000) Intraprotoplasmic and wall-localised formation of arabinoxylan-bound diferulates and larger ferulate-coupling products in maize cell-suspension cultures. *Planta* **211**, 679–692.

Funk, C., Ralph, J., Steinhart, H. and Bunzel, M., (2005) Isolation and structural characterisation of 8-O-4/8-O-4-and 8-8/8-O-4-coupled dehydrotriferulic acids from maize. *Phytochemistry* **66**, 363–371.

Garcia, E., Filisetti, T.M.C.C., Udaeta, J.E.M. and Lajolo, F.M. (1998) Hard-to-cook beans (Phaseolus vulgaris): involvement of phenolic compounds and pectates. *Journal of Agricultural and Food Chemistry* **46** (6), 2110–2116.

Gaspar, Y.M., Johnson, K.L., McKenna, J.A., Bacic, A. and Scholtz, C.T. (2001) The complex structures of arabinogalactan-proteins and the journey towards a function. *Plant Molecular Biology* **47**, 161–176.

Ingham, L.M., Parker, M.L. and Waldron, K.W. (1998) Peroxidase: changes in soluble and bound forms during maturation and ripening of apples. *Physiologia Plantarum* **102**, 93–100.

Ishii, T., Matsungaga, T., pellerin, P., O'Neil, M.A., Darvill, A. and Albersheim, P. (1999) The plant cell wall polysaccharide rhamnogalactorunan II self-assembles into a covalently cross linked dimmer. *Journal of Biological Chemistry* **274**, 13098–13104.

Iwai, H., Masako, N., Ishii, T. and Satoh, S. (2002) A pectin glucuronyltransferase gene is essential for intercellular attachment in the plant meristem. *Proceedings of the National Academy of Sciences of the United States of America* **99** (25), 16319–16324.

Jarvis, M.C. (1998) Intercellular separation forces generated by intracellular pressure. *Plant, Cell and Environment* **21**, 1307–1310.

Jarvis, M.C., Briggs, S.P.H. and Knox, J.P. (2003) Intercellular adhesion and cell separation in plants. *Plant, Cell and Environment* **26**, 977–989.

Kim, J.B. and Carpita, N.C. (1992) Changes in esterification of the uronic acid groups of cell-wall polysaccharides during elongation of maize coleoptiles. *Plant Physiology* **98** (2), 646–653.

Knox, J.P., Linstead, P.J., King, J., Cooper, C. and Roberts, K. (1990) Pectin esterification is spatially regulated both within cell walls and between developing tissues of root apices. *Planta* **181**, 512–521.

Kollofel, C. and Linssen, P.W.T. (1984) The formation of intercellular spaces in the cotyledons of developing and germinating pea seeds. *Protoplasma* **12**, 12–19.

Lecain, S., Ng, A., Parker, M.L., Smith, A.C. and Waldron, K.W. (1999) Modification of cell-wall polymers of onion waste – Part I. Effect of pressure-cooking. *Carbohydrate Polymers* **38** (1), 59–67.

Liners, F. and Van Custem, P. (1992) Distribution of pectic polysaccharides throughout walls of suspension-cultured carrot: an immunocytochemical study. *Protoplasma* **170**, 10–21.

Liu, K. (1995) Cellular, biological and physicochemical basis for the hard-to-cook defect in legume seeds. *Critical Reviews in Food Science and Nutrition* **35** (4), 263–298.

MacKinnon, I.M., Jardine, W.G. and O'Kennedy, N. (2002) Pectic methyl and non methyl esters in potato cell walls. *Journal of Agricultural and Food Chemistry* **50**, 342–346.

Martin-Cabrejas, M.A., Waldron, K.W. and Selvendran, R.R. (1994) Changes in Spanish pear during ripening. *Journal of Plant Physiology* **144**, 541–548.

Matoh, T., Takasaki, M., Takabe, K. and Kobayashi, M. (1998) Immunochemistry of rhamnogalacturonan II in cell wall of higher plants. *Plant and Cell Physiology* **39**, 483–491.

Mattson, S. (1946) The cookability of yellow peas: a colloid-chemical and biological study. *Acta Agricultura Sueden* **2**, 185–187.

McCartney, L. and Knox, J.P. (2002) Regulation of pectic polysaccharide domains in relation to cell development and cell properties in the pea testa. *Journal of Experimental Botany* **53** (369), 707–713.

Menges, M. and Murray, J.A.H. (2002) Synchronous Arabidopsis suspension cultures for analysis of cell-cycle gene activity. *Plant Journal* **30** (2), 203–212.

Menges, M., Hennig, L., Gruissem, W. and Murray, J. A.H. (2003) Genome-wide gene expression in an Arabidopsis cell suspension. *Plant Molecular Biology* **53**, 423–442.

Micheli, F. (2001) Pectin methylesterase: cell wall enzymes with important roles in plant physiology. *Trends in Plant Science* **6** (9), 474–419.

Morris, E.R., Powell, D.A., Gidley, M.J. and Rees, D.A. (1982) Conformations and interactions of pectins. *Journal of Molecular Biology* **155**, 507–516.

Morris, V.J., Ring, S.G., MacDougall, A.J. and Wilson, R.H. (2003) Biophysical characterisation of plant cell walls. In: *The Plant Cell Wall* (ed. Rose, J.K.C.). Blackwell Publishing, Oxford, and CRC Press, Boca Raton.

Ng, A., Greenshields, R.N. and Waldron, K.W. (1997) Oxidative cross-linking of corn-bran hemicellulose: formation of ferulic acid dehydrodimers. *Carbohydrate Research* **303**, 459–462.

Ng, A. and Waldron, K.W. (1997a) Effect of steaming on cell wall chemistry of potatoes (Solanum tuberosum cv. Bintje) in relation to firmness. *Journal of Agricultural and Food Chemistry* **45**, 3411–3418.

Ng, A. and Waldron, K.W. (1997b) Effect of cooking and pre-cooking on cell-wall chemistry in relation to firmness of carrot tissues. *Journal of the Science of Food and Agriculture* **73**, 503–512.

Ng, A., Harvey, A.J., Parker, M.L., Smith, A.C. and Waldron, K.W. (1998) Effect of oxidative coupling on the thermal study of texture and cell wall chemistry of beet root (Beta vulgaris). *Journal of Agricultural and Food Chemistry* **46**, 3365–3370.

O'Neill, M.A., Ishii, T., Albersheim, P. and Darvill, A. (2004) Rhamnogalacturonan II: structure and function of a borate cross-linked cell wall pectic polysaccharide. *Annual Review of Plant Biology* **55**, 109–39.

O'Neill, M.A., Warrenfeltz, D., Kates, K., Pellerin, P., Doco, T., Darvil, A.G. and Albersheim, P. (1996) Rhamnogalactorunan II, a pectic polysaccharide in the walls of growing plant cell, forms a dimer that is covalently linked by a borate diester. *Journal of Biological Chemistry* **274**, 13098–13104.

Orfilla, C., Seymour, G.B., Willats, W.G.T., Huxham, I.M., Jarvis, M.C., Dover, C.J., Thompson, A.J. and Knox, J.P. (2001) Altered middle lamella homogalacturonan and disrupted deposition of (1–5)-a-L-Arabinan in the pericarp of Cnr, a ripening mutant of tomato. *Plant Physiology* **126**, 210–221.

Parker, M.L. and Waldron, K.W. (1995) Texture of Chinese Waterchestnut: involvement of cell-wall phenolics. *Journal of the Science of Food and Agriculture* **68**, 337–346.

Parker. M.L., Ng, A., Smith, A.C. and Waldron, K.W. (2000), Esterified phenolics of the cell walls of chufa (Cyperus esculentus L.) tubers and their role in texture. *Journal of Agricultural and Food Chemistry* **48** (12), 6284–6291.

Parker, C.C., Parker, M.L., Smith, A.C. and Waldron K.W. (2001) Pectin distribution at the surface of potato parenchyma cells in relation to cell-cell adhesion. *Journal of Agricultural and Food Chemistry* **49**, 4364–4371.

Parker, C.C., Parker, M.L., Smith, A.C. and Waldron K.W. (2003) Thermal stability of texture in Chinese water chestnut may be dependent on 8,8′-diferulic acid (aryltetralyn form). *Journal of Agricultural and Food Chemistry* **51**, 2034–2039.

Ralph, J., Quideau, S., Grabber, J.H. and Hatfield, R.D. (1994) Identification and synthesis of new ferulic acid dehydrodimers present in grass cell walls. *Journal of the Chemical Society-Perkin Transactions* **1** (23), 3485–3498.

Ridley, B., O'Neil, M.A. and Mohnen, D. (2001) Pectins: structure, biosynthesis, and oligosaccharide-related signalling. *Phytochemistry* **57**, 929–967.

Roberts, J.A., Whitelaw, C.A. and Gonzalez-Carranza, Z.H. (2000) Cell separation processes in plants – models, mechanisms and manipulation. *Annals of Botany* **86**, 223–235.

Roberts, J.A., Elliot, K.A. and Gonzalez-Carranza, Z.H. (2002) Abscission, dehiscence, and other cell separation processes. *Annual Review of Plant Biology* **53**, 131–158.

Rodriguez-Arcos, R.C., Smith, A.C. and Waldron, K.W. (2002) Mechanical properties of green asparagus. *Journal of the Science of Food and Agriculture* **82** (3), 293–300.

Rodriguez-Arcos, R.C., Smith, A.C. and Waldron, K.W. (2004a) Effect of storage on wall-bound phenolics in green asparagus. *Journal of Agricultural and Food Chemistry* **50** (11), 3197–3203.

Rodriguez-Arcos, R.C., Smith, A.C. and Waldron, K.W. (2004b) Ferulic acid crosslinks in asparagus cell walls in relation to texture. *Journal of Agricultural and Food Chemistry* **52** (15), 4740–4750.

Rouau, X., Chynier, V., Surget, A., Gloux, D., Barron, C., Meudec, E., Loius-Montero, J. and Criton, M. (2003) A dehydrotrimer of ferulic acid from maize bran. *Phytochemistry* **63**, 899–903.

Round, A.N. (1999) *Atomic Force Microscopy of Plant Cell Wall Polysaccharides*. PhD Thesis, University of East Anglia, Norwich, UK.

Roy, S., Janeau, A. and Vian, B. (1994) Analytical detection of calcium ions and immunocytochemical visualisation of homogalacturonic acid sequences in the cell walls of apple fruit. *Plant Physiology and Biochemistry* **32**, 633–640.

Roy, S., Gillen, G., Conway, W.S., Watada, A.E. and Wergin, W.P. (1995) Uses of secondary ion mass spectrometry to image ^{44}calcium uptake in the cell walls of apple fruit. *Plant Physiology and Biochemistry* **32**, 633–640.

Shevell, D.E., Kunkel, T. and Chua, N.-H. (2000) Cell wall alterations in the Arabidopsis emb309 mutant. *The Plant Cell* **12**, 2047–2059.

Shomer, I., Paster, N., Lindner, P. and Vasilver, R. (1990) The role of cell-wall structure in the hard-to-cook phenomenon in beans (*Phaseolus vulgaris* L). *Food Structure* **9** (2), 139–149.

Smallwood, M., Beven, A., Donavan, N., Neil, S.J., Peart, J., Roberts, K. and Knox, J.P. (1994) Localization of cell wall proteins in relation to the developmental anatomy of the carrot root apex. *Plant Journal* **5**, 237–246.

Stanley, D.W. (1992) A possible role for condensed tannins in bean hardening. *Food Research International* **25** (3), 187–192.

Thibault, J.-F. (1986) Some physicochemical properties of sugar-beet pectins modified by oxidative cross-linking. *Carbohydrate Research* **155**, 183–192.

Thompson, J.E. and Fry, S.C. (2000) Evidence for covalent linkage between xyloglucan and acidic pectins in suspension-cultured rose cells. *Planta* **211**, 275–286.

Tijskens, L.M.M., Rodis, P.S., Hertog, M.L.A.T., Waldron, K.W., Ingham, L., Proxenia, N. and VanDijk, C. (1997a) Activity of peroxidase during blanching of peaches, carrots and potatoes. *Journal of Food Engineering* **34**, 355–370.

Tijskens, L.M.M., Waldron, K.W., Ng, A., Ingham, L. and VanDijk, C. (1997b) The kinetics of pectin methyl esterase in potatoes and carrots during blanching. *Journal of Food Engineering* **34**, 371–385.

Van Buren, J.P. (1979) The chemistry of texture in fruits and vegetables. *Journal of Textural Studies* **10**, 1–23.

Van Buren, J.P. and Peck, N.H. (1982) Effect of K-fertilisation and addition of salts on the texture of canned snap bean pods. *Journal of Food Science* **47** (1), 311–313.

Verma, D.P.S. (2001) Cytokinesis and building of the cell plate in plants. *Annual Review in Plant Physiology and Plant Molecular Biology* **52**, 751–784.

Waldron, K.W., Smith, A.C., Parr, A.J., Ng, A., Parker, M.L. (1997) New approaches to understanding and controlling cell separation in relation to fruit and vegetable texture. *Trends in Food Science and Technology* **8**, 213–221.

Waldron, K.W. and Selvendran, R.R. (1990a) Composition of the cell-walls of different asparagus (asparagus officinalis) tissues. *Physiologia Plantarum* **80**, 568–575.

Waldron, K.W. and Selvendran, R.R. (1990b) Effect of maturation and storage on asparagus (asparagus officinalis) cell wall composition. *Physiolgia Plantarum* **80**, 576–583.

Waldron, K.W. and Selvendran, R.R. (1992) Cell wall changes in immature asparagus stem tissue after excision. *Phytochemistry* **31**, 1931–1940.

Waldron, K.W., Parker, M.L. and Smith, A.C. (2003) Plant cell walls and food quality. *Comprehensive Reviews in Food Science and Food Safety* **2**, 101–119.

Wende, G., Waldron, K.W., Smith, A.C. and Brett, C.T. (2000) Tissue-specific developmental changes in cell-wall ferulate and dehydrodiferulates in sugar beet. *Phytochemistry* **55** (2), 103–110.

Willats, W.G.T., McCartney, L., Mackie, W. and Knox, J.P. (2001a) Pectin: cell biology and prospects for functional analysis. *Plant Molecular Biology* **47**, 9–27.

Willats, W.G.T., Orfila, C., Limberg, G., Buchholt, H.C., Van Alebeek, G.W.M., Voragen, A.G.J., Marcus, S.E., Christensen, T.M.I.E., Mikkelsen, J.D., Murray, B.S. and Knox, J.P. (2001b) Modulation of the degree and pattern of methyl esterfication of pectic homoglacturonan in plant cell walls—implications for pectin methyl esterase action, matrix properties, and cell adhesion. *Journal of Biological Chemistry* **276**, 19404–19413.

Williams, M.N.Y., Freshour, G., Darvill, A.G., Albersheim, P. and Hahn, M.G. (1996) An antibody Fab selected from a recombinant phage display library detects deesterified pectic polysaccharide rhamnogalactorunan II in plant cells. *Plant Cell* **8**, 673–685.

Index